NEURAL NETWORKS AND SYSTOLIC ARRAY DESIGN

SERIES IN MACHINE PERCEPTION AND ARTIFICIAL INTELLIGENCE*

Vol. 34: Advances in Handwriting Recognition
(Ed. *S.-W. Lee*)

Vol. 35: Vision Interface — Real World Applications of Computer Vision
(Eds. *M. Cheriet and Y.-H. Yang*)

Vol. 36: Wavelet Theory and Its Application to Pattern Recognition
(*Y. Y. Tang, L. H. Yang, J. Liu and H. Ma*)

Vol. 37: Image Processing for the Food Industry
(*E. R. Davies*)

Vol. 38: New Approaches to Fuzzy Modeling and Control — Design and Analysis
(*M. Margaliot and G. Langholz*)

Vol. 39: Artificial Intelligence Techniques in Breast Cancer Diagnosis and Prognosis
(Eds. *A. Jain, A. Jain, S. Jain and L. Jain*)

Vol. 40: Texture Analysis in Machine Vision
(Ed. *M. K. Pietikäinen*)

Vol. 41: Neuro-Fuzzy Pattern Recognition
(Eds. *H. Bunke and A. Kandel*)

Vol. 42: Invariants for Pattern Recognition and Classification
(Ed. *M. A. Rodrigues*)

Vol. 43: Agent Engineering
(Eds. *Jiming Liu, Ning Zhong, Yuan Y. Tang and Patrick S. P. Wang*)

Vol. 44: Multispectral Image Processing and Pattern Recognition
(Eds. *J. Shen, P. S. P. Wang and T. Zhang*)

Vol. 45: Hidden Markov Models: Applications in Computer Vision
(Eds. *H. Bunke and T. Caelli*)

Vol. 46: Syntactic Pattern Recognition for Seismic Oil Exploration
(*K. Y. Huang*)

Vol. 47: Hybrid Methods in Pattern Recognition
(Eds. *H. Bunke and A. Kandel*)

Vol. 48: Multimodal Interface for Human-Machine Communications
(Eds. *P. C. Yuen, Y. Y. Tang and P. S. P. Wang*)

Vol. 49: Neural Networks and Systolic Array Design
(Eds. *D. Zhang and S. K. Pal*)

*For the complete list of titles in this series, please write to the Publisher.

Series in Machine Perception and Artificial Intelligence – Vol. 49

NEURAL NETWORKS AND SYSTOLIC ARRAY DESIGN

Editors

David Zhang

Hong Kong Polytechnic University, Hong Kong

Sankar K. Pal

The Indian Statistical Institute, Calcutta, India

World Scientific

New Jersey • London • Singapore • Hong Kong

Published by

World Scientific Publishing Co. Pte. Ltd.
P O Box 128, Farrer Road, Singapore 912805
USA office: Suite 1B, 1060 Main Street, River Edge, NJ 07661
UK office: 57 Shelton Street, Covent Garden, London WC2H 9HE

British Library Cataloguing-in-Publication Data
A catalogue record for this book is available from the British Library.

NEURAL NETWORKS AND SYSTOLIC ARRAY DESIGN

ISBN 981-02-4840-7

Printed in Singapore by Uto-Print

Preface

In recent years, Artificial Neural Networks (ANNs) have become a subject of very extensive research, since they are envisioned as an alternative to solving problems of certain tasks such as pattern recognition, vision and speech recognition, where conventional artificial intelligence techniques had been unable to tackle efficiently. Neural networks are biologically inspired; they are composed of elements that perform in a manner analogous to the most elementary functions of biological neurons. However, as simulations of large neural networks on a sequential computer frequently require days and weeks of computations, and the long computational time had been a critical obstacle to the progress in neural network research, extensive effort is being devoted to the parallel implementation of neural networks. Systolic Arrays (SAs) offer some of the best solutions to the parallel implementation of neural networks. It can overcome the communication problems generated by the highly interconnected neurons, and can exploit the massive parallelism inherent in the problem. Moreover, since the computation model of neural networks can be represented by a series of matrix by vector multiplications, the classical systolic algorithms can be used to implement them.

SA design differs from that of the conventional von Neumann computer in its highly pipelined computation. Once a data item is brought out from the memory it can be used effectively at each cell it passes while being pumped from cell to cell along the array. This is especially appealing for a wide class of compute-bound computations, where multiple operations are performed on each data item in a repetitive manner. This avoids the classic memory access bottleneck problem commonly incurred in von Neumann machines. In addition, SA has a spatial locality and a temporal locality. It manifests a locally communicative interconnection structure, and there is at least one unit-time delay allocated so that signal transactions from one processing element to the next can be completed. When a large number of processors work together, communication becomes a significant problem. Therefore, regular and local communication in systolic arrays is advantageous.

ANNs and SAs have many similar features. The communication only between neighbor neurons in neural networks is similar to the spatial/temporal locality of systolic arrays. The highly interconnected structures of neural networks can be mapped onto the regular structure of a systolic array, and the massive parallelism can be exploited by the large number of processing elements of systolic arrays. Since SAs can overcome the communication problems generated by highly interconnected neurons, and can exploit the massive parallelism inherent in the problem, there have been several research efforts on developing systolic algorithms and systolic array structures to implement various ANNs. Attempts are made to develop a mapping approach, in which the systolic algorithms are partitioned and mapped onto the parallel computers, and a VLSI approach, in which VLSI systolic arrays dedicated to specific models are designed.

This book provides a collection of sixteen chapters containing tutorial articles and new material describing, in a unified way, the basic concepts, theories and characteristic features of integrating and formulating different facets of ANNs and SAs, with recent developments and significant applications. The articles, written by various experts from around the world, demonstrate the varied ways this integration can be made to design methodologies, algorithms and architectures, and implementations for ANN applications efficiently. With the exception of the first chapter's tutorial introduction to ANNs and SAs, each chapter provides exhaustive information on developments in the respective areas, keeping a cohesive character with other chapters. The approaches typically include the mapping of systolic algorithms for neural networks onto parallel computers and the design of VLSI systolic arrays dedicated to one or two specific models, besides describing some available systems and applications.

They also provide a balanced mixture of both theory and implementation.

This book, which is unique in character, is useful to graduate students and researchers in computer science, electrical engineering, systems science and information technology not only as a reference, but also as a text book for some parts of the curriculum. The researchers and practitioners in industry and R&D laboratories working in the fields of system design, VLSI, parallel processing, control, pattern recognition, neural networks and vision will also benefit.

We would like to take this opportunity to thank all the contributors for agreeing to write for the book. We owe a vote of thanks to Ian Seldrup and Alan Pui of World Scientific, Singapore, for taking the initiative in bringing the volume out. The technical/software support provided by Adams Kong, Michael Wong, Henry Ko and Yunzhi Lin are also acknowledged. The project was initiated when Prof. S.K. Pal was working as a Visiting Professor in the department of computing, Hong Kong Polytechnic University, Hong Kong, during November-December 2000.

David Zhang and Sankar K. Pal
December 2001

Contents

CHAPTER 1

NEURAL NETWORKS AND SYSTOLIC ARRAYS: MODELS AND INTEGRATIONS

D. Zhang
The Centre for Multimedia Signal Processing and Department of Computing
Hong Kong Polytechnic University, Kowloon, Hong Kong
E-mail: csdzhang@comp.polyu.edu.hk

S.K. Pal
Machine Intelligence Unit, Indian Statistical Institute,
203, B.T. Road, Calcutta 700035, India
E-mail: sankar@isical.ac.in

This chapter briefly provides a tutorial introduction to artificial neural networks (ANN) and systolic arrays (SA). The existing ANN and SA models, architectures, implementation techniques and their integration are reviewed, respectively. Also, all chapters are briefly outlined to show our organization of the book.

1.1. Introduction

Artificial neural networks (ANN) are massively parallel interconnected networks of simple (usually adaptive) nodes which are intended to interact with objects of the real world in the same way as biological nervous systems do [1]. The interest in these networks is due to the general opinion that they are able to perform some complicated and creative tasks, such as pattern recognition, similar to the way they are performed by human brains [2-4]. The implementations of these tasks by traditional computing methods have only reached relatively low performances in some limited aspects or environments. Nevertheless, as neural systems show some properties, like association, generalization, parallel searching, fault tolerance, and adaptation to changes in the environment, which are analogous to human brain properties, they promise improved results.

The usage of ANN for pattern recognition may be traced back to the perceptron models originated by Rosenblatt [2]. The perceptron models used the concept of reward and punishment. In late 1960s, the progress in ANN models slowed down due to the limited capabilities of the early single layer perceptron models. In the mid-1970s and early 1980s, with the availability of enhanced computing power the progress in the development of ANN models accelerated. Researchers were able to model and test their theories about the functioning of the brain.

Today a number of well-developed theories and models of ANNs are available [5-22]. These networks consist of a large number of simple processing elements called nodes that represent the neurons. The nodes are interconnected by the synaptic connections. These models are capable of learning and making decisions; and are suitable for a variety of pattern recognition tasks.

Pattern recognition techniques can be grouped into two classes: supervised and unsupervised techniques. In supervised methods, certain number of samples is available for each category and these samples are used to train the classifier. In the case of unsupervised classification no training samples are available, and the network learns by detecting the similarity between the input patterns.

ANNs consist of parallel distributed processing (PDP) models. The PDP models are well described in the work of Rumelhart and McCelland [23]. The functional synthesis of these models consists of establishing a relationship between the several inputs and one or more outputs. In ANN, the nodes are connected to each other by the synaptic connections or the links. There is an associated synaptic strength or a weight with each connection. During the learning, the weights that represent the knowledge stored in the network are updated. The layered ANNs consist of two or several layers of nodes and each layer contains several nodes. The observed feature vector is presented to the input nodes. The input values may represent the probability that the discrete feature is present. Each possible decision or outcome can be represented by a node in the output layer.

Since ANNs have become a subject of great interest due to their massive parallelism, modularity and stability, research into ANNs has not only embraced modeling and algorithm, but also dealt with corresponding architectures and implementations [24-25].

There have been a variety of implementation approaches for ANN, including software and hardware, electronic and optical implementations [25-30]. The real promise for pattern recognition applications of ANNs currently lies in specialized hardware, in particular VLSI implementations [31-35].

Typically, the high interconnectivity and relatively low precision needed for signals in ANNs are well tailored to an analog approach. With increasing ANN size, however, designing large integrated analog circuits, based on VLSI implementation, is a difficult task. On the other hand, digital processing, which appears to be inferior to analog processing in terms of computational density, has advantages of flexibility in terms of programming for a variety of architectures and learning strategies, as well as simplifying the task of memory retention. There has been a strong development tendency to implement ANNs in VLSI using digital or mixed technologies [31,36].

It has already been proved that systolic arrays (SA) are a good approach to ANN implementation and suitable for VLSI design [20,37-40]. SA architecture in parallel processing came as a product of certain environment, the means and the background knowledge for its realization. The needs can be outlined as the ever-increasing tendency for faster computations, especially in areas like real-time signal processing and large-scale scientific computation. The means could be provided by the remarkable advances in VLSI technology and computer aided design (CAD).

Ever since Kung proposed the systolic architecture in 1978, its elegant solution to demanding problem and its potential performance have attracted great attention [41]. SA can be applied to many fields, such as matrix arithmetic, signal processing, image processing, solution of differential equations, data structure, graph algorithms, computer aided design (CAD), and so forth [37-43]. Many systolic algorithms have been designed for a great diversity of areas. Few problems resist attack from SA. Efforts have been made to implement SA in a VLSI chip, in a programmable processor array, or in a reconfigurable SA. While SAs were originally used for fixed or special purpose architecture, SA concept has been extended to general-purpose Single Instruction Multiple Data (SIMD) and Multiple Instruction Multiple Data (MIMD) architectures.

This chapter is organized as follows: Section 1.2 outlines some basic ANN models. Typical ANN architectures and corresponding systolic array implementations in VLSI are reviewed in Sections 1.3 and 1.4, respectively. In Section 1.5, each chapter is simply summarized to draw our organization of the book.

1.2. Neural Network Models

ANNs are information processing systems. Due to the variety of fields of interest and applications, a broad range of ANN models has been emerged. In all these fields, the term "neural networks" is characterized by a combination of adaptive learning algorithms and parallel distributed implementations. Although ANNs are biologically motivated, their resemblance to the brain models is not straightforward. Since the basic concepts of biological neural networks provide a common ground for understanding the ANN models, we will first introduce basic biological neural networks in order to appreciate the differences and similarities with ANN.

1.2.1. *Biological Neural Networks*

As we know, the von Neumann computer performs poorly on certain tasks, such as pattern recognition that humans handle routinely. It is interesting to compare the human brain with a serial modern von Neumann computer from the information processing aspects. Although the neuron's switching time (a few milliseconds) is about a million times slower than modern computer elements, they have a thousand-fold greater connectivity than today's supercomputers. Furthermore, the brain is very powerfully efficient – it consumes less than 100 watts – by contrast a supercomputer may dissipate 10^5 watts.

The basic element of neural networks of a brain is a neuron. The neurons consist of four basic parts: cell body, synapses, axons, and dendrites. The cell body essentially sums the membrane potential provided by the synapses. The synapses provide an output. Axons are the connections between the neurons that carry charge, and the dendrites are the branch-like structures, which provide the sensory input to a cell body.

In fact, the synapse represents the junction between an axon and a dendrite. How two or more neurons interact remains largely mysterious, and complexities of different neurons vary greatly. Generally speaking, a neuron sends its output to other neurons via its axon. An axon carries information through a series of action potentials, or waves of current, which depends on the neuron's voltage potential. More precisely, the membrane generates the action potential and propagates down the axon and its branches, where axonal insulators restore and amplify the signal as it propagates, until it arrives at a synaptic junction.

There are several important features obtained from biological neural networks, which apply to all ANNs:

1) Each neuron acts independently of all others – each neuron's output relies only on its constantly available inputs from the abutting connections.
2) Each neuron relies only on local information – the information that is provided by the adjoining connections.
3) The large number of connections provides a large amount of redundancy and facilitates a distributed representation.

First two features allow neural networks to operate efficiently in parallel. The last one provides neural networks with inherent fault-tolerance.

1.2.2. *Artificial Neural Networks*

ANNs mimic the functioning of the neural networks of a brain. ANNs consist of a large number of simple processing elements, called nodes or artificial neurons, that interact with other nodes using weighted connections or synaptic connections or links. These artificial neurons were designed to mimic the characteristics of the biological neurons. A neuron accepts a set of inputs, each representing the output of another neuron. Each input is multiplied by a corresponding weight, analogous to a synaptic strength, and all of the weighted inputs are then summed to determine the state or active level of the neuron. ANNs are characterized by a few key properties: the network topology, which is specified by the interconnection pattern of the neurons; the recall procedure, which is specified by the output function of the neurons according to the input; and the training/learning procedure, which establishes the values of the weights.

Information processing takes place through the interaction between the nodes. Each node is associated with an activation value $\varphi_j(t)$. The activation value passes through an activation function $f(\varphi_j)$ to provide an actual output $y_j(t)$. These outputs pass through the unidirectional synaptic connections. There is an associated number, w_{ij}, — called the weight or the connection strength, that determines the amount of effect node i can have on node j. For each node all the inputs are combined, and the total input, along with the current activation, determines the new activation value.

Usually ANNs consist of a number of layers and nodes in each layer. The most general model assumes the complete interconnections between all the nodes, and resolves the cases of the nonconnected nodes (i, j) by setting the weights $w_{ij} = 0$. The networks can be synchronous or asynchronous. The synchronous networks are controlled by clock pulses; whereas in asynchronous networks the nodes respond instantaneously to the incoming inputs.

The connections between the nodes can be bi-directional or unidirectional. The activation function and the activation values to be used in the network are often restricted in the range [0, 1]. In the case of discrete values, the activation values can take only two values, 0 or 1. The number of activation functions can be used to define the propagation law in the network.

The linear transfer function produces a linearly modulated output from the input x as

$$f(x) = \alpha x \tag{1-1}$$

where x ranges over the real numbers and α is a positive scalar. If $\alpha = 1$, it is equivalent to removing the transfer function completely.

The step transfer function is described as

$$f(x) = \begin{cases} \beta & \text{if } x \geq \theta \\ -\delta & \text{if } x < \theta \end{cases} \tag{1-2}$$

where β and δ are positive scalars, and θ is a predefined value. The ramp transfer function is a combination of the linear and step transfer functions. It is defined as

$$f(x) = \begin{cases} \gamma & \text{if } x \geq \gamma \\ x & \text{if } |x| < \gamma \\ -\gamma & \text{if } x \leq -\gamma \end{cases} \tag{1-3}$$

where γ is the saturation value for the transfer function. The most common sigmoid function provides an output value from 0 to 1. It is defined as

$$f(x) = \frac{1}{1+e^{-\alpha x}} \tag{1-4}$$

where $\alpha > 0$. The Gaussian transfer function is

$$f(x) = \exp(-x^2/v) \tag{1-5}$$

where $\gamma > 0$ is the predefined variance.

ANNs are used for a variety of applications because they are capable of learning, which is a basic requirement of intelligence. Knowledge is represented by the weights in the network and the weights are updated or modified during the learning process. The dynamic modification of these weights allows the network to memorize information or to learn. The networks can learn in supervised or unsupervised modes. In the case of supervised mode, during the learning phase the network is presented with the pair of input and output patterns, the actual output is compared with the desired output and the weights are adjusted so that the error between the actual output and the desired output is minimized. During the searching (i.e., the decision-making) phase, the network maps an unknown input pattern to a known output pattern. Self-organizing networks learn in an unsupervised mode. With each input pattern the weights are modified.

A number of learning algorithms are available for updating the weights. These include Kohonen learning, Competitive learning, ART (Adaptive Resource Theory), Neocognitron, and so on. Most of the learning algorithms are based on the Hebbian rule, which states that when the nodes *i* and *j* are simultaneously excited, the strength of the connection between them is increased. The extension of the same rule is that the strength of the connection between the two nodes, in proportion to the product of their simultaneous activation, is adjusted, that is

$$\Delta w_{ij} = \lambda y_i y_j \tag{1-6}$$

where λ is a constant, and y_i and y_j are activation of the two nodes, and Δw_{ij} represents the change in the weights [23].

1.3. Typical ANN Architectures

One may assume that the neural models developed in computational neuroscience at a high level could be directly implemented in silicon. This assumption is false, however, since the technology, the physical devices and the circuits severely limit the performance of integrated neural networks. As a result, a need for the development of a special ANN architecture has arisen.

The complexity of ANN does not stem from the complexity of its nodes but rather from the multitude of ways in which a large collection of these nodes can interact. Therefore neural computation is an emergent property of the network that is only vaguely evident in any single device. It is generally assumed that the performance of ANNs arises from the collective behavior of many primitives, highly interconnected processing units. ANN models also reflect highly parallel, regular and modular architectures based on matrices that make them attractive for VLSI techniques [45-47].

The basic algorithm is the matrix vector multiplication where the input vector is multiplied by the weights stored in the network, and the products are summed up and evaluated by an activation function. This leads to the simple architecture of a matrix. Learning is accomplished by modifying the variable weights. There are different algorithms for adaptation, which are competitive or application specific. The learning of correlation between the input and output vectors is generally slow in such networks [28].

A feedback of the output to the input yields a more complex behavior of the network. The behavior of such a network can be characterized by minimizing an "energy" or cost function. This energy associated with each network state is a global quantity not felt by an individual node. This concept can be illustrated graphically by a contour map of the energy function in the N-dimensional state space, where N is the number of nodes in the network. Starting at any initial state, the network state is always climbing "downhill" on an energy-terrain, coming to a local valley and stopping for certain types of networks [48-49]. This process may often be very time consuming.

One of the goals is to realize networks that are more dynamic, e.g., they should have the ability of one shot learning and of evaluating the input information with regard to match and mismatch. These features come up in multiple matrices that are interacting. Hartmann [50] presented an architecture of a closed loop antagonistic network, called *CLAN*. It makes the one shot learning by a global feedback via the network NN_F to the output network NN_0 and it detects the input information by two networks NN_1 and

NN_2 which are working in an antagonistic way so that it evaluates the match and the mismatch between the input information and the stored patterns.

An essential research area in future will be the use of ANNs in combination with fuzzy logic to implement knowledge based systems [51]. Rules can be written into a fuzzy controller. If we supplement this controller by an ANN the whole system gets the important feature of learning. In such a neuro-fuzzy system self-organizing networks can take the function of the fuzzifier and the defuzzifier which process the input and output data. Such a micro expert system may find a broad application range. A general-purpose network will be too slow and too complex in these cases so the application specific network is strongly demanded. Another way of integration of fuzzy logic with ANN is performed under the heading *fuzzy neural networks* where fuzzy sets can be used both at input, output and during learning so that the application domain of the ANN gets augmented [5,11,13] because of its capability of handling different kinds of input and of generating linguistic rules.

1.4. Systolic Array Implementation in VLSI

Research into ANNs has not only embraced modeling and algorithms, but has also dealt with corresponding architectures and implementations [24-25]. ANN applications suitable for implementation in VLSI [52-53] cover a wide spectrum, from dedicated feedforward networks for real-time feature detection to general-purpose engines for exploring learning algorithms. There have been a variety of implementation approaches for ANNs, including software and hardware, electronic and optical implementations [26-30]. The real promise for ANNs currently lies in specialized hardware, in particular VLSI implementations [31-35].

The implementation of a neural algorithm into silicon as special-purpose hardware is an important step in the realization of ANN by integrated circuits [36,54]. Each node and synapse takes a dedicated piece of silicon and can be implemented in analog or digital circuit technique.

Most of the silicon neural networks are implemented in MOS or CMOS technology. They offer the opportunity to use the analog technique within a node and the digital technique for the information exchange between the nodes similar to biological systems. In this case the errors coming from the low precision of analog circuits are limited to the nodes themselves.

Typically, the high interconnectivity and relatively low precision needed for signals in ANN are well tailored to an analog approach. With increasing ANN size, however, designing large integrated analog circuits, based on VLSI implementation, is a difficult task. On the other hand, digital processing, which appears to be inferior to analog processing in terms of computational density, has advantages of flexibility in terms of programming for a variety of architectures and learning strategies, as well as simplifying the task of memory retention. There is a strong development tendency to implement ANNs in VLSI using digital technologies [31,36].

1.4.1. *Digital Implementation*

In the digital implementations, well-developed digital techniques are exploited. Unlike analog implementations, digital implementations have concentrated on the architecture design. Basic distributed architectures are analyzed with reference to the performances they allow when adapted to support the logical connectivity requirements of ANNs. On the other hand, the specific solutions geared to support neural structures are proposed by a number of authors with the aim of optimizing speed performances at the same time as silicon compactness, ease of design and possibly flexibility in adapting to the mapping of different networks. A set of techniques can be developed for estimating chip area, performance, and power consumption in the early stage of design to facilitate architectural exploration. The capacity and performance estimation techniques for digital architectures and implementation strategies for very large networks have been discussed in [8]. This method defines convergence time, efficiency and complexity as some measures for the evaluation of digital neural hardware architectures.

Some mapping methods of an ANN to an array of custom processors have been studied [28,55,56]. The mapping could be optimized for the training phase of the back-propagation algorithm. Some typical array structures are bus-coupling, ring and mesh topologies. It shows that a much finer grain regular processor system can be realized with the mesh topology. The best match function of associative memories can be done in a bit-serial fashion with a highly pipelines processor running above 100 MHz. Another implementation of associative memory including the Hebb-like learning process has also been presented.

1.4.2. *Systolic Array (SA) Implementation*

SAs are arrays of processors that are connected to a small number of nearest neighbours in a mesh-like topology. Processors perform a sequence of operations on data that flows between them. Generally the operations will be the same in each processor, with each processor performing an operation (or small number of operations) on a data item and then passing it on to its neighbour. This makes possible to implement specialized algorithms on a VLSI chip using multiple, regularly connected processing elements (PE's) to exploit the great potential of pipelining and multi-processing.

In fact, SAs are a form of SIMD parallel computers that involve large arrays of simple processors and use vector pipelines to flow data across the array. The introduction of SAs in the late 1970s had an enormous impact on the area of special purpose computing. One of the many advantages of this approach is that each input data item can be used a number of times once it is accessed, and thus, a high computation throughput can be achieved with only modest bandwidth. Other advantages include modular expandability, simple and regular data and control flows, and use of simple and uniform cells.

Systolic arrays have been classified into (semi-) SAs with global data communications and (pure) SAs without global data communications [41]. In *semi-SAs*, a data item accessed from memory is broadcast to and used by a number of possibly nonidentical cells concurrently. Although this approach is potentially faster than SA without data broadcast, providing (or collecting) a data item to (or from) all the cells in each cycle requires the use of a global bus that may eventually slow down the processing speed as the number of cells increases. On the other hand, a *pure SA* eliminates the use of broadcast buses and implements the algorithm in pipelines extending in different directions. Several data items flowing along different pipes with the same or different rates may meet and interact. The PE's operate synchronously, that is, each data item must stay in a PE for one and only one clock cycle, and all the necessary operands to be processed by a PE in each computational step must arrive at this PE simultaneously. This mode of pipelining is referred to as *systolic processing*.

In general, SAs are somewhat inflexible to implement because they must be algorithmically specialized. Studies have been made to design reconfigurable interconnections for the pure SA approach in order to provide the flexibility needed while retaining the benefits of uniformity and locality

[57]. Another approach is based on the mapping of different algorithms onto a fixed computer architecture [58].

The functioning of a systolic array depends on:

a) the structure of the individual cells that are replicated across the array;
b) how the data is moved across the array (a function of how they are connected); and
c) how the data is fed into the array.

So far, SAs have been designed to perform a range of important but repetitive tasks, including data compression; discrete Cosine transforms; Fourier transformations; genetic algorithms; image processing and detection systems; matrix inversion; matrix multiplication; shortest-path problems; signal processing; tridiagonal banded matrix multiplication; and sorting [58-67].

1.4.3. *SA and ANN*

As mentioned before, simulations of large neural networks on a sequential computer frequently require days and weeks of computations, and the long computational time had been a critical obstacle for the progress in neural network research. Accordingly, extensive efforts are being devoted to the parallel implementation of neural networks. Systolic array (SA) is one of the best solutions to the parallel implementation of neural networks. It can overcome the communication problems generated by the highly interconnected neurons, and can exploit the massive parallelism inherent in the problem. Moreover, since the computation model of neural networks can be represented by a series of matrix by vector multiplications, the classical systolic algorithms can be used to implement them.

SA design differs from that of the conventional von Neumann computer in its highly pipelined computation. Once a data item is brought out from the memory it can be used effectively at each cell it passes while being pumped from cell to cell along the array. This is especially appealing for a wide class of compute-bound computations, where multiple operations are performed on each data item in a repetitive manner. This avoids the classic memory access bottleneck problem commonly incurred in von Neumann machines.

SA has a spatial locality and a temporal locality. It manifests a locally communicative interconnection structure, and there is at least one unit-time delay allocated so that signal transactions from one processing element to the next can be completed. When a large number of processors work together, communication becomes a significant problem. Therefore, regular and local communication in systolic array is advantageous.

ANN and SA have many similar features. The communication only between neighbor neurons in neural networks is similar to the spatial/ temporal locality of systolic arrays. The highly interconnected structures of neural networks can be mapped onto the regular structure of systolic array, and the massive parallelism can be exploited by the large number of processing elements of systolic arrays. The propagation of each neuron's state to the neighbor neurons corresponds to the pipelined computation/ communication. Since the operations required in neural network computation are simple and primitive, and the results of the operations are propagated through the network, the computation can be executed synchronously. Moreover, the fact that the input and output are presented from outside the world of neural network, and the input and output are performed only at the array boundaries is another resemblance between them.

Since SA can overcome the communication problems generated by the highly interconnected neurons, and can exploit the massive parallelism inherent in the problem, there have been several research efforts on developing systolic algorithms and systolic array structures to implement various neural networks. Attempts are made to develop -- a mapping approach, in which the systolic algorithms are partitioned and mapped onto the parallel computers, and a VLSI approach, in which VLSI systolic arrays dedicated to specific models are designed.

1.5. Book Perspective

In Chapter 2, a robust method is developed for determining and quantifying components in a composite spectrum. The proposed method is supported by a linear recurrent neural network based on the Hopfield model (HRNN). A systolic array for its implementation is proposed. In order to describe the systolic structure, the authors define a methodology based on the dependence graph. The proposed systolic structure is divided into two parts, one rectangular array for implementing the weight matrix determination, and a linear array for obtaining the threshold vector. Both structures are used for realizing the iterative process of the neural network. These structures allow us to reduce the computational cost form $O(N^2)$ to $O(N)$ so that the weight matrix is obtained, and $O(N^3)$ to $O(N)$ in order to realize the iterative process.

In Chapter 3, the authors discuss the application of systolic arrays to speed up the performance of a new algorithm that performs unsupervised analysis of remote sensing images with high dimensionality. The methodology is based on mathematical morphology, which is a non-linear image analysis and pattern recognition technique. The proposed method, which integrates spectral and spatial information in the analysis process, is fully described in this chapter, along with its hardware implementation by systolic arrays. A comparison between the hardware implementation and the software-based implementation is addressed.

Chapter 4 focuses mainly on the architecture of the ANN machine because that only very seldom systems, so far can be considered as massively parallel and, hence, exploit the huge intrinsic parallelism of ANN models. A prototype SIMD computer with 400 processing elements (PEs), expandable to 1,600 PEs, has been designed, built and tested. Then, the performance of the machine is analyzed using the Kohomen ANN model. Finally, a critical analysis is made how the dame general approach could be kept for future systems, requiring important steps toward generality and better ease-of-use.

Mixed-signal VLSI design provides a tradeoff between the fast and compact but fixed analog implementations of neuro/fuzzy feedforward algorithms and the programmable but area and power consuming digital counterparts. In Chapter 5, a sequential study of such systems is performed. Basic blocks for sequential mixed-signal neural and fuzzy computing are proposed, and two sequential example processors are described. Feedback from the designed processors and sub-circuits are allowed considering the technology constraints for analysis and extension to different s sequential degrees.

In Chapter 6, the authors firstly introduce Cerebellar Model Articulation Controller (CMAC) neural network. Secondly, an enhanced model, named the higher-order CMAC, is defined to illustrate its excellent approximation capability function. Next, they present a cost-efficient systolic array architecture to implement the higher-order CMAC with digital hardware. Finally, a case study, used to solve the color calibration problem, is developed to demonstrate how to realize higher-order CMAC with the presented systolic array architecture.

In Chapter 7, the Quadrant Interlocking Factorization (QIF) procedure is implemented on Systolic and Wavefront Array Processors. At first, the fast direct hardware implementation of the QIF procedure, for the solution of a set of linear equations on a special-purpose pipelined + shared memory architecture with multi-layered meshes, is investigated. The proposed Multi-layered DEWavefront Array Processor complex (M-DEWAP) consists of three square meshes interconnected via shared memory circuits, while pipelined techniques are adopted. The potential of a block generalized WZ factorization procedure for direct hardware implementation on a special-purpose network of Wavefront Array Processor Modules (WAPms) is then investigated. The definition of a wavefront modules is given and the development of an Eigenvalue Evaluator Engine (E^3) based on this module is discussed. Finally, a new SA is introduced for the solution of linear system equations occurring in scientific engineering calculations.

Hyperspectral image sensor developments on the study of the Earth's surface give way to images with higher spectral and spatial resolutions. In fact, the higher the resolution, the greater the size of these images. The use of these sensors by space-borne satellite systems will provide an enormous and continuous flow of data with constraints placed on onboard storage, and data transmission bandwidth. New algorithms and computer capabilities will be necessary for the classification, compression and pre-processing of these images. In Chapter 8, an SOM (self organizing map) algorithm for hyperspectral classification implemented by one systolic array is presented to provide real-time hyperspectral compression facilities.

In Chapter 9, a new optimizing method is presented to reach both goals, adequate hidden structure and learning time reduction. This approach is based on the idea of selecting the optimum hidden nodes of the network during the training process, avoiding the retraining phase of the reduced-size network. The features of the proposed method favor the use of systolic array for its implementation. Exploiting the benefits of pipelined processing, it presents a systolic architecture that implements the optimizing and learning algorithm described in this chapter.

Chapter 10 presents a system design methodology for fuzzy clustering neural networks (FCN). This methodology emphasizes coordination between FCN model definition, architectural description, and systolic implementation. Two mapping strategies both from FCN model to system architecture and from the given architecture to systolic arrays are then described. There also includes a comparison with FCM architecture with respect to hardware complexity. Some mapping strategies and typical systolic structures, based on which the FCN architecture can be systematically implemented by the corresponding SAs, have been provided.

In Chapter 11, a methodology for parallel time-delay window computing is developed by considering the features and characteristics of such neural speech recognition systems. A model for time-delay window computing and its corresponding architecture definition are described. Then, two kinds of processing stages used in pipelined architecture and their building elements are explained. Some mapping strategies from window computing model into systolic array structures are defined. Three typical speech recognition applications and their performance analysis by parallel window computing are finally given.

The aim of Chapter 12 is to provide a review of different approaches to biosignal classification using adaptive logic networks (ALNs), compare their performance on a challenging medical data set, and draw conclusions on their applicability to realistic biomedical engineering problems. Various adaptive logic network architectures were built and evaluated based on the relational approach for ALNs and focused on the development and control of an integrated intelligent classification engine for biosignal engineering. The fundamental theme of representing knowledge in hardware ANN was investigated with relation to constructing ALN based classifiers. The rules for fast and efficient design of intelligent engines based on ALN classifiers are also discussed. Finally, a new neurochip design strategy is introduced which combines ANN approaches with the machine learning objective of imitating human intelligence.

In Chapter 13, the author introduces optimal architectures for designing multiplierless multi-layer feedforward ANN suitable for various combinations of discrete/continuous input to discrete/continuous output mappings. For illustration, the details of one design for discrete input to discrete output mapping and one design for continuous input to continuous output mapping are presented. Results show that each of these multiplierless multi-layer ANN can retain nearly identical recall performance of the corresponding continuous-weight network, while having reduced hardware cost and increased computational speed in digital implementation.

Chapter 14 describes a real time intelligent decision making system, called RAPID implementable in VLSI. The architecture is unique in that the entire system can be implemented as a linear SA. The system consists of a back-propagation based ANN that can learn and adapt to a dynamically changing environment and a fuzzy expert system for decision making. It has shown that the integrated system performs better than the ANN and the fuzzy expert approaches.

Chapter 15 presents a detailed technical description of a large-scale evolvable hardware system for evolving compartmentally modeled neural circuits directly in silicon at high speed. The core of the system is a three-dimensional array of reconfigurable logic with 1.2 million fine-grained function units and 1.2 Gbyte distributed memory. An application example is developed, describing an evolution of compartmentally modeled neural networks, based on a substrate of three-dimensional cellular automata, and a simulation of a hardware-based 75-million neuron artificial brain in real time.

Last chapter discusses the need for implementing ANNs on parallel architectures, in particular the back-propagation learning algorithm since multiplayer feed-forward ANNs with back-propagation learning is one of the most widely used model. Based on the ANN model and the parallel architecture, two different methods for mapping the network model and its learning algorithm to the architecture are developed. Both methods achieve *0(PL)* time complexity for *P* number of input patterns which are asymptotic optimal for RMESH (Reconfigurable Mesh Model) architecture.

References

[1] T. Kohonen. An introduction to neural computing. *Neural Networks* 1, pages 3-16, 1988.

[2] R. Rosenblatt. *Principles of neurodynamics: perceptrons and the theory of brain mechanism*, Spartan Books, 1962.

[3] S. Grossberg. *Neural networks and natural intelligence*, MIT Press, 1988.

[4] P. G. J. Lisboa (ed.). *Neural networks: current applications*, Chapman & Hall, 1992.

[5] C. C. Klimasauskas. Neural networks: a short course. *PC AI*, pages 26-30, 1988.

[6] E. Gelenbe (ed.). *Neural networks advances and applications*, North-Holland, 1991.

[7] S. K. Pal and S. Mitra. *Neuro-Fuzzy Pattern Recognition: Methods in Soft Computing*, N.Y., John Wiley, 1999.

[8] J. B. Burr. Digital neural network implementation. *Neural Networks, Concepts, Applications, and Implementations*, Prentice- Hall, pages 237-285, 1991.

[9] S. K. Pal, A. Ghosh and M. K. Kundu (eds.). *Soft Computing for Image Processing*, Heidelberg, Physica Verlag, (Springer), 2000.

[10] H. P. Graf, L. D. Jackel and W. E. Hubbard. VLSI implementation of a neural network model. *Computer*, pages 41-49, 1988.

[11] S. K. Pal and S. Mitra. Multi-layer Perceptron, Fuzzy Sets and Classification. *IEEE Trans. Neural Networks*, 3, pages 683-697, 1992.

[12] Y. H. Pao. *Adaptive pattern recognition and neural networks*, Adison-Wesley, 1989

[13] S. K. Pal, S. Mitra and P. Mitra. Rough Fuzzy MLP : Modular Evolution, Rule Generation and Evaluation. *IEEE Trans. Knowledge and Data Engineering*, (accepted).

[14] J. Basak and S. K. Pal. X-tron : An Incremental Connectionist Model for Category Perception. IEEE Trans. *Neural Networks*, 6, pages 1091-1108, 1995.

[15] J. Basak and S. K. Pal. PsyCOP: A Psychologically Motivated Connectionist System for Object Perception. *IEEE Trans. Neural Networks*, 6, pages 1337-1354, 1995.

[16] R. P. Lippmann. Review of neural networks for speech recognition. *Neural Computing* 1, pages 1-38, 1989.

[17] A. Waibel et al.. Phoneme recognition using time-delay neural networks. *IEEE Trans. ASSP*, 37(3), pages 328-339, 1989.

[18] K. J. Lang, and A. H. Waibel. A time-delay neural networks architecture for isolated word recognition. *Neural Networks* 3, pages 23-43, 1990.

[19] L. Fausett. *Fundamentals of neural networks: architectures, algorithms, and applications*, Prentice-Hall, 1994.

[20] S. Y. Kung. *Digital neural networks*, Prentice-Hall, 1993.

[21] L. P. J. Veelenturf. *Analysis and applications of artificial neural*, Prentice-Hall, 1995.

[22] D. W. Patterson. *Artificial neural networks: theory and applications*, Prentice-Hall, 1996.

[23] D. E. Rumelhart and J. L. McCelland. *Parallel distributed processing I*, MIT Press, 1986.

[24] P. Antognetti and V. Milutinovic (eds.). *Neural networks: concepts, applications, and implementations*, Prentice Hall, 1991.

[25] E. Sanchez-Sinencio and C. Lau (eds.). *Artificial neural networks: paradigms, applications, and hardware implementations*. IEEE Press, 1992.

[26] R. Eckmiller and C. V. D. Malsburg. *Neural computers*, Springer Verlag, 1988.

[27] J. M. J. Murre. Transputers and neural networks: an analysis of implementation constraints and performance. *IEEE Trans. Neural Networks* 4(2), pages 284-292, 1993.

[28] J. Ghosh and K. Hwang. Mapping neural networks onto message-passing multicomputers. *Journal of Parallel and Distributed Computing*, April, 1989.

[29] R. H. Nielsen. Performance limits of optical, electro-optical, and electronic neurocomputers. *Proc. of the SPIE*, 634, pages 277-306, 1986.

[30] B. W. Lee and B. J. Sheu. *Hardware annealing in electronic neural networks*, Kluwer, 1991.

[31] B. A. White and M. I. Elmasry. The digi-neucognitron: a digital neocognitron neural network model for VLSI. *IEEE Trans. Neural Networks*, 3(1), pages 73-84, 1992.

[32] S. Satyanarayana, et al.. A reconfigurable VLSI neural network. *IEEE Journal of Solid-State Circuits*, 27, pages 67-81, 1992.

[33] Y. He, U. Cilingiroglu and E. S. Sinencio. A high-density and low-power charge-based hamming network.. *IEEE Trans. VLSI Systems* 1(1), pages 56-62, 1993.

[34] C. Mead and M. Ismail (eds.). *Analog VLSI implementation of neural systems*, Kluwer, 1989.

[35] J. Choi and B. J. Cheu. VLSI design of compact and high-precision analog neural network processors. *IJCNN'92 II*, pages 637-641, 1992.

[36] U. Ramacher and U. Rucker (eds.), *VLSI design of neural networks*, Kluwer, 1991.

[37] J. J. Navarro, J. M. Llaberia, and M. Valero. Partitioning: An essential step in mapping algorithm into systolic array processors. *IEEE Computer*, pages 77-89, 1987.

[38] D. Zhang and M. I. Elmasry. Mapping neural networks onto systolic arrays. *The Journal: Neural, Parallel & Scientific Computations* 4(3), pages 341-352, 1996.

[39] S. Y. Kung. Parallel architecture for artificial neural nets. *Proc. of Intl' Conf. on Systolic Arrays*, pages 163-174, 1988.

[40] S. Y. Kung. VLSI array processors. *IEEE ASSP Magazine* 2(3), pages 4-22, 1985.

[41] H. T. Kung. Why systolic architecture?. *IEEE Trans. on Computer*, pages 37-46, 1983.

[42] D. I. Moldovan. On the design of algorithms for VLSI systolic array. *Proc. IEEE* 71(1), pages 113-120, 1983.

[43] D. Zhang. *Parallel VLSI Neural System Designs*, Springer-Verlag, Singapore, 1999.

[44] A. Prieto (ed.). Artificial neural networks. *IWANN' 91*, Springer-Verlag, 1991.

[45] J. A. Vlontzos and S. Y. Kung. Digital neural network architecture and implementation. *VLSI Design of Neural Networks*. pages 205-227, June, 1990.

[46] R. Mason, W. Robertson, and D. Pincock. An hierarchical VLSI neural network architecture. *IEEE Journal Solid-State Circuits* 27(1), pages 106-108, 1992.

[47] R. L. Harvey, P. N. DiCaprio and K. G. Heinemann. A neural network architecture for general image recognition. *Lincoln Laboratory Journal* 4(2), pages 189-207, 1991.

[48] J. J. Hopfield and D. W. Tank. Neural computation of decisions in optimization problems. *Biological Cybernetics* 52, pages 141-152, 1985.

[49] R. P. Lippmann. An introduction to computing with neural nets. *IEEE ASSP Magazin* 4, pages 4-22, 1987.

[50] G. Hartmann. The closed loop antagonistic network (CLAN). *Advanced neural computers*, R. Eckmiller (ed), North-Holland, Amsterdam, pages 279-285, 1990.

[51] W. Eppler. Implementation of fuzzy production systems with neural networks. *Parallel Processing in Neural Systems and Computers*, R. Eckmiller, et al. (eds), North-Holland, Amsterdam, pages 247-252, 1990.

[52] C. Mead and L. Conway. *Introduction to VLSI systems, Addison-Wesly,* 1980.

[53] D. A. Pucknell. *Basic VLSI design*, Kamran Eshraghian, 1994.

[54] M. Sami and J. C. Daguerre (eds.). *Silicon architectures for neural nets*, North-Holland, 1990.

[55] K. Funahashi. On the approximate realization of continuous mappings by neural networks. *Neural Networks* 2, pages 183-192, 1989.

[56] H. White. Connectionist nonparametric regression: multilayer feedforward networks can learn arbitrary mappings. *Neural Networks* 3, pages 535-550, 1990.

[57] L. Snyder. Introduction to the configurable, highly parallel computer. *IEEE Computer* 15, pages 47-56, 1982.

[58] D. Zhang and S. K. Pal. A fuzzy clustering neural networks (FCNs) system design methodology. *IEEE Trans. on Neural Networks* 11(5), pages 1174 – 1177, 2000.

[59] H. Amin, et al.. Two-ring systolic array network for artificial neural networks. *IEE Proceedings- Circuits, Devices and Systems* 146(5), pages 225 –230, 1999.

[60] W. Eppler, et al.. Neural chip SAND/1 for real time pattern recognition. *IEEE Trans. on Nuclear Science* 45(4), pages 1819 –1823, 1998.

[61] H. Amin, et al.. Efficient two-dimensional systolic array architecture for multilayer neural network. *Electronics Letters* 33, 24, 2055 –2056, 1997.

[62] J. S. Ker. et al.. Systolic implementation of higher-order CMAC and its application in colour calibration. *IEE Proceedings- Circuits, Devices and Systems* 144(3), pages 129 –137, 1997.

[63] J. Dehaene, M. Moonen and J. Vandewalle. Continuous-time recursive least-squares estimation, adaptive neural networks and systolic arrays. *IEEE Trans. on Circuits and Systems I: Fundamental Theory and Applications* 42(2), pages 116 –119, 1995.

[64] D. Naylor, S. Jones and D. Myers. Backpropagation in linear arrays-a performance analysis and optimization. *IEEE Trans. on Neural Networks* 6(3), pages 583 –595, 1995.

[65] S. R. Jones, K. M. Sammut and J. Hunter. Learning in linear systolic neural network engines: analysis and implementation. *IEEE Trans. on Neural Networks* 5(4), pages 584 –593, 1994.

[66] D. Naylor and S. Jones. A performance model for multilayer neural networks in linear arrays. *IEEE Trans. on Parallel and Distributed Systems* 5(12), pages 1322 –1328, 1994.

[67] J. M. Moreno, et al.. An analog systolic neural processing architecture. *IEEE Micro* 14(3), pages 51 –59, 1994.

CHAPTER 2

SYSTOLIC ARRAY METHODOLOGY FOR A NEURAL MODEL TO SOLVE THE MIXTURE PROBLEM

R.M. Pérez, P. Martínez, A. Plaza and P.L. Aguilar

Departamento de Informática, Universidad de Extremadura
Campus Universitario s/n 10071 Cáceres, Spain
Email : rosapere,pablomar,aplaza,paguilar@unex.es

In this chapter we present a robust method for determining and quantifying components in a composite spectrum; we assume that the patterns of the individual spectra are known in advance. The proposed method is supported by a linear recurrent neural network based on the Hopfield model (HRNN). The HRNN has very important characteristics in the implementation of low- complexity VLSI structures, since only multiplying and adding operations are required for an inversion process. We propose a systolic array for its implementation. In order to describe the systolic structure, we use a methodology based on the dependence graph. We suggest a sequential algorithm that verifies the unique assignment rule and the locality of the data dependences. The proposed systolic structure is divided into two parts, one rectangular array for implementing the weight matrix determination, and a linear array for obtaining the Threshold vector. Both structures are used for realising the iterative process of the Neural Network. These structures allow us to reduce the computational cost from o(N2) to o(N) so that the weight matrix is obtained, and o(N3) to o(N) in order to realise the iterative process.

2.1. Introduction

In most signal processing applications such as Remote Sensing, Environmental Control, and so forth, it is necessary to determine and quantify the components in a composite spectrum, which is obtained from a given mixture of elements. In hyperspectral analysis, this issue is known as the determination of end-members and abundances, while in general spectroscopy, it is called the *Mixture Problem*.

Formally, this development may be considered as follows: Under the assumption that we know the spectra of K elements (*Basic References*), we must determine the unknown composition of a *cocktail* of the mentioned elements by using a radiation spectrum obtained from this mixture.

The main goal of this chapter is to explore the possibility of applying a Neural Networks Methodology to achieve a reliable, robust and efficient solution to this problem, based on the inherent parallelism of neural networks.

Adams et al. [1] studied this problem in the context of Surface Mineralogy Prospection, and Lawton [2] proposed conventional digital algorithms for its solution. These approaches are fairly slow because serial computation is implied.

A method based on Optical Neural Networks was presented by Barnard and Cassasent [3]. The main drawback in this approach is the lack of uniqueness for the proposed solution.

The possibility of using the Multiple Regression Theory, thus enabling an optimal solution in terms of uniqueness, has been developed by Díaz et al. [4]. This approach is based on the use of the Pseudo-Inverse Matrix, supported by a Linear Associative Memory, built by implementing Pyle's algorithm [5].

The general solution method proposed here is based on the Hopfield Recurrent Neural Network (HRNN). It is a flexible, efficient and robust approach aimed to solve the problem. The Gradient Method for error reduction is applied to ensure the convergence of the algorithm. The use of this model is fully justified, since the spectrum formation in the *Mixture Problem* is essentially a linear process [6].

One of the most remarkable properties deduced from the algorithmic structure of this method, is related to the possibility of implementing low-complexity VLSI structures. These are suitable to support the algorithm, since multiplications and additions are only required for programming an inversion process [7]. These operations can be implemented through basic processor arrays. This approach matches the VLSI system design principle, according to which, modular, pipelined, and parallel architectures are exploited, and the communication hardware is reduced to local interconnections in most cases [8].

A typical example of VLSI parallel/pipelined architectures is observed in systolic arrays. These introduce many advantages; for instance, the exploitation of pipelining is very natural in networks set up regularly and locally. This saves the cost associated with communication. A second benefit is the good balance between computation and communication, which is critical to the effectiveness of array computing [8].

In the following sections, a method developed from the HRNN-based (Hopfield model) linear recurrent neural network is explained. Such a development aims to solve the mixture problem and its implementation by means of systolic arrays.

2.2. Algorithmic Method

In order to describe the algorithmic method proposed, we must previously consider that a Composite Spectrum $\boldsymbol{y}$ can be seen as an *N-dimensional* vector. This vector is built by sorting the emission intensities associated with each energy channel vs. the channel number, where N is the total number of energy channels:

$$y = [y_1, y_2, \ldots, y_N]^T \qquad y_n \geq 0 \qquad 1 \leq n \leq N \tag{2-1}$$

Here, y_n is the intensity measured as the number of photons whose energy is comprised in the *n-th* energy channel interval.

As a result, a *Reference Spectrum* is a spectrum of the same nature, but produced by an *Individual Source*. We denote these spectra as $\boldsymbol{r_k}$, with $1 \leq k \leq K$. The set of K reference vectors is named the *Reference Set*, and it must be evaluated in advance. For the sake of compactness, the *Reference Set* is referred to as the *Reference Matrix* $\boldsymbol{R}$ composed by the reference column vectors:

$$R = [r_1, r_2, \ldots, r_K] \tag{2-2}$$

In a general sense, the set of Composite Spectra is the range of all possible spectra that may be produced by a linear combination of all elements belonging to the *Reference Set*. When the *Reference Set* is composed by linearly independent K vectors, the result is a *K-dimensional Vector Space*, integrated by all the vectors $\boldsymbol{y}$, and explained as:

$$y = Rc = \sum_{i=1}^{K} c_i r_i \tag{2-3}$$

where $\boldsymbol{c}$ is the *Contribution Vector*, which is defined as:

$$c = [c_1, c_2, \dots, c_K]^T \qquad c_k \geq 0 \qquad 1 \leq k \leq K \tag{2-4}$$

and where every contribution $\boldsymbol{c_i}$ is a function of the relative intensities of the Composite and Reference Spectra.

Our goal is to estimate $\boldsymbol{c}$, as we assume that $\boldsymbol{R}$ and $\boldsymbol{y}$ are known. For the mixture described by an estimation of $\boldsymbol{c}$, called $\boldsymbol{c'}$, the difference between the measured spectral vector $\boldsymbol{y}$ and its re-constructed version $\boldsymbol{y'}$:

$$\varepsilon = y - y' = y - \sum_{i=1}^{K} c'_i \, r_i \tag{2-5}$$

is called *Estimation Error,* which gives us a measure of how well the estimation of $\boldsymbol{c}$ has been accomplished. This error is exploited to optimise the estimation process by means of a *Least Mean Square (LMS)* minimization procedure. In relation to $\boldsymbol{c}$, this is laid out to minimise the *Measure Function F(c),* being defined as:

$$F(\boldsymbol{c}) = \|\varepsilon\|^2 = \|y - \boldsymbol{Rc'}\|^2 \tag{2-6}$$

To solve this problem, we apply an iterative process, supported by the *Linear Hopfield Minimization Procedure*, that is basically a progressive refinement of the *Contribution Vector*:

$$c'(t+1) = c'(t) + \Delta c'(t) = c'(t) + \lambda R^T [y - Rc'(t)] \tag{2-7}$$

where we have used an estimation of the *Gradient of F(c)* in relation to $\boldsymbol{c}$, exposed as:

$$\nabla_c F(\boldsymbol{c}) = -2\boldsymbol{R}^T [y - \boldsymbol{Rc}] \tag{2-8}$$

In compact notation, we can Eq. (2-7) as:

$$\boldsymbol{c(t+1)} = \lambda \boldsymbol{q} + \boldsymbol{Pc(t)} \qquad \text{with} \qquad q = R^T y \qquad P = I - \lambda R^T R$$

$$\text{where } 1 \le i, j \le K \qquad p_{ij} = -\lambda \sum_{p=1}^{N} r_{pi} r_{pj} \qquad i \ne j$$

$$p_{ii} = 1 - \lambda \sum_{p=1}^{N} r_{pi} r_{pi} \; q_i = \sum_{p=1}^{N} r_{pi} y_p \qquad (2\text{-}9)$$

in which λ is a parameter dependent on the trace of $\mathbf{R^T R}$. This controls the speed of convergence, whereas $\boldsymbol{p_{ij}}$ denotes the weight from the *i-th* node to the *j-th* node. This method simply requires multiplying and adding operations to solve the *Mixture Problem*. The HRNN structure proposed according to this algorithm can be seen in Fig. 2-1 and Fig. 2-2.

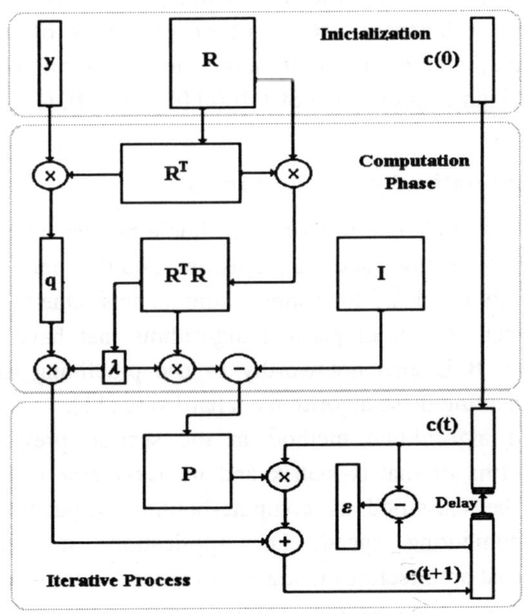

Fig. 2-1. HRNN Algorithm.

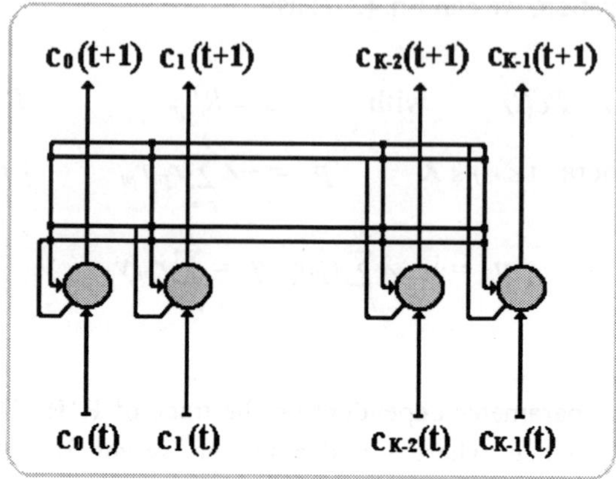

Fig. 2-2. HRNN Iterative Process.

The HRNN presents significant features, such as a reasonable computational cost, robustness versus increasing levels of noise in the composite spectrum, and an inherent ability to prevail in quite uneven mixtures, as in the high disproportion of 1:1000 [6], [7], [9], [10].

2.3. VLSI Implementation by Systolic Arrays

VLSI technology has made one thing clear: Simple regular interconnections lead to cheap implementations and high densities, and this implies both high performance and low overhead for support components. Due to these traits, an interesting design is that of parallel algorithms that have simple and regular data flows. It is also noteworthy to use pipelining as a general method for implementing these algorithms in hardware [11].

The proposed algorithmic method in the section previous can be implemented in terms of matrix-matrix and matrix-vector multiplications. These tasks can be classified as compute-bound computations, thereby requiring high computing speed. For applications involving matrix calculations, low-cost architectures that are specialized in terms of algorithm and high-performance, such as the systolic array processors, have been conceived [12].

In the systolic background, VLSI devices are formed by arrays of interconnected processing cells having a high degree of modularity. Each processor operates on a string of data that flows regularly over the network [13].

In order to derive the systolic arrays for the proposed algorithm, we follow a systematic mapping methodology based on dependence graphs. The mapping procedure is summarised into key tasks [13]: To pipeline all the variables in the algorithm; to find the set of data dependence vectors; to identify a valid transformation for the data dependence vectors and the index set; and to map the algorithm onto the hardware.

In the following parts of this section, previous concepts are applied to obtain systolic array structures for the HRNN. We consider three phases: First, the determination of the rectangular systolic array for the weight matrix configuration, to which the reference spectra set ***R*** data is applied; secondly, a linear systolic array is set to achieve the threshold of the neural network, for which the transpose of the matrix ***R*** and the composition vector ***y*** are developed; finally, the structure for the iterative process is designed by exploiting the systolic system described in the previous phases.

2.3.1. *Weight Matrix Determination*

The HRNN displays a hebbian learning, since the reference set is developed in order to come up with the weight matrix. Proceeding from (9), the following sequential algorithm is devised to compute the values associated with the net connections:

```
For i=1 to K do
        For j=1 to K do
                P_ij=0;    P_ii = 1
                For m=1 to N do
                        P_ij=P_ij+(-λ)RT_jm R_mj
                Endfor
        Endfor
Endfor
```

where RT is the matrix whose rows have the reference spectra, i.e., $RT_{ij} = R_{ji}$.

Table 2-1 provides the set index for this algorithm. Each index element is shown as a three-fold tuple (i,j,m). Note that for both index elements (i,j,1) and (i,j,2), the same value of P(i,j) is used; in other words, the value P(i,j) can be piped in the m direction. Similarly, the values RT(i,m) and R(m,j) can be piped in the j and i directions respectively.

Table 2-1. Index set of the weight matrix algorithm.

i,j,m	P(i,j)=	P(i,j)	RT(i,m)	*R(m,j)
1,1,1	1,1	1,1	1,1	1,1
1,1,2	1,1	1,1	1,2	2,1
...	...	...	...	...
1,1,N	1,1	1,1	1,N	N,1
1,2,1	1,2	1,2	1,1	1,2
...	...	...	...	...
1,2,N	1,2	1,2	1,N	N,2
...	...	...	...	...
1,K,1	1,K	1,K	1,1	1,K
...	...	...	...	...
1,K,N	1,K	1,K	1,N	N,K
2,1,1	2,1	2,1	2,1	2,1
...	...	...	...	...
2,K,N	2,K	2,K	2,N	N,K
...	...	...	...	...
K,1,1	K,1	K,1	K,1	N,1
...	...	...	...	...
K,K,N	K,K	K,K	K,N	N,K

The data set of dependence vectors can be identified by equating indexes of all the possible pairs of generated and exploited variables

$d_P=(001)^T$ $d_{RT}=(010)^T$ $d_R=(100)^T$

The dependence matrix can be expressed as:

$$D=\begin{pmatrix}0&0&1\\0&1&0\\1&0&0\end{pmatrix}$$

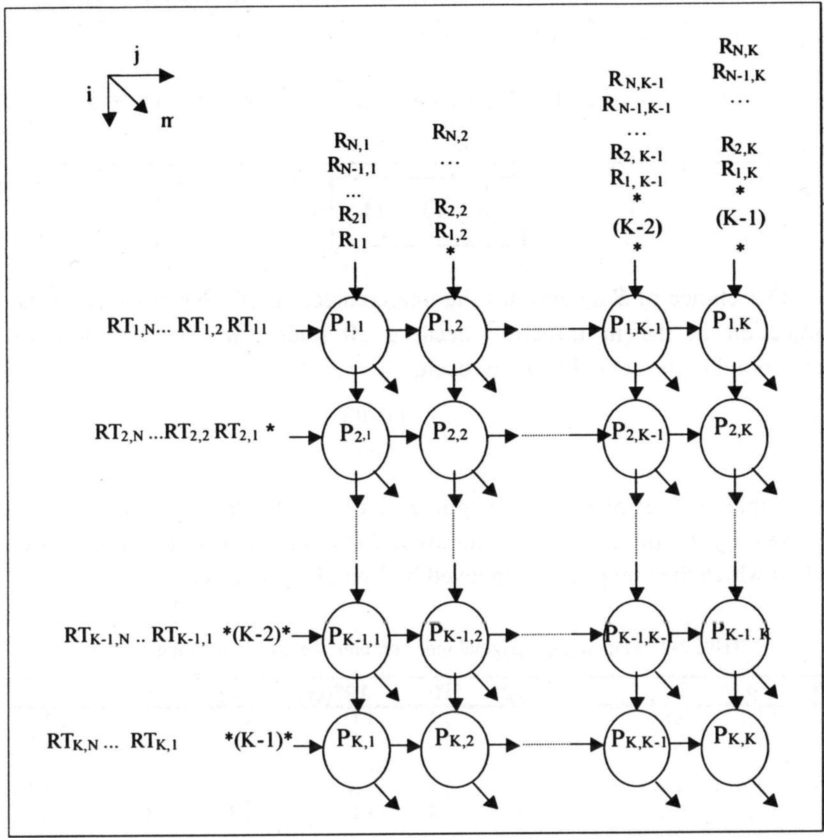

Fig. 2-3. Dependence graph for the obtaining the weight matrix **P**.

Fig. 2-3 shows the dependence graph for $\boldsymbol{R^T R}$. This graph is mapped onto the m direction, for which a transformation $\boldsymbol{T}$ is sought; this change must be a bijection and monotonic increase of the form:

$$T = \begin{pmatrix} t_{11} & t_{12} & t_{13} \\ t_{21} & t_{22} & t_{23} \\ t_{31} & t_{32} & t_{33} \end{pmatrix} = \begin{pmatrix} \pi \\ S \end{pmatrix}$$

The nodes on an equi-temporal hyperplane cannot be projected to the same PE [12]. This must be done as a valid order, and as a reduction of the turnaround. For such aims, we chose π as:

$$\pi * d_p = (t_{11}\, t_{12}\, t_{13}) * (0\ 0\ 1)^T > 0 \rightarrow t_{13} > 0 \rightarrow t_{13} = 1$$

$$\pi * d_{RT} = (t_{11}\, t_{12}\, t_{13}) * (0\ 1\ 0)^T > 0 \rightarrow t_{12} > 0 \rightarrow t_{12} = 1$$

$$\pi = (1\ 1\ 1)$$

Our choice of ***S*** determines the interconnection of the processors. It is a projection in the m direction because all nodes in this direction will correspond to the same PE. As a result,

$$S = \begin{pmatrix} 010 \\ 100 \end{pmatrix}$$

Table 2-2 displays the mapping between original and transformed indexes (by ***T***). In this Table, columns ***πxI*** and ***SxI*** indicate the time and the cell in which the computation indexed by I will be processed.

Table 2-2. Mapping the original index set onto the transformed index set.

T	i,j,p	T(I)	P(i,j)=	P(i,j)	-λRT(i,p)	*R(p,j)	πI	SI
T	1,1,1	3,1,1	1,1	1,1	1,1	1,1	3	1,1
T	1,1,2	4,1,1	1,1	1,1	1,2	2,1	4	1,1
...	...	...	...	...	...	...	...	...
T	1,1,N	N+1,1,1	1,1	1,1	1,N	N,1	N+2	1,1
T	1,2,1	4,2,1	1,2	1,2	1,1	1,2	4	2,1
...	...	...	...	...	...	...	...	...
T	1,2,N	N+3,2,1	1,2	1,2	1,N	N,2	N+3	2,1
...	...	...	...	...	...	...	...	...
T	1,K,1	K+2,K,1	1,K	1,K	1,1	1,K	K+2	K,1
...	...	...	...	...	...	...	...	...
T	1,K,N	N+K+1,K,1	1,K	1,K	1,N	N,K	N+K+1	K,1
T	2,1,1	4,1,2	2,1	2,1	2,1	2,1	4	1,2
...	...	...	...	...	...	...	...	...
T	2,K,N	N+K+2,K,2	2,K	2,K	2,N	N,K	N+K+1	K,2
...	...	...	...	...	...	...	...	...
T	K,1,1	K+2,1,K	K,1	K,1	K,1	N,1	K+2	1,K
...	...	...	...	...	...	...	...	...
T	K,K,N	N+2K,K,K	K,K	K,K	K,N	N,K	2K+N	K,K

Therefore, to obtain the weight matrix, KxK PEs are needed, while required computation time is 2K+N. The interconnection between those processors is defined as:

$$Sd_i = \begin{bmatrix} x \\ y \end{bmatrix}$$

where x and y refer to the movement of the variable along the directions j and i respectively. For our case,

$$S \cdot \mathrm{d}_P = \begin{pmatrix} 0 & 1 & 0 \\ 1 & 0 & 0 \end{pmatrix} \bullet \begin{pmatrix} 0 \\ 0 \\ 1 \end{pmatrix} = \begin{pmatrix} 0 \\ 0 \end{pmatrix} \qquad S \cdot \mathrm{d}_{RT} = \begin{pmatrix} 0 & 1 & 0 \\ 1 & 0 & 0 \end{pmatrix} \bullet \begin{pmatrix} 0 \\ 1 \\ 0 \end{pmatrix} = \begin{pmatrix} 1 \\ 0 \end{pmatrix}$$

$$S \cdot \mathrm{d}_R = \begin{pmatrix} 0 & 1 & 0 \\ 1 & 0 & 0 \end{pmatrix} \bullet \begin{pmatrix} 1 \\ 0 \\ 0 \end{pmatrix} = \begin{pmatrix} 0 \\ 1 \end{pmatrix}$$

thus mean that variables P do not travel in any direction, and are updated in terms of time; and that RT and R move with a speed of one grid per unit time along the direction *j* and *i* respectively.

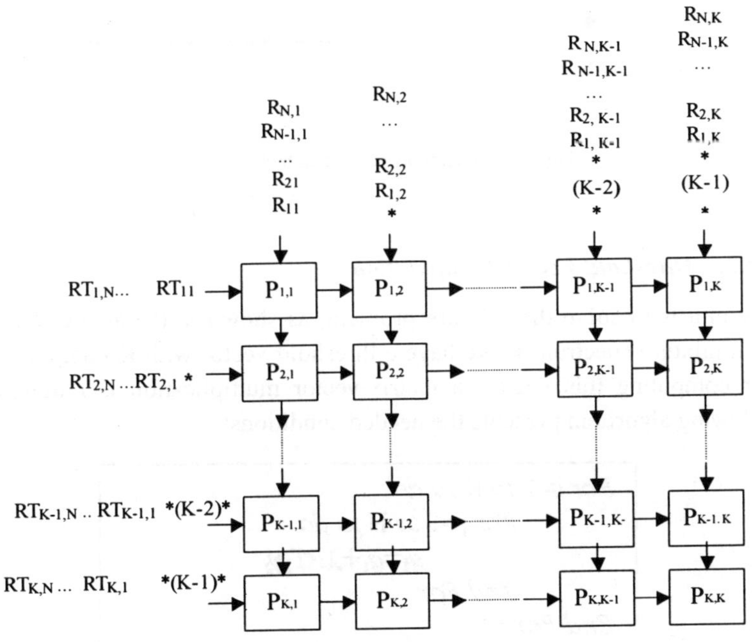

Fig. 2-4. Systolic array for the weight matrix **P**.

Fig. 2-4 reproduces interconnections between PEs. As can be observed, this architecture has a rectangular array design. The elements of matrix ***RT*** are fed from the left side of the array, while matrix R is pipelined into the array from the upper edge.

At time 3 (first clock cycle of computing) $R_{11}$and RT_{11} enter PE_{11}, which contains the variable P_{11} -- at previous times all P_{ij} are initialised as $P_{ii}=1$; $P_{ij}=0$ and the value of the learning parameter λ are stored in each PE. These perform multiplying and adding operations, and they accumulate the partial results, as is shown in Fig. 2-5. The process continues until all the values are computed.

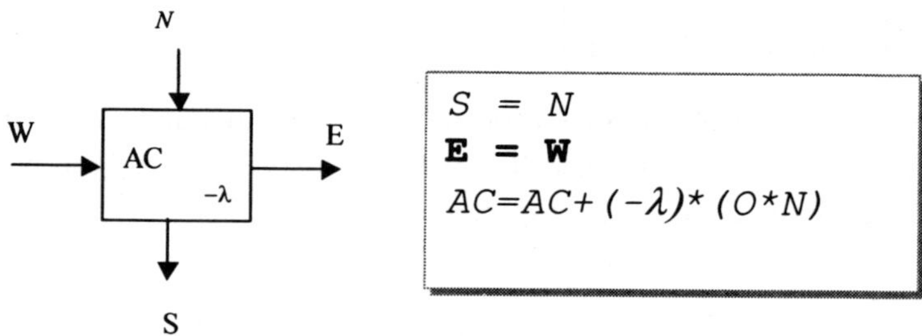

Fig. 2-5. Operations performed by Fig. 2-4 PEs.

2.3.2. *Threshold Vector Computation*

The aim is to solve the mixture problem, as shown in (9) above. Thus, for each mixture spectrum ***y***, we have a threshold vector with K components, ***q***. For computing this vector, a matrix-vector multiplication is required; the following algorithm presents the needed conditions:

```
For i=1 to K to do
        For j=1 to N to do
                q_i=q_i+λRT_ij y_j
        End For
End For
```

where λ is the convergence parameter.

Table 2-3 supplies the set index for this algorithm. In this Table, each index element is shown as a two tuple (i,j). Based on this, data dependence can be deduced in a similar manner to the previous case.

Table 2-3. Index set of threshold vector.

i,j	q(i)=	q(i)	+RT(i,j)	*y(j)
1,1	1	1	1,1	1
1,1	1	1	1,2	2
...	...	...	...	...
1,N	1	1	1,N	N
2,1	2	2	2,1	1
...	...	...	...	..
2,	2	2	2,N	N
...	...	...	...	...
K,1	K	K	K,1	1
...	...	...	...	...
K,N	K	K	K,N	N

Thus, the data dependence vectors are:

q(i,j)=q(i,j-1)	$D_q=(01)^T$
y(i,j)=y(i-1,j)	$D_y=(10)^T$

So, the matrix dependence is:

$$D=\begin{pmatrix}0 & 1\\ 1 & 0\end{pmatrix}$$

If the transformation T is selected as:

$$T=\begin{pmatrix}t_{11} & t_{12}\\ t_{21} & t_{22}\end{pmatrix}=\begin{pmatrix}\pi\\ S\end{pmatrix} \qquad T=\begin{pmatrix}1 & 1\\ 1 & 0\end{pmatrix}$$

Then:

$$S\cdot d_q=(1\ \ 0)\bullet\begin{pmatrix}0\\ 1\end{pmatrix}=(0) \qquad S\cdot d_y=(1\ \ 0)\bullet\begin{pmatrix}1\\ 0\end{pmatrix}=(1)$$

It means that variable ***q*** does not travel in any direction and is updated in time; and that variable ***y*** moves to the right with a speed of one per time unit. Table 2-4 presents the mapping between original and transformed indexes.

Table 2-4. Mapping the original index set onto the transformed index set.

I=(i,j)	T(I)	q(i)=	q(i)	+RT(i,j)	*y(j)	πI	SI
1,1	2,1	1	1	1,1	1	2	1
1,2	3,1	1	1	1,2	2	3	1
...	...	...	...	...	...	...	...
1,N	N+1,1	1	1	1,N	N	N+1	1
2,1	3,2	2	2	2,1	1	3	2
...	...	...	...	...	..	...	...
2,N	N+2,2	2	2	2,N	N	N+2	2
...	....	...	...	...	...	...	...
K,1	K+1,K	K	K	K,1	1	K+1	K
...	...	...	...	...	...	...	...
K,N	K+N,K	K	K	K,N	N	K+N	K

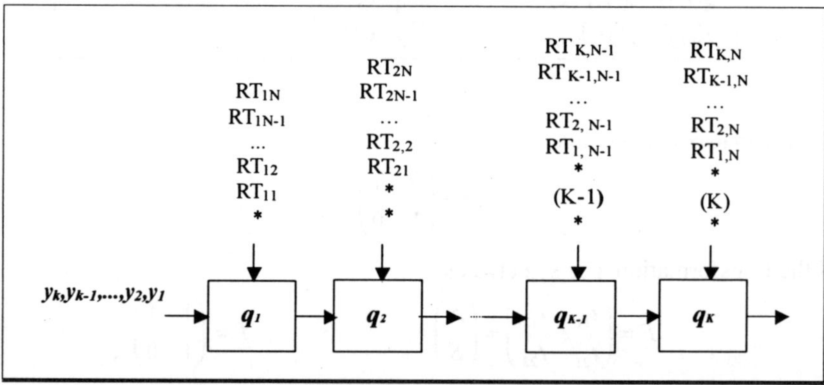

Fig. 2-6. Required interconnections among PE's for the systolic array in the computation of the threshold vector q.

Fig. 2-6 reproduces the interconnections between the PEs. As can be examined, this architecture has a linear array design. The elements of vector y are fed from the left side of the array and transmitted among neighbouring

PEs. The arithmetic processing capabilities of each PE should include multiplication and accumulation operations (see Fig. 2-7).

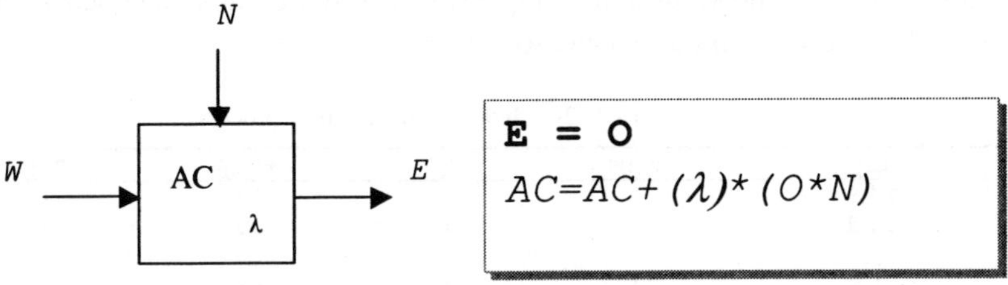

Fig. 2-7. Operations performed by PEs in Fig. 2-6.

2.3.3. *Contribution Achievement*

The previous sub-sections have shown how the weight matrix and threshold vector can be computed by means of systolic structures. Next, we should be able to solve the mixture problem with the iterative process (Fig. 2-2) given for (9), and expressed as:

$$c(t+1) = \left(\sum_{j=1}^{N} p_{ij} c_j(t) + q_i \right) \tag{2-10}$$

where ***c(t+1)*** and ***c(t)*** represent two consecutive network states. Therefore, when the neural model converges, the contribution of components in a mixture spectrum is obtained and each c_i determines the proportion of the correspondent reference spectrum $\boldsymbol{r}_i$ in the mixture. The algorithm describing the process given in (10) can be expressed as:

```
For t=1 to t_max do
        For i=1 to K do
                c'_i(t)=q_i
                For j=1 to K do
                        c'_i(t)=c'_i(t)+P_ij c_j(t)
                Endfor
                c_i(t+1)=c'_i(t)
        Endfor
Endfor
```

where **c'** represents the net state at the ***(t+1)*** time

Table 2-5 represents the set index for this algorithm. In this Table, each index element is shown as a three tuple (t,i,j). Based on this, data dependence can be located in a similar manner to the previous case.

Table 2-5. Index for the iterative process.

t,i,j	c'(i)=	c'(i)	+P(i,j)	*c(j)
1,1,1	1	1	1,1	1
1,1,2	1	1	1,2	2
...	...	...	...	...
1,1,K	1	1	1,K	K
...	...	...	...	...
1,K,1	K	K	K,1	1
...	...	...	...	...
1,K,K	K	K	K,K	K
2,1,1	1	1	1,1	1
2,1,2	1	1	1,2	2
...	...	...	...	..
2,K,1	K	K	K,1	1
...	...	...	...	...
2,K,K	K	K	K,K	K
t_{max},1,1	1	1	1,1	1
t_{max},1,2	1	1	1,2	2
...		...	...	...
t_{max},1,K	K	K	1,K	K
t_{max},2,1	2	2	2,1	1
...	...	...	...	..
t_{max},2,K	2	2	2,K	K
...	...	...	...	...
t_{max},K,1	K	K	K,1	1
...	...	...	...	...
t_{max},K,K	K	K	K,K	K

According to Table 2-5, data dependence can be obtained as:

In order to account for the previous derivations, the π and ***S*** values are chosen as:

$c'(t,i,j)=c'(t,i,j-1)$	$d_{c'}=(0\ 0\ 1)^T$
$c(t,i,j)=c(t,i-1,j)$	$d_c=(0\ 1\ 0)^T$
$P(t,i,j)=P(t-1,i,j)$	$d_P=(1\ 0\ 0)^T$

$$\pi = (111); \qquad S = \begin{pmatrix} 010 \\ 001 \end{pmatrix} \qquad \Rightarrow T = \begin{pmatrix} 111 \\ 010 \\ 001 \end{pmatrix}$$

Through this selection, we can apply the designed rectangular array, which calculated and stored one weight in each PE. According to the selected matrix $\boldsymbol{T}$, the variable: $\boldsymbol{P}$ does not travel in any direction, $\boldsymbol{c}$ and $\boldsymbol{c'}$ move with a speed of one grid per unit time downwards and rightwards respectively. Note that the first vaule of c_i' is q_i, thus implying that the results of the designed linear array must be used, inputted and stored into the first column of PEs.

Table 2-6 yields the mapping between original and transformed indexes. The Table demonstrates that the involved cycle number to solve the problem is t_{max} +2K.

Fig. 2-8 captures the interconnections between the PEs. Extra links between the last column (PE_{iK})and the first row (PE_{Ki}) have been added for realising the iterative process at the time t . This graph includes a multiplexer to select among threshold vector q_i (at the inicialisation cycle), 0 (at the resting cycles), and a convergence control circuit.

The array output is obtained sequentially. To feed the upper input, a demultiplexer must be used, and a module K counter is also needed as the control of this circuit. The diagram for this is shown in Fig. 2-9.

The implementation of the convergence control can be seen in Fig. 2-10. This control is formed by two adders and one comparator. Its output is related to the network stability condition.

Finally, the arithmetic processing capabilities of each PE should include multiplying and accumulating operations (see Fig. 2-11).

2.3.4. *Design Summary*

It seems a good idea, at this stage, to sum up the proposed systolic methodology: The HRNN model can be mapped onto a systolic structure formed by a rectangular array of KxK PEs and a linear array of K PEs. The PEs should be programmable due to the varying functionalities involved. For their description, we have distinguished data movements and PE design requirements in both cases.

Table 2-6. Mapping between original and transformed indexed.

I=(t,i,j)	T(I)	c'(i)=	C'(i)	+p(i,j)	*c(j)	πI	SI
1,1,1	3,1,1	1	1	1,1	1	3	1,1
1,1,2	4,1,2	1	1	1,2	2	4	1,2
...	...	...	...	...	...	...	...
1,1,K	K+2,K,N	1	1	1,K	K	K+2	K,N
...	...	...	...	...	...	...	...
1,K,1	K*2,K,1	K	K	K,1	1	K+2	K,1
...	...	...	...	...	...	...	...
1,K,K	2K+1,K,K	K	K	K,K	K	2K+1	K,K
2,1,1	4,1,1	1	1	1,1	1	4	1,1
2,1,2	5,1,2	1	1	1,2	2	5	1,2
...	...	...	...	...	..	...	...
2,K,1	K+3,K,1	K	K	K,1	1	K+3	K,1
...	...	...	...	...	...	...	...
2,K,K	2K+2,K,K	K	K	K,K	K	2K+2	K,K
...	...	...	...	...	...	...	...
t_{max},1,1	t_{max} +2,1,1	1	1	1,1	1	t_{max} +2	1,1
t_{max},1,2	t_{max} +31,2	1	1	1,2	2	t_{max} +3	1,2
...	...		...	...	...	...	...
t_{max},1,K	t_{max} +K+11,K	K	K	1,K	K	t_{max} +K+1	1,K
t_{max},2,1	t_{max} +3,2,1	2	2	2,1	1	t_{max} +3	2,1
...	...	...	...	...	..	...	...
t_{max},2,K	t_{max} +K+2,2,K	2	2	2,K	K	t_{max} +K+2	2,K
...	...	...	...	...	...	...	...
t_{max},K,1	t_{max} +2K+1,K,1	K	K	K,1	1	t_{max} +K+1	K,1
...	...	...	...	...	...	...	...
t_{max},K,K	t_{max} +2K,K,K	K	K	K,K	K	t_{max} +2K	K,K

Data movements within the linear systolic design

The data movements comprise:

- ✓ The data y_j are handled sequentially in the first PE, in the original order. They are subsequently propagated downwards to all other PEs.
- ✓ The transpose of the reference spectra matrix, ***RT*** (row by row) in the PEs.
- ✓ When the value y arrives at the ith PE (from the top), it is multiplied by the stored data RTij to yield a partial sum qi to be accumulated in the same PE.
- ✓ The PE changes mode to be ready for the iterative process phase.

Processor element design requirements within the linear systolic design

The processor elements include the following key components:

a) Each PE should store a row of the transposition of the reference spectra matrix ***RT***, i.e., a column of matrix ***R***.

b) Data are transmitted in one direction between two neighbouring PEs.

c) Each PE should support all the arithmetic processing capabilities, including the multiplying and accumulating operations.

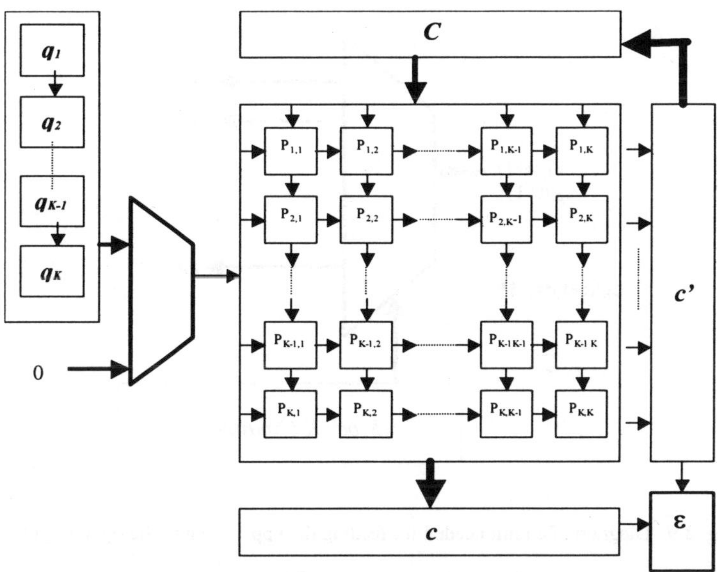

Fig. 2-8. Systolic structure for the iterative process.

Data movements within the rectangular systolic design

In this case, we have distinguished two main steps: First, the weight matrix computation, and, secondly, the iterative process.

<u>Weight Computation</u>

The data movements comprise:

- ✓ The reference spectra ***R***, as input into the first row of the PE array -- one column for each PE. They are subsequently propagated downwards to all PEs on the same column.
- ✓ The transposition of the reference spectra matrix ***RT***, as input on the first column of the PE array -- one row for each PE. They are subsequently propagated rightwards to all PEs on the same row.
- ✓ When the values R_{ji} and RT_{ij} arrive at the ijth PE (from the top and the right positions respectively), both are multiplied to yield a partial sum P_{ij} that is accumulated in the same EP. This value will be used during the iterative process phase.

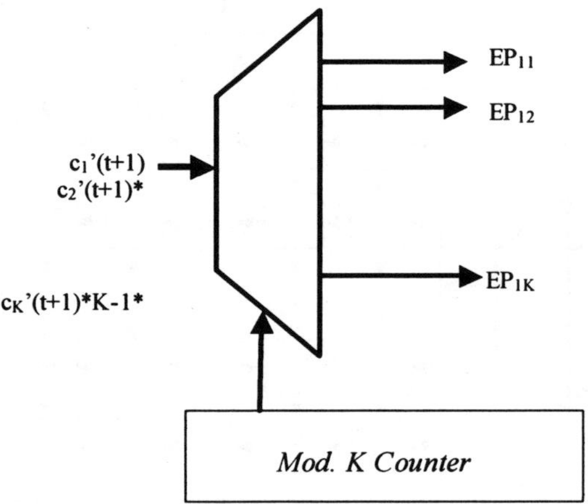

Fig. 2-9. Diagram of circuit needed for feeding the upper input to the right output.

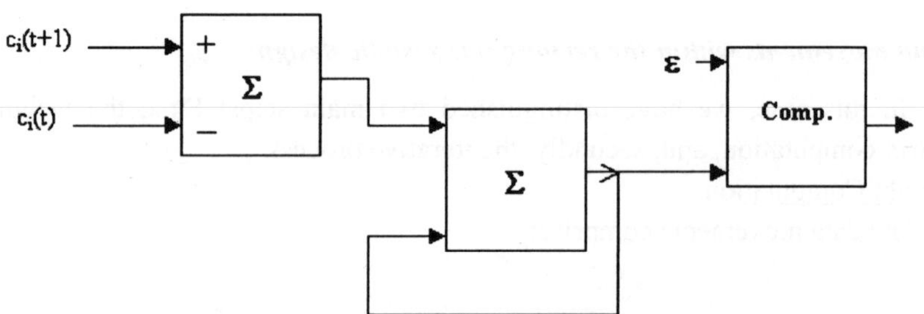

Fig. 2-10. Diagram of the control convergence process.

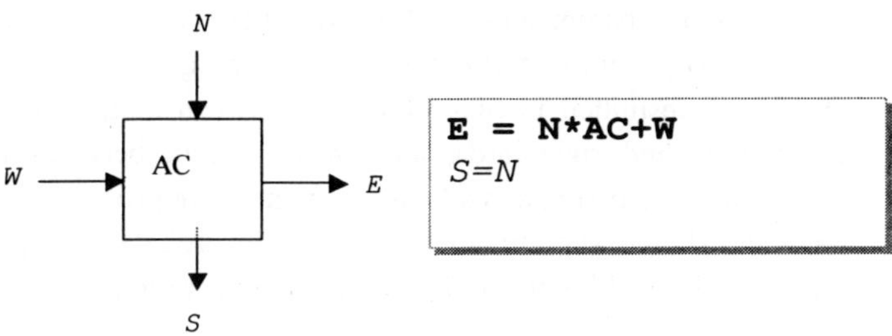

Fig. 2-11. Operations performed by PEs in Fig. 2-8.

Iterative process

In this case, the data movements comprise:

a) At the time t, the contribution vector ***c(t)*** as input on the first row of the PE array -- one component $c_j(t)$ for each *1j*th PE. They are subsequently propagated downwards to all PEs on the same column.

b) The threshold vector, previously calculated, **q**, as input on the first column of the PE array -- one component q_i for i1th PE. They are stored on all PEs on this first column.

c) When the values q_i and $c_1(t)$ arrive at the *i1*th PE (from the right and the top sides respectively), the products of the P_{i1} are computed with c1, and from λ with qi both results are added and routed to the *i2*th PE. When the respective values arrive at the rest of *ij*th PEs, they are multiplied to yield a partial sum to be propagated rightwards to all PEs on the same row.

d) As soon as the final partial sum is computed at the last PE of each row, PE_{iK}, can be routed to the corresponding PE_{1i} on each column.

Processor element design requirements within the rectangular systolic design

In this case, the processor elements comprise the following key components:

a) Each PE_{ij} should store a row and a column of the reference matrix ***R***, plus a row of the transposition reference matrix ***RT***, and the value of the learning parameter $-\lambda$. Moreover, it is necessary that each PE has a memory cell for storing the computed weight P_{ij}.

b) Data are transmitted in two directions between neighbouring PEs, downwards and rightwards, and an extra link between the last column and the first row is added for the iterative process.

c) Each PE should support all the arithmetic processing capabilities including the multiplying and accumulating operations.

2.4. Conclusions

In this chapter, a recursive Neural Network has been introduced to solve the Mixture Problem, based on the Hopfield Model. This model finds the composition of the mixture by operating with the spectra of the Reference Set. One interesting property, deduced from the algorithmic structure of the method, is related to the low-complexity VLSI structures, amenable for supporting the algorithm, since only multiplications and additions are required for programming an inversion process.

The main goal of this chapter has been to demonstrate how this model can be implemented through a systolic array. The proposed methodology is applied as a linear structure for determining the net thresholds, and as a rectangular structure of programmable processor elements to obtain the net weight matrix and to implement the iterative process.

The linear array permits us to reduce the sequential computation time of K*N order to K+N clock cycles. This is an important advantage because the dimension of the used spectra is given by 1024 channels. In the weight determination phase, we need K*K*N cycles for a sequential computation while the systolic structure proposed reduces it to 2K+N. Finally, for realising the iterative process, the reduction is from $K*K*t_{max}$ (where t_{max} is the maximum iteration number) to $2K+t_{max}$.

The proposed algorithmic method can be applied to deal with several issues in spectroscopy, such as IR and visible spectrum decomposition. In addition, applications can be developed for hyperspectral unmixing through remote sensing on the Earth's surface. The reduction of computation time is an important benefit if a response is wished in real time.

Acknowledgements

The authors wish to express their gratitude to Dr. Alejandro Curado Fuentes for his linguistic revision of this chapter.

References

[1] J. Adams, P. Johnson, M. Smith and T. George. A Semi-Empirical Method for Analysis of the Reflectance Spectra of Binary Mineral Mixtures. *Journal of Geophysic Res.*, 88, pages 3557–3561, 1983.

[2] W. Lawton and M. Martin. The Advanced Mixture Problem-Principles and Algorithms. *Technical Report IOM384*, Jet Propulsion Laboratory, 1985.

[3] E. Barnard and P. Cassasent, Optical Neural Net for Classifying Imaging Spectrometer, *Applied Optics*, 18(15), pages 3129-3133, Aug. 1989.

[4] J. C. Díaz, P. Aguayo, P. Gómez, V. Rodellar and P. Olmos. An Associative Memory to Solve the Mixture Problem in Composite Spectra. *Proc. of the 35th Midwest Symposium on Circuits and System*, Washington DC, pages 422–428, August 1992.

[5] V. Rodellar, M. Hermida, A. Díaz, A. Lorenzo, P. Gómez, P. Aguayo, J. C. Díaz and R. W. Newcomb. A VLSI Arithmetic Unit for a Signal Processing Neural Network. *Proc. of the 35th Midwest Symposium on Circuits and System*, Washington DC, pages 891–894, Aug. 1992.

[6] R. M. Pérez, P. Martínez, A. Silva and P. L. Aguilar. Influence of the Fixed Point Format on the Accuracy of the Neural Network Solution to the Mixture Problem. *Proc. of the 3rd Advanced Training Course: Mixed Design of Integrated Circuits and Systems*, Lodz, Poland, pages 413–418, May 1999.

[7] R. M. Pérez, P. Martínez, L. Martínez, J. C. Díaz, V. Rodellar and P. Gómez , A Hopfield Neural Network to Solve the Mixture Problem. *Proc. of the VI Spanish Symposium on Pattern Recognition and Image Analysis*, Córdoba, Spain, pages 744–745, April 1995.

[8] S. Y. Kung. *Digital Neural Networks*. Prentice-Hall, Inc., pages 85–91, 1993.

[9] R. M. Pérez, P. Martínez, A. Silva and P. L. Aguilar. Adaptive Algorithm to Solve the Mixture Problem with a Neural Networks Methodology. *Proc. of 3rd. International Workshop In Image/Signal Processing: Advances In Computational Intelligence, IWISP'96*, Manchester, United Kingdom, pages 133–136, Nov. 1996.

[10] R. M. Pérez, P. Martínez, A. Silva and P. L. Aguilar. An Adaptative Solution to the Mixture Problem with Drift Spectra. *Proc of the Third International Conference on Signal Processing*, Beijing, China, Oct. 1996.

[11] S. Y. Kung. VLSI Array Processors, *IEEE ASSP Magazine*, 2(3), pages 4–22, July 1985.

[12] J. J. Navarro, J. M. Llaberia, and M. Valero. Partitioning: An Essential Step in Mapping Algorithms Into Systolic Array Processors. *IEEE Computer*, pages 77–89, July 1987.

[13] D. I. Moldovan, On the Design of Algorithms for VLSI Systolic Arrays. *Proceedings of the IEEE*, 1(1), Jan. 1983.

CHAPTER 3

MORPHOLOGICAL ENDMEMBER IDENTIFICATION AND ITS SYSTOLIC ARRAY DESIGN

P.L. Aguilar, A. Plaza, P. Martínez, R.M. Pérez
Departamento de Informática, Universidad de Extremadura
Campus Universitario s/n 10071 Cáceres, Spain
E-mail: paguilar,aplaza,pablomar,rosapere@unex.es

Systolic arrays can be used in many different applications in order to improve performance. In particular, digital image processing algorithms are suitable to be implemented by systolic structures, since basic image manipulation operations are usually repetitive and can be mapped into a rectangular systolic structure. In this chapter we discuss the application of systolic arrays to speed up the performance of a new algorithm that performs unsupervised analysis of remote sensing images with high dimensionality (hyperspectral images). The methodology is based on mathematical morphology, which is a non-linear image analysis and pattern recognition technique that has proved to be especially well suited to segment images with irregular and complex shapes, but has rarely been applied to the classification/segmentation of hyperspectral images. The proposed method, which integrates spectral and spatial information in the analysis process, is fully described in this chapter, along with its hardware implementation by systolic arrays. A comparison between the hardware implementation and the software-based implementation is addressed.

3.1. Introduction

High-performance special-purpose computer systems are used to meet specific application requirements. As hardware cost decreases exponentially and the scale of integration in digital circuits is ever growing, the development of those circuits is obvious for special-purpose applications that demand large processing requirements [1].

Systolic arrays benefit from advances in semiconductor technology to provide adequate results in the case of applications requiring optimal throughput.

In terms of configuration, systolic arrays derive from other array-like structures, such as iterative arrays, cellular automata and processor arrays. These architectures are based on regular and modular formations that match the computational needs of many algorithms.

There exist several applications which include algorithms and processing modules appropriate for implementation by systolic arrays, such as digital image and signal processing, pattern recognition, linear and dynamic programming, etc. In fact, many of these require special-purpose implementations when used in real-time environments.

When systolic arrays were originally proposed, they were intended for applications requesting optimal throughput and large processing bandwidth, usually at the cost of response time. These applications can usually be supported by array-like structures consisting of a few types of single-processing units.

This chapter introduces an example of the applications mentioned above. In particular, we propose a systolic array structure that speeds up performance in a brand-new methodology; this framework serves to conduct an unsupervised analysis of multispectral or hyperspectral images. The use of systolic arrays in this specific application is very appropriate and illustrative of a practical case, as it demonstrates how basic image handling tasks can be mapped onto a rectangular systolic structure that takes advantage of the parallel processing capabilities provided by VLSI/ULSI technologies. We thus describe this methodology and its corresponding implementation by working with systolic arrays.

3.2. Overview of the Proposed Application

During the past decade, a number of airborne and satellite hyperspectral sensors have been developed or improved for remote sensing applications [2], [3], [4]. Image spectrometry enables the detection of materials, objects and regions in a particular scene with great accuracy. Hyperspectral data typically consist of hundreds of thousands of spectra, and, as a result, the analysis of this information is a key issue [5].

A very useful and commonly accepted approach to analyze hyperspectral data has been the identification of the purest spectra or "endmembers" [6]. Some researchers have taken up the chore of constructing spectral libraries of pure elements that can be matched with every spectrum in a hyperspectral image in order to classify the scene [7]. This processing method is suitable when pure materials, contained in the library, are at ground level, but in real-world situations, only the strongest features are matched, mainly due to the fact that the materials are spatially or intimately mixed. In this sense, the most common technique is to determine endmember spectra directly from the image. Once the individual endmembers have been identified, several methods can be used to map their spatial distributions, associations and abundances [8], [9].

A wide variety of methodologies has been proposed in the literature in order to find endmembers in data cubes [10]. One of the most successful approaches is the Pixel Purity IndexTM (PPI) algorithm [11], which finds the "purest" pixels in the scene through a series of repeated projections onto randomly oriented lines in the N-dimensional space. These potential endmember spectra are loaded into an N-dimensional scatterplot and rotated in real time until a trained analyst selects extremities within the data cloud; these are likely to match scene endmembers. The procedure, based on the geometry of convex sets, is well accepted (made available through a commercial software system called ENVI®, standing for Environment for Visualizing Images), but it is time-consuming and highly interactive.

Several methods for autonomous extraction of endmembers have been recently proposed in the literature. The N-FINDR algorithm [10] finds the simplex with a maximum volume to be enclosed within all the points of the data cloud. The ORASIS algorithm [12] uses a process called Exemplar Selection to reduce the data set by rejecting any "redundant" spectra (this requires the calculation of the angle between spectral vectors). The Iterative Error Analysis (IEA) approach [13] performs a series of constrained sieving tasks and chooses as endmembers those pixels that minimize errors in the unmixed image. In addition to these physical methods, there are several statistics-driven approaches that rely on clustering algorithms [14]. The major drawback of all previous methods is that they only consider the spectral information contained in data cubes. Spatial information has not been fully exploited yet, specially in unsupervised classification. The integration of both spatial and spectral information is becoming more relevant as the sensors used in spaceborne platforms tend to increase the spatial resolution [15].

Mathematical morphology theory [16] is a non-linear technique widely used for image analysis and pattern recognition. Although it is namely befitting for segmentation of binary or grayscale images that have irregular and complex shapes, its application in the classification/segmentation of multispectral or hyperspectral images has been rare [17], [18]. In this chapter, we discuss a brand-new automated methodology to find endmembers in hyperspectral data cubes by applying mathematical morphology.

3.3. Methodology

In this section, we provide some basic concepts about mathematical morphology theory; we also describe a schema to extend morphological operators to the hyperspectral domain. Finally, we propose an automated methodology, based on morphology, to extract endmembers from data cubes.

3.3.1. *Classical Mathematical Morphology*

Mathematical morphology theory [16] has become a productive tool for image analysis and pattern recognition. In binary morphology, images are represented as sets, in which foreground pixels are members of a set ***X*** and background pixels belong to the complementary set ***XC***. The two basic operations of mathematical morphology consist in the transformation of an image by another set ***K***, known as the structuring element. The shape and size of the structuring element determine the spatial characteristics of the resulting image. The two basic morphology operations, dilation and erosion, are defined respectively as follows:

$$X \oplus K = \{s \mid \mathrm{K}_{\mathrm{s}} \cap \mathrm{X} \neq \varnothing\} \tag{3-1}$$

$$X \otimes K = \{s \mid K_s \subseteq \mathrm{X}\} \tag{3-2}$$

Erosion and dilation are said to be dual regarding complementation. We define the opening of ***X*** with respect to ***K*** as the following set:

$$X_K = (X \otimes K) \oplus K \tag{3-3}$$

Similarly, we attribute the closing of ***X*** regarding ***K*** to the following set:

$$X^K = (X \oplus K) \otimes K \tag{3-4}$$

The opening of ***X*** by ***K*** is thus defined in terms of an erosion followed by a dilation. Similarly, the closing of ***X*** by ***K*** is defined in terms of a dilation followed by an erosion.

The principles of mathematical morphology have also been extended to the grayscale image case [19]. Grayscale images are described as a grey level function *f(x,y)* on the points of the Euclidean 2-space. Grayscale dilation, erosion, opening and closing are respectively described as follows:

$$(f \oplus k)(x,y) = \underset{(s,t)\in \mathrm{k}}{Max}\{f(x-s, \mathrm{y}-\mathrm{t}) + \mathrm{k(s, t)}\} \tag{3-5}$$

$$(f \otimes k)(x,y) = \underset{(s,t)\in \mathrm{k}}{Min}\{f(x+s, \mathrm{y}+\mathrm{t}) - \mathrm{k(s, t)}\} \tag{3-6}$$

$$f_k(x, \mathrm{y}) = ((f \otimes k) \oplus k)(x,y) \tag{3-7}$$

$$f^k(x, \mathrm{y}) = ((f \oplus k) \otimes k)(x,y) \tag{3-8}$$

The expressions for grayscale dilation and erosion are markedly similar to the convolution integral that is often encountered in digital image processing. Here, adding and differentiation replace multiplication and minimization, whereas maximization replaces addition.

3.3.2. *Extending Mathematical Morphology to the N-Dimensional Space*

Mathematical morphology precision requires an algebraic structure ***T*** (complex lattice) so that:

- ***T*** is induced by a (partial) ordering relation.
- For any family of elements in ***T***, there exists a major member as small as possible, called supremum, and a minor member as big as possible called the infimum [18].

In hyperspectral images, these two properties are absent because there is no natural means for a complete arrangement of multivariate pixels. Two main strategies have been considered in the solution of this problem as a result:

The first one consists of processing each channel of the hyperspectral image separately. This marginal approach is not satisfactory since it fails to account for the existing correlation among individual channels.

The second tactic is based on a sheer vector approach to process all the hyperspectral channels at the same time. This strategy requires the definition of a vector-driven organizational task that determines the supremum and the infimum of any family of N-dimensional vectors.

The definition of a vector ordering relation can be done as follows. Let $\boldsymbol{x_1}, \boldsymbol{x_2}, \ldots, \boldsymbol{x_n}$ be the hyperspectral image pixels within a filtering window that represents the structuring element of a morphological operation. We can define a measurement of the dissimilarity between two of those pixels, $\boldsymbol{x_a}$ and $\boldsymbol{x_b}$ by:

$$dist(x_a, \mathrm{x_b}) = \cos^{-1}\left(\frac{x_a \cdot x_b}{\|x_a\| \cdot \|x_b\|}\right) \quad (3\text{-}9)$$

This is the angular distance function, expressed in radians. The following scalar quantity can be calculated to measure the global distance between a particular pixel $\boldsymbol{x_i}$ and a set $\boldsymbol{x_j}$, j=1..m of neighboring pixels.

$$d_i = \sum_{j=1}^{m} dist(x_i, \mathrm{x_j}) \quad (3\text{-}10)$$

The supremum and infimum of a set of $\boldsymbol{x_i}$, i=1..n hyperspectral pixels are respectively defined as:

$$\sup(x_i) = h = \arg\max_i d_i \quad (3\text{-}11)$$

$$\inf(x_i) = l = \arg\min_i d_i \quad (3\text{-}12)$$

Thus, it is easy to determine basic morphology operations such as erosion and dilation by maximum and minimum operations, as indicated in the subsection above. While $\boldsymbol{h}$ is the most singular pixel (spectra-based) in a spatial neighborhood, $\boldsymbol{l}$ is nothing else than the median of the pixels in the neighborhood, according to the classical vector median definition [18].

3.3.3. *Autonomous Morphological Endmember Extraction (AMEE)*

We propose a morphology-based automated algorithm to extract endmembers from hyperspectral images. The input for this process is the full hyperspectral image cube, with no previous dimensionality reduction or pre-processing. The procedure must examine the full dataset to find those pure pixels that can be used to describe the various mixed pixels in the scene. This is done through the following steps.

Let ***X*** be the original image and ***K*** a structuring element. We propose the following morphological operator to identify endmembers:

$$X\Phi K = dist\left(X, X^{K}\right) = dist\left(X, \left(X \oplus K\right) \oplus K\right) \tag{3-13}$$

This operator works as follows: First, a hyperspectral closing operation is applied to the original image; then, a measurement of the dissimilarity between each pixel of the original image and the correspondent pixel in the closed image is calculated in terms of the angular distance function.

It is important to emphasize that Φ extracts spectral information on a pixel-by-pixel scale, much like any other existing methods, but the process is driven by spatial information. As a result, spatial and spectral responses are simultaneously considered in the analysis.

Let ***K*** be a square-shaped, 3x3 structuring element. Fig. 3-1 illustrates the working procedure of the operator when applied to a target pixel ***E*** in the original image, and using ***K*** as a structuring element. First, sheer spectra-based hyperspectral pixels (***Q***) in a spatial neighborhood of the target pixel are selected after projecting each pixel against all the spatial neighbors. This procedure is repeated for every pixel in ***X***, leading to a new image ***X⊕K***. From this image, only the pixels that provide a good representation of their neighbors are selected by the minimum operator, which in our case is equivalent to a vector median filtering. The pixels in **(*X*⊕*K*)⊕*K*** match the definition of an endmember, i.e. a pure spectra pixel that can be used to describe several mixed pixels in the scene. Finally, dissimilarity between ***E***, corresponding to the target pixel in the original image, and ***V***, the pixel selected by the operator, is calculated to check whether the target pixel is likely to be an endmember.

The described operation is repeated by working with structuring elements that have an increasing size, so that detailed information of the spatial context associated with each hyperspectral pixel is obtained.

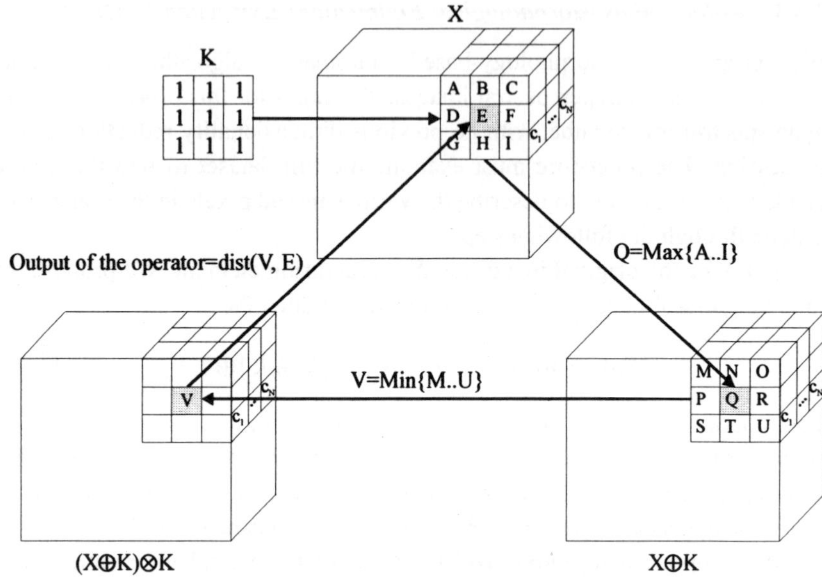

Fig. 3-1. Working procedure of the Φ hyperspectral morphological operator.

The overall result of applying a sequence of morphological operators to the original image is a spatial/spectral signature related to each pixel of the scene. From this information, a 2-dimensional grayscale rule image is generated by following the basic idea of the Pixel Purity Index algorithm: Those pixels that have been repeatedly selected by the operator during the process are more likely to be declared "pure". This image is automatically thresholded, and the resulting regions are developed by applying a spatial-spectral technique, as described in [20], to obtain the final endmembers. In our research, we focus on the hardware implementation of morphological eroding and dilating operations by means of systolic arrays, indicating the efficiency of these structures in the hardware optimization of a specific special-purpose software algorithm.

3.4. Systolic Array Implementation

The algorithm described above is characterized by the repetitive use of very simple operations with the input data. This fact implies that systolic arrays may provide an efficient VLSI implementation of this methodology. We also give a detailed account of the operations of AMEE algorithms involved in systolic array design:

- Calculation of the spectral dissimilarity between two particular pixels in the hyperspectral image (i.e. x^i and x^j) by means of the angular distance function, expressed in Eq (3-5).
- Extraction of the purest spectra pixel from a set of observations. In order to perform this operation, as mentioned in Section 3.3, the definition of a vector ordering relation to determine the *supremum* and the *infimum* of any family of N-dimensional vectors is required. It is important to emphasize that each vector is associated with a particular hyperspectral pixel, and the family of vectors corresponds to a spatial neighborhood defined by the structuring element of the morphological operation (see Section 3.3). The calculation of the *supremum* and *infimum* can be done as follows:
    - ✓ Select one vector x^i related to a hyperspectral pixel.
    - ✓ Compute the angular distance between x^i and all the other vectors in the family, accumulating the partial results obtained (see Eq (3-6)). This calculation provides the cumulative distance between x^i and the rest of the vectors in the family.
    - ✓ Repeat Steps 1 and 2 with all the pixels in the spatial neighborhood. Therefore, we will have an accumulated distance value d_i per pixel.
    - ✓ According to Eq (3-7), determine the vectors having extreme values of d_i.

In the next subsections, we present a graph-dependent methodology to calculate the angular distance and the *supremum* and *infimum* by using systolic arrays.

3.4.1. *Rectangular Systolic Array to Calculate the Angular Distance*

Let X be a matrix containing all the pixels that belong to a certain spatial neighborhood, as defined by a structuring element.

$$X = \begin{bmatrix} [x_1^1 & x_2^1 & \cdots & x_N^1] \\ [x_1^2 & x_2^2 & \cdots & x_N^2] \\ & \vdots & & \\ [x_N^M & x_N^M & \cdots & x_N^M] \end{bmatrix}$$

The following pseudo-code algorithm is used to calculate the angular distance. *M* is the total number of pixels in the neighborhood. *Prod(i)* accumulates the outer product between pixel x^i and the rest of pixels in the neighborhood. *Norm_Xi* and *Norm_Xj* are the norms of pixels x^i and x^j, respectively. Finally, *dist(i)* is the accumulated distance of pixel x^i in relation to all the other pixels from the neighborhood.

Table 3-1. Set of indexes of the algorithm to calculate the outer product and vector norms.

i j k	Prod(i)	Norm_Xi(i)	Norm_Xj(i)	Xi (i,k)	Xj (j k)
1 1 1	1	1	1	(1,1)	(1,1)
1 1 2	1	1	1	(1,2)	(1,2)
1 2 1	1	1	1	(1,1)	(2,1)
1 2 2	1	1	1	(1,2)	(2,2)
......	.....	.....	.....	.....	.....
1 N 1	1	1	1	(1,1)	(1,1)
......	.....	.....	.....	.....	.....
1 N N	1	1	1	(1,N)	(N,N)
.....	.....	.....	.....	.....	.....
2 1 2	2	2	2	(2,2)	(1,2)
2 2 1	2	2	2	(2,1)	(2,1)
2 2 2	2	2	2	(2,2)	(2,2)
.....	.....	.....	.....	.....	.....
M,1,1	M	M	M	(1,1)	(1,1)
.....	.....	.....	.....	.....	.....
M,N,N	M	M	M	(M,N)	(N,N

First, we proceed to the design of the systolic structure, corresponding to the first three iterations of this algorithm. From that initial design, we perform other operations needed in order to obtain the angular distance. Table 3-1 shows the set of indexes for this algorithm. In this table, each index element is shown as a three-tuple (*i,j,m*). Note that for both index elements (*i,j,1*) and (*i,j,2*), the same value of *Prod(i)* is used, i.e. *P(i)* can be piped onto *k* direction *k*. Similarly, values Xi(*i,k*) and Xj(*j,k*) can be piped onto directions *j* and *i*, respectively.

```
Begin
  For i = 1 to M
    For j = 1 to M
      For k = 1 to N
        Prod(i) = Prod (i) + x (i,k) * x (j,k)
        Norm_Xi (i) = Norm_Xa (i) + x (i,k) * x (i,k)
        Norm_Xj (i) = Norm_Xb (i) + x (j,k) * x (j,k)
      End For
    End For
  End For
  For i = 1 to M
    d (i) = Prod (i) / (Norm_Xi (i) * Norm_Xj (i) )
    dist (i) = cos^-1 (d (i) )
  End For
End
```

Data-dependence vectors are found by equating indexes of all possible pairs of generated and used variables.

$d_1^T = (i, j, k) - (i-1, j, k) = (1, 0, 0)$ to x(j,k)
$d_2^T = (i, j, k) - (i, j-1, k) = (0, 1, 0)$ to x(i,k)
$d_3T = (i, j, k) - (i, j, k-1) = (0, 0, 1)$ to prod(i) and norm_Xi, and norm_Xj

Fig. 3-2 shows the graph of dependences for the described algorithm variables. In this figure, we can observe the behaviour and movement of the variables: x^i moves from left to right, x^i moves from top to bottom, and the rest of the variables move onto direction j.

In order to map this algorithm with loops onto a systolic array, the approaches proposed by Kuhn [21] and Moldovan [22] are followed. The loop indexes of our algorithm are transformed into new loop indexes by a bijective and monotonic function ***T***, which allows parallelism and pipelining, while preserving the dependences of the original algorithm.

$$T = \begin{bmatrix} \Pi \\ S \end{bmatrix} \qquad T = \begin{bmatrix} t_{11} & t_{12} & t_{13} \\ t_{21} & t_{22} & t_{23} \\ t_{31} & t_{32} & t_{33} \end{bmatrix}$$

where

$$\Pi = \lfloor t_{11} \quad t_{12} \quad t_{13} \rfloor \qquad S = \begin{bmatrix} t_{21} & t_{22} & t_{23} \\ t_{31} & t_{32} & t_{33} \end{bmatrix}$$

In order to achieve a valid arrangement, and, at the same time, reduce the turnaround time (i.e. $\pi^* d_i > 0$), the mapping function π is selected as:

$$\Pi = \begin{bmatrix} 1 & 1 & 1 \end{bmatrix}$$

Our choice of $\boldsymbol{S}$ determines the interconnection of the processors. We select $\boldsymbol{S}$ as:

$$S = \begin{bmatrix} 0 & 1 & 0 \\ 1 & 0 & 0 \end{bmatrix}$$

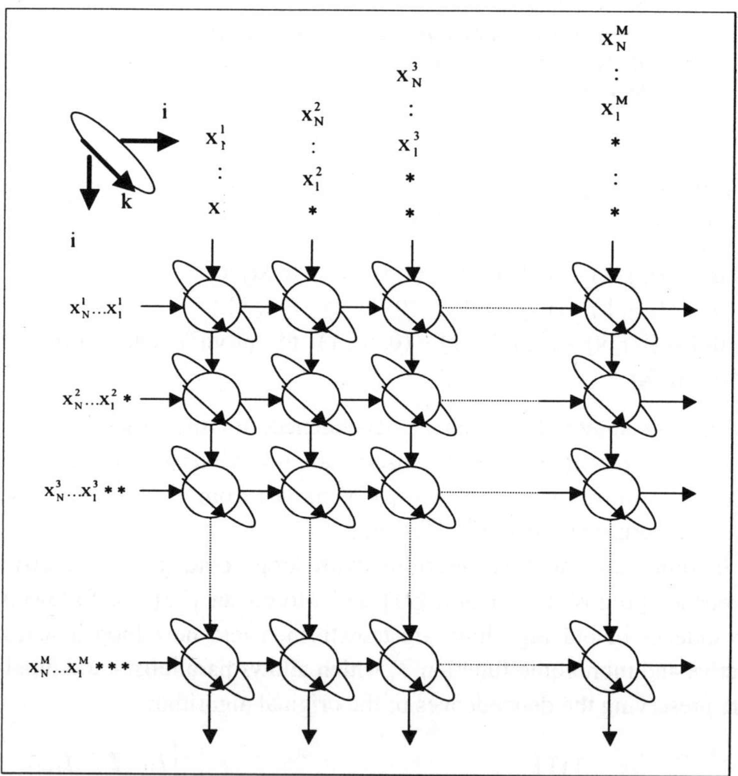

Fig. 3-2. Dependence graph for the obtaining the necessary computation the angular distance.

The advantage of this selection is that a projection onto direction k is performed. Thus, all the nodes aligned in direction k will be processed by the same processor element or PE. The interconnection function between processors is defined by:

$$Sd_i = \begin{bmatrix} x \\ y \end{bmatrix}$$

where x and y refer to the movement of the variable along directions j and i, respectively. In our case:

$$S{\cdot}d_1 = \begin{pmatrix} 0 & 1 & 0 \\ 1 & 0 & 0 \end{pmatrix} \bullet \begin{pmatrix} 1 \\ 0 \\ 0 \end{pmatrix} = \begin{pmatrix} 0 \\ 1 \end{pmatrix} \qquad S{\cdot}d_{12} = \begin{pmatrix} 0 & 1 & 0 \\ 1 & 0 & 0 \end{pmatrix} \bullet \begin{pmatrix} 0 \\ 1 \\ 0 \end{pmatrix} = \begin{pmatrix} 1 \\ 0 \end{pmatrix} S{\cdot}d_3 = \begin{pmatrix} 0 & 1 & 0 \\ 1 & 0 & 0 \end{pmatrix} \bullet \begin{pmatrix} 0 \\ 0 \\ 1 \end{pmatrix} = \begin{pmatrix} 0 \\ 0 \end{pmatrix}$$

Table 3-2. Mapping the original index set onto the transformed index set.

I=(i j k)	T*I	Computations	Π*I	S*I
(1 1 1)	(3 1 1)	$Prod_{1\ 1}=x_1^1*x_1^1$ $Norm_X1_1=x_1^1*x_1^1$ $Norm_X1_1=x_1^1*x_1^1$	3	(1 1)
(1 1 2)	(4 1 1)	$Prod_{1\ 1}=Prod_{1\ 1}+x_2^1*x_2^1$ $Norm_X1_1=Norm_X1_1+x_2^1*x_2^1$ $Norm_X1_1=Norm_X1_1\ x_2^1*x_2^1$	4	(1 1)
......	......	......	.....	.....
(1 1 N)	(N+2 1 1)	$Prod_{1\ 1}=Prod_{1\ 1}+x_N^1*x_N^1$ $Norm_X1_1=Norm_X1_1+x_N^1*x_N^1$ $Norm_X1_1=Norm_X1_1\ x_N^1*x_N^1$	N+2	(1,1)

Table 3-2 represents the mapping functions between the original and transformed (by T) indexes. In this table, columns πxI and SxI indicate the time and the cell processing the computation indexed by I.

From Table 3-2, we can conclude that a rectangular array formed by MxM processor elements (PEs) is needed. The number of clock cycles required to perform all the operations involved in the full calculation is 2M+N. Thus, in cycle N+3, the processing unit PE11 will have calculated the product between x_t and its own, as well as its norm. One cycle afterwards, PE12 and PE21 will have performed the same operations for pixels x_1 and x_2. In this train of thought, we can conclude that by the cycle 1+N+M, the product between x_t and all the pixels of the spatial neighborhood will have been performed, and after 2M+N cycles, the operations associated with PEMM will have been finished.

Table 3-2 (cont.). Mapping the original index set onto the transformed index set.

I=(i j k)	T*I	Computations	Π*I	S*I
(1 2 1)	(4 2 1)	$Prod_{21}=x_1^1*x_1^2$ Norm_X1_2= Norm_X1_2+ $x_1^1*x_1^1$ Norm_X2_1= Norm_X2_1 $x_1^2*x_1^2$	4	(2 1)
(1 2 2)	(5 2 1)	$Prod_{21}=x_2^1*x_2^2$ Norm_X1_2= Norm_X1_2+ $x_2^1*x_2^1$ Norm_X2_1= Norm_X2_1 $x_2^2*x_2^2$	5	(2 1)
......	......	......	.....	.....
(1 2 N)	(N+3 2 1)	$Prod_{21}= Prod_{21}+x_N^1*x_N^2$ Norm_X1_2= Norm_X1_2+ $x_N^1*x_N^1$ Norm_X2_1= Norm_X2_2 $x_N^2*x_N^2$	N+3	(2 1)
(2 1 1)	(4 1 2)	$Prod_{12}=x_1^2*x_1^1$ Norm_X2_1= $x_1^2*x_1^2$ Norm_X1_2= $x_1^1*x_1^1$	4	(1 2)
(2 1 2)	(5 1 2)	$Prod_{12}= Prod_{12}+x_2^2*x_2^1$ Norm_X2_1= Norm_X2_1+ $x_2^2*x_2^2$ Norm_X1_2= Norm_X1_2 $x_k^1*x_k^1$	5	(1 2)
......	......	......	.....	.....
(2 j k)	(2+j+k j 2)	$Prod_{j2}= Prod_{j2}+x_k^j*x_k^2$ Norm_X2_j= Norm_X2_j+ $x_k^j*x_k^j$ Norm_Xj_2= Norm_Xj_2 $x_k^2*x_k^2$	2+j+k	(j 2)
......	......	......	.....	.....
(i j,k)	(i+j+k k i)	$Prod_{ji}= Prod_{ji}+x_k^j*x_k^i$ Norm_Xi_j= Norm_Xi_j+ $x_k^j*x_k^j$ Norm_Xj_i= Norm_Xj_i $x_k^i*x_k^i$	i+j+k	(k i)
......	......	......	.....	.....
(M M N)	(2M+N M M)	$Prod_{MM}= Prod_{MM}+x_K^M x_K^M$ Norm_XM_M= Norm_XM_M+ $x_K^M*x_K^M$ Norm_XM_M= Norm_XM_M+$x_K^M*x_K^M$	2M+N	(M M)

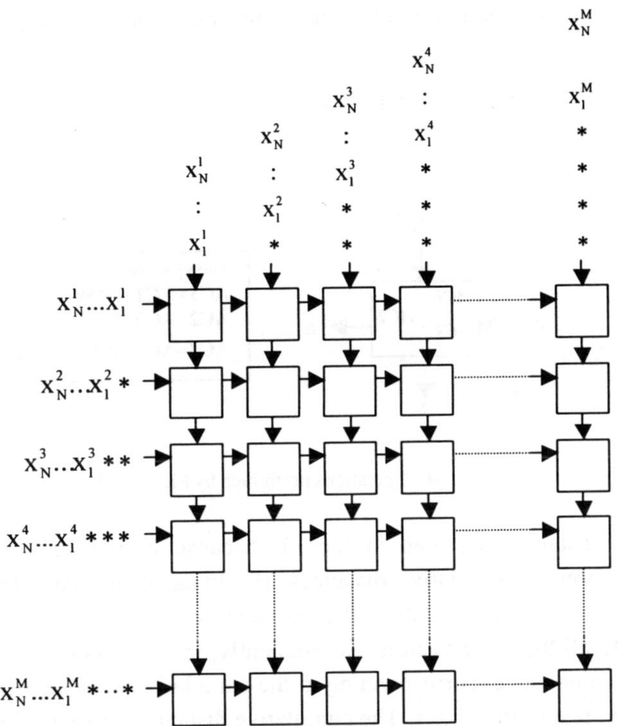

Fig. 3-3. Required interconnections between PEs to the systolic array for computing the outer-product and the norms of pixels.

Fig. 3-3 represents the required interconnections between the PEs. As shown in this figure, the resulting architecture has a rectangular array design. The pixels fed from left to right and top to bottom are transmitted between neighbouring PEs.

The capabilities of arithmetic processing related to each PE should include multiplication and accumulation operations (see Fig. 3-4). It is also required that each PE contains three accumulator cells to store the outer product and the norms of the involved pixels.

It is important to emphasize that AC1 refers to *Prod(i)*, AC2 to *norm_*x^i and AC3 to *norm_*x^j. As described above, in cycle N+2, the processor element PE11 performs the operations associated with the calculation of variable *d(i)* and the angular distance; this is one of the objectives of our algorithm. In subsequent cycles, the same operation will be performed for other PEs.

$$AC1 = \cos^{-1}\left(\frac{AC1}{AC2 * AC3}\right)$$

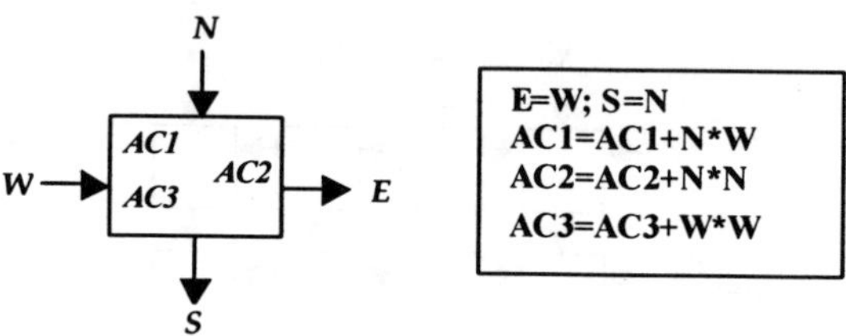

Fig. 3-4. Operations performed by Fig. 3-3 PEs.

This distance is stored in the PE because it will be needed in the determination of cumulative distances. If adding all the AC1 by rows and columns, we obtain as output the cumulative distance of each vector in relation to all the other vectors. Subsequently, the outputs of PE_{1M} and PE_{M1} have the same value, corresponding to the cumulative distance between pixel x1 and the rest of the pixels. The cumulative distance between x2 and the rest of the pixels can be obtained by the outputs of PE_{2M} and PE_{M2}, and so on. In order to achieve the calculation of cumulative distances, it is necessary that, during the M cycles that follow the determination of the angular distance, each PE performs the operations:

$$S = N + AC1 \qquad\qquad E = W + AC1$$

It is important to note that, in this case, the input to the extreme left and at the very top of the initial elements must be 0. Thus, in order to obtain all the cumulative distances, M clock cycles must be added to the 2M+N cycles expressed in Table 3.2. From the previous exposition of ideas, we thus derive that, during those M cycles, the PEs perform operations that differ from those described in Fig. 3-4. Therefore, programmable PEs are required to obtain different functionalities at different clock cycles.

3.4.2. *Linear Systolic Array to obtain Supremum and Infimum of Cumulative Distances*

By following the procedure described in section 3, after obtaining the cumulative distances at each pixel of the neighborhood, we must obtain the pixels with the maximum and minimum associated cumulative distance values. The following simple algorithm can be used to realize such a computation.

```
Begin
        Infimum=∞
        Supremun=0
        For i = 1 to M
                If dist_acum(i) < infimum
                        Then Infimum = dist_acum(i)
                Endif
        Endfor
        For j=1 to M
                If dist_acum(j) >supremum
                        Then Supremun = dist_acum(j)
                Endif
                End For
End
```

In order to implement the algorithm, we propose the use of two linear arrays, one to calculate the supremum and another to calculate the infimum. Each one will have M processor elements respectively distributed in unique horizontal and vertical directions.

Fig. 3-5 represents the required interconnections between the PEs. As shown in this figure, the architecture has a linear array design. The supremum (or infimum) initial value is fed from the leftmost (or upmost) side of the array, and is transmitted between neighboring PEs.

The only operation performed at each PE is a comparison, and the resulting value is transmitted to the closest PE on the right or at the bottom. After M cycles, the last PE provides the maximum (or minimum) of the angular distances. The operations supported at each PE are shown in Fig. 3-6, where a) corresponds to the horizontal linear array and b) to the vertical array.

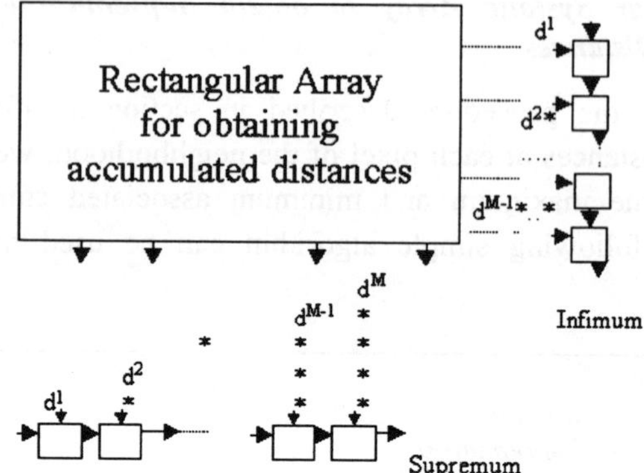

Fig. 3-5. Required interconnections between PEs for calculating the supremum and infimum values.

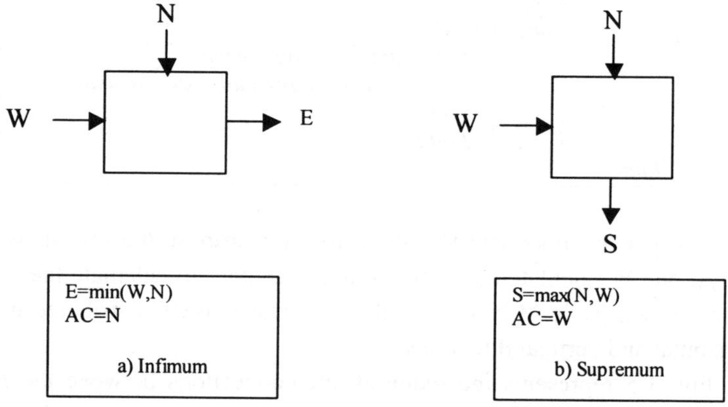

Fig. 3-6. Operations performed by Fig. 3-5 PEs of linear array.

The pixel associated with the maximum and minimum value of the cumulative distance is determined by a simple procedure: By providing feedback to the outputs of the linear arrays and comparing the outputs corresponding to each pixel, stored at each PE. The elements PEi of the horizontal array and PEj of the vertical array which are able to detect the matching, will indicate the desired extreme vectors $x^i\, x^j$. Those vectors have been stored in elements PE_{iM} and PE_{Mj} of the rectangular array when *Prod(i)* and *Prod(j)* were calculated. Fig. 3-7 shows a diagram of the systolic array final design, including feedback connections between the PEs as well as the operations performed with each processing element.

3.5. Summary of Design

We deem that, at this stage, summing up the described systolic array is suitable. The AMEE model can be mapped onto a systolic structure formed by a rectangular array of MxM PEs and two linear arrays, each one consisting of M PEs. The PEs should be programmable due to the variable functionalities involved. In order to describe the design, we distinguish between data movements and PE design requirements.

Data movements into the rectangular systolic design

In this implementation, we have distinguished two main steps: First, the angular distance, and, second, cumulative distance computations. Next, we offer a detailed description of the steps required to perform each operation.

Angular Distance Computation

Data movements comprise:

- ✓ Pixel vector set ***X***, as input into the first row of the PE array, -- one column for each PE. They are subsequently propagated downwards to all PEs on the same column. This same set serves as input for the first column of the PE array. They are subsequently propagated rightward to all PEs on the same row.
- ✓ When the values of pixel vectors x_k^j and x_k^i arrive at the *ij*th PE (from the top and right respectively) both are multiplied by themselves and by it, giving way to partial sums *Prodi_j(k)*, *Norm_xi(k)* and *Norm_xj(k)*. These values are accumulated in the same EP in order to compute the angular distance, which will be used in determining the cumulative distance.

Cumulative distance determination

In this case, data movements comprise the following steps:

- ✓ Once angular distances are calculated, a null (0) value is introduced in the PEs on the first row and first columns of the array, following a time-pipelined scheme (i.e., 0-value is introduced in the first PE at time t, in the second PE at time t+1, and so forth).

- ✓ Each PE_{ij} (or PE_{ji}) performs the addition between the value received from its neighboring PE at the left (PE_{i-1j} or PE_{ij-1}) and the stored distance, sending the resulting value to its neighboring PE at the right (PE_{ij+1}) or bottom (PE_{i+1j}).
- ✓ As soon as the data have arrived at the last PE of each row or column (PE_{iM}, or PE_{Mj}), the data can be routed to the linear array in order to determine the extreme distance values.

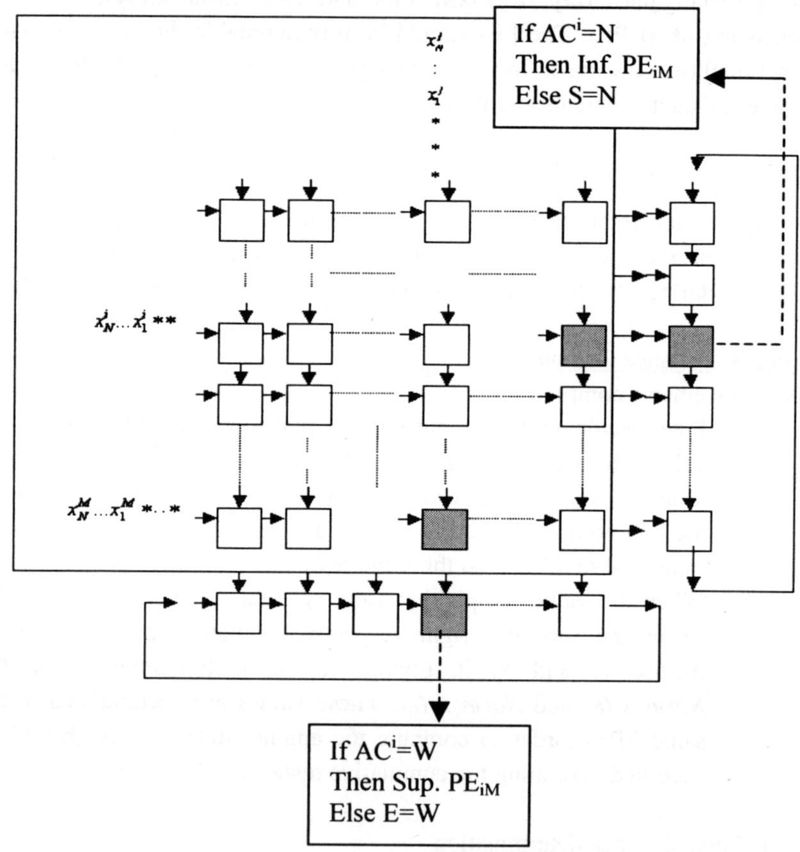

Fig. 3-7. Final systolic design scheme. PEs related to the calculation of extreme pixels are outlined.

Processor element design requirements into the rectangular systolic design

In this case, the processor elements comprise the following key components:

a) Each PE_{ij} should store a row of the set of pixel vectors $\boldsymbol{X}$, and the last element of each row and column must be able to store the pixel that arrives in a pipelined way from the left. Moreover, it is necessary that each PE incorporates three memory cells for storing the required computing.

b) Data are transmitted in two directions between neighbouring PEs, downwards and rightwards.

c) Each PE should support all arithmetic processing capabilities, including multiplication, addition, and division operations. However, for distance computation there is one non-linear function involved: $\cos^{-1}$.

3.6. Conclusions

This chapter has mainly introduced a brand-new automated approach for the unsupervised analysis of multispectral or hyperspectral image scenes that relies on mathematical morphology concepts. The extension of classic morphological operations to the hyperspectral domain enables the integration of spectral and spatial information in the analysis process and, thus, the incorporation of image processing in the field of hyperspectral analysis. In particular, a systolic array structure is proposed to speed up the performance of this new methodology, conducting an unsupervised analysis of multispectral or hyperspectral images.

The proposed methodology is applied as a rectangular structure for determining the accumulative distances, and as a linear structure of programmable processor elements to obtain the extreme values of these distances for the extraction of endmembers in the image.

The rectangular array allows us to reduce the sequential computational time of the MxMxN order to 2M+N clock cycles. This is an important advantage because of the high dimension of the used data. In the extreme distance determination phase, the supremun and infimum are simultaneously obtained.

In relation to the neural network for hyperspectral analysis, these pixels corresponding to the endmembers can be used as training patterns of unsupervised neural networks, providing a robust and efficient solution to the abundances determination problem (hyperspectral unmixing).

Acknowledgements

This work has been conceived under the funding project *Aplicación de las imágenes hiperespectrales a la vigilancia de recursos naturales* (TIC 2000-0739-C04-3), Ministerio de Educación y Ciencia (Spain).We also wish to thank Dr. Alejandro Curado, from our university Department of English, for his linguistic revision of this chapter.

References

[1] H. T. Kung. Why Systolic Architectures?, *IEEE Trans. Computer*, pages 37-46, Jan. 1983.

[2] R. O. Greeen. *AVIRIS Earth Science Workshop Proceedings.* Available at http://makalu.jpl.nasa.gov/, 1998-2000.

[3] G. Vane, R. O. Green, T. G. Chrien, H. T. Enmark, E. G. Hansen and W. M. Porter. The Airborne Visible/Infrared Imaging Spectrometer (AVIRIS). *Remote Sensing of Environment,* 44, pages 127-143, 1993.

[4] F. A. Kruse and J. W. Boardman. Fifteen Years of Hyperspectral Data: Northern Grapevine Mountain. *Summaries of the Eighth JPL Airborne Earth Science Workshop*, Nevada,1999.

[5] D. Landgrebe,.On Progress Toward Information Extraction Methods for Hyperspectral Data, *Proc. SPIE* . San Diego, California, August-1997.

[6] F. A. Kruse. Spectral Identification of Image Endmembers Determined from AVIRIS Data. *Summaries of the Seventh JPL Airborne Earth Science Workshop*, 1998.

[7] D. A. Roberts, P. Dennison, S. Ustin, E. Reith and M. Morais. Development of a Regionally Specific Library for the Santa Monica Mountains Using High Resolution AVIRIS Data. *Summaries of the Eighth JPL Airborne Earth Science Workshop*, 1999.

[8] J. W. Boardman and F. A. Kruse. Automated Spectral Analysis: A Geological Example Using AVIRIS Data, Northern Grapevine Mountains, Nevada, *Proc. 10th Thematic Conference*, Geologic Remote Sensing, San Antonio, Texas, May 1994.

[9] F. A. Kruse and J. W. Boardman. Fifteen Years of Hyperspectral Data: Northern Grapevine Mountains. *Summaries of the Eighth JPL Airborne Earth Science Workshop*, Nevada, Pasadena, California, 1999.

[10] E. M. Winter and M. E. Winter. Autonomous Hyperspectral End-member Determination Methods. *Part of the EUROPTO Conference on Sensors,*

Systems and Next-Generation Satellites, Proc. SPIE, Florence, Italy, Sep. 1999.

[11] J. W. Boardman, F. A. Kruse and R. O. Green. Mapping Target Signatures via Partial Unmixing of AVIRIS Data. *Summaries of the Fifth JPL Airborne Earth Science Workshop*, Pasadena, California, 1995.

[12] P. Palmadesso, J. Antoniades, M. Baumback, J. Bowles and L. J. Rickard. Use of Filter Vectors and Fast Convex Set Methods in Hyperspectral Analysis. *Proceedings of SPIE*, 1999.
Available at http://nemo.nrl.navy.mil/public/publications.html.

[13] T. Szeredi, K. Staenz and R. Neville. Automated Endmembers Selection: Part I Theory. *Remote Sensing of Environment*, 1999.

[14] S. Beaven, L. E. Hoff, and E. M. Winter. Comparison of SEM and Linear Unmixing Approaches for Classification of Spectral Data. *Proc. SPIE*, 1999.

[15] L. O. Jiménez and J. Rivera-Medina, On the Integration of Spatial and Spectral Information in Unsupervised Classification for Multispectral and Hyperspectral Data. *Part of the EUROPTO Conference on Sensors, Systems and Next-Generation Satellites, Proc. SPIE*, Florence, Italy, Sep. 1999.

[16] J. Serra. *Image Analysis and Mathematical Morphology*. Academic Press, London, 1982.

[17] P. Soille. Morphological Partitioning of Multispectral Images. *Journal of Electronic Imaging*, 5,(3), pages 252-265, July 1996.

[18] P. Lambert and J. Chanussot. Extending Mathematical Morphology to Color Image Processing. *CGIP'2000*, Saint-Etienne, France, 2000.

[19] S. R. Sternberg. Greyscale Morphology. *Computer Vision Graphics and Image Processing*, 35, pages 283-305.

[20] A. Plaza, P. Martínez, J. A. Gualtieri and R. M. Pérez. Spatial/Spectral Endmember Extraction from AVIRIS Hyperspectral Data Using Mathematical Morphology. *Summaries of the JPL/AVIRIS Workshop*, Pasadena, California, 2001.

[21] R. H. Kunh. *Optimization and Interconnection Complexity for Parallel Processors, Single Stage Networks and Decision Trees*. Ph. D. Dissertation, Department of Computer Science, University of Illinois, Urbana-Champaign, 1980.

[22] D. I. Moldovan. On the Design of Algorithms for VLSI Systolic Array. *Proc. IEEE*, 71 (1), pages 113-120, 1983.

CHAPTER 4

MANTRA I:
A SYSTOLIC ARRAY FOR NEURAL COMPUTATION

M.A. Viredaz

Compaq, Western Research Laboratory
250 University Ave., Palo Alto, CA 94301, USA
E-mail: viredaz@computer.org

P. Ienne

Swiss Federal Institute of Technology Lausanne
Processor Architecture Laboratory
IN-F Ecublens, CH – 1015 Lausanne, Switzerland
E-mail: paolo.ienne@epfl.ch

In the last decade, the need has arisen for medium-cost dedicated computers for *artificial neural network (ANN)* models. Several machines have been proposed. However, only very seldom can systems be considered as massively parallel and, hence, exploit the huge intrinsic parallelism of ANN models. The *MANTRA I* machine addresses this issue by targeting synapse-level parallelism on a bidimensional systolic array, based on a custom VLSI circuit called *GENES IV*. A prototype SIMD computer with 400 *processing elements (PEs)*, expandable to 1600 PEs, has been designed, built, tested, and analyzed.

This chapter focuses mainly on the architecture of the machine. The performance of the machine is analyzed using the Kohonen ANN model. Finally, a critical analysis suggests how the same general approach could be kept for future systems, requiring important steps toward generality and better ease-of-use.

4.1. Introduction

Research on *artificial neural networks (ANNs)* and computer science have coexisted for almost six decades. In the 80's, the ANN domain entered a new era with the development of models such as J. Hopfield's ***recurrent networks***, T. Kohonen's ***self-organizing feature maps***, and the ***back-propagation rule***[3]. In the late 80's and early 90's, besides super-computers, no platform was at the same time versatile enough to implement several ANN models and fast enough to be used on large problems. However, the need for moderate-cost systems, offering the power of super-computers for this class of algorithms, was clearly identified.

A survey of ***multi-model neural computers***[1,2,9,10,11,12,13,14] shows that most use only a moderate amount of parallelism. However, ANN models contain a huge intrinsic parallelism, that makes them well suited for massively parallel implementations. This field has been left almost unexplored, and the primary goal of the MANTRA I project is to study dedicated architectures that feature generality within the ANN field and cost-effective performance by using of massive parallelism[16]. Super-computers and general-purpose massively-parallel systems are not considered because of their typically high price.

A VLSI integrated circuit, named GENES IV[6], has been designed as a building block to implement a square systolic array. It exploits synapse-level parallelism (i.e., one real or virtual ***processing element (PE)*** is allocated per synapse or neural connection), while most other systems implement only neuron-level parallelism (i.e., one PE per neuron). This original array topology has a number of advantages. First, thanks to the much finer parallelism grain and the resulting massively parallel architecture, a much higher throughput is achievable. Another advantage of synapse-level parallelism is that, since synapses are much more numerous than neurons, a larger number of PEs can be effectively exploited on smaller problems, thus extending the domain of applicability towards ANNs of realistic sizes. Finally, the data bandwidth required by the array is proportional to its perimeter (i.e., $O(\sqrt{N})$) and not to the number of processing units (i.e., $O(N)$) as in almost all other unidimensional systems. Ring- or bus-based systems, for finer parallelism, often have to avoid unrealistic bandwidths by means of large and expensive storage inside the processing units. Internal storage is reduced to a minimum in the GENES IV PE.

The MANTRA I machine[15,16] is based on a mesh of up to 40×40 PEs. While most existing ANN systems are dedicated to a particular application and/or model, the MANTRA I machine does not hard-wire any algorithm, but provides basic primitives to implement the target models. The parallel or SIMD module of the machine is controlled by a microprocessor-based unit, the control module. Several features have been implemented to allow the computation of arbitrarily large problems and to maximize the hardware utilization rate. The first prototype became operational in 1993.

The architecture of the systolic array and of the machine are presented in Secs. 4.2 and 4.3. Programming and performance are discussed in the next two sections. Finally, a thorough analysis, leading to requirements for a more general and efficient system, is given in Sec. 4.6. A description of ANN models is beyond the scope of this chapter. The interested reader is referred to any general introductory book[3]. Since all examples presented here are based on Kohonen's self-organizing feature maps, a brief overview of this algorithm is provided in an appendix at the end of the present chapter.

4.2. The GENES IV Systolic Array

In this section a novel design of the *Generic Element for Neuro-Emulator Systolic (GENES)* array, called *GENES IV*[6], is presented.

4.2.1. *Principle*

Only a few basic operations — efficiently implementable on a systolic array — are necessary to implement widely used ANN models like feed-forward networks using the back-propagation rule — as well as similar simpler algorithms: the Perceptron, Adaline, and delta or least mean square (LMS) rules — Hopfield's recurrent networks, and Kohonen's self-organizing feature maps. The complexity of some operations grows with the number of neurons m (i.e., $O(m)$), while the complexity of the others scale with the number of synaptic connections (i.e., $O(m \cdot n)$ or $O(m^2)$, where n is the number of inputs). The bidimensional GENES IV array is intented for the latter category.

Each PE implements a synaptic connection, storing the corresponding weight $W_{i,j}$. PEs are connected to form a square mesh, as shown in Fig. 4-1. Beside the *no operation (NOP)* command, the GENES IV instruction set is composed of six operations, shown in Table 4-1. Each

Table 4-1. Operations of the GENES IV array (non-transpose mode).

Instruction	PE operation	Array global result
NOP	$h_{i,j} := h_{i,j-1}$	$\vec{\mathbf{O}}^h := \vec{\mathbf{I}}^h$
Matrix-vector product	$h_{i,j} := h_{i,j-1} + W_{i,j} \cdot v_{i-1,j}$	$\vec{\mathbf{O}}^h := \vec{\mathbf{I}}^h + \mathbf{W} \cdot \vec{\mathbf{I}}^v$
Square Euclidean distance	$h_{i,j} := h_{i,j-1} + (v_{i-1,j} - W_{i,j})^2$	$O_i^h := I_i^h + \lvert\vec{\mathbf{I}}^v - \mathbf{W}_i^{\mathrm{T}}\rvert^2$
Minimum element	$h_{i,j} := \begin{cases} h_{i,j-1} & \text{if } h_{i,j-1} \leq v_{i-1,j} \\ N_{\max} & \text{otherwise} \end{cases}$	$O_i^h := \begin{cases} I_i^h & \text{if } I_i^h \leq \min_j (I_j^v) \\ N_{\max} & \text{otherwise} \end{cases}$
Maximum element	$h_{i,j} := \begin{cases} h_{i,j-1} & \text{if } h_{i,j-1} \geq v_{i-1,j} \\ N_{\min} & \text{otherwise} \end{cases}$	$O_i^h := \begin{cases} I_i^h & \text{if } I_i^h \geq \max_j (I_j^v) \\ N_{\min} & \text{otherwise} \end{cases}$
Hebbian learning rule	$W_{i,j} := W_{i,j} + h_{i,j-1} \cdot v_{i-1,j}$	$\mathbf{W} := \mathbf{W} + \vec{\mathbf{I}}^h \cdot \vec{\mathbf{I}}^{v\,\mathrm{T}}$
Kohonen learning rule	$W_{i,j} := W_{i,j} + h_{i,j-1} \cdot (v_{i-1,j} - W_{i,j})$	$\mathbf{W}_i := \mathbf{W}_i + I_i^h \cdot (\vec{\mathbf{I}}^{v\,\mathrm{T}} - \mathbf{W}_i)$

operation takes up to three operands: the matrix **W** and the two input vectors $\vec{\mathbf{I}}^h$ and $\vec{\mathbf{I}}^v$. Its result is either one of the two output vectors $\vec{\mathbf{O}}^h$ and $\vec{\mathbf{O}}^v$ (***matrix-vector product***, ***squared Euclidean distance***, and search for the ***minimum*** and ***maximum elements*** of a vector), or an update of the matrix **W** (***Hebbian*** and ***Kohonen learning rules***).

All operations can be performed on the transpose of the stored matrix, by exchanging the roles of rows and columns. This is mainly used with the matrix-vector product to implement the back-propagation rule.

Virtual matrices larger than the physical array can be decomposed into sub-matrices and processed iteratively. The term $\vec{\mathbf{I}}^h$, that appears in the matrix-vector product and squared Euclidean distance, has been added for that purpose.

To avoid emptying the array before each operation change, and refilling it afterwards, as required in a "conventional" SIMD system, each PE is locally controlled by the ***systolic instruction flow*** shown in Fig. 4-2.

4.2.2. *Implementation*

The actual implementation of the GENES IV array, illustrated in Fig. 4-3, differs from the functional structure presented in Sec. 4.2.1. Input and output ports are located on the north-west to south-east diagonal to "diagonalize" the input vectors and "undiagonalize" the output vectors. The architecture of a GENES IV PE is presented in Fig. 4-4. Inputs and weights are coded as 16-bit two's complement integers (except for a 33-bit intermediate weight representation used for some operations), while outputs are 40 bits wide (including an overflow bit). In order to reduce the silicon area and pin count, the computation and communication are bit-serial. Therefore, the size of the widest register determines the duration of an operation (i.e., 40 clock cycles), called *macro-cycle*.

A 2×2 PE VLSI chip, shown in Fig. 4-5, has been implemented with standard cells in a $1\,\mu$m CMOS technology. It has been successfully tested at 20 MHz.

4.3. The MANTRA I Machine

Fig 4-6 presents the architecture of the *MANTRA I* machine[15,16]. It consists of a *parallel* or *SIMD module* and a *control module.*

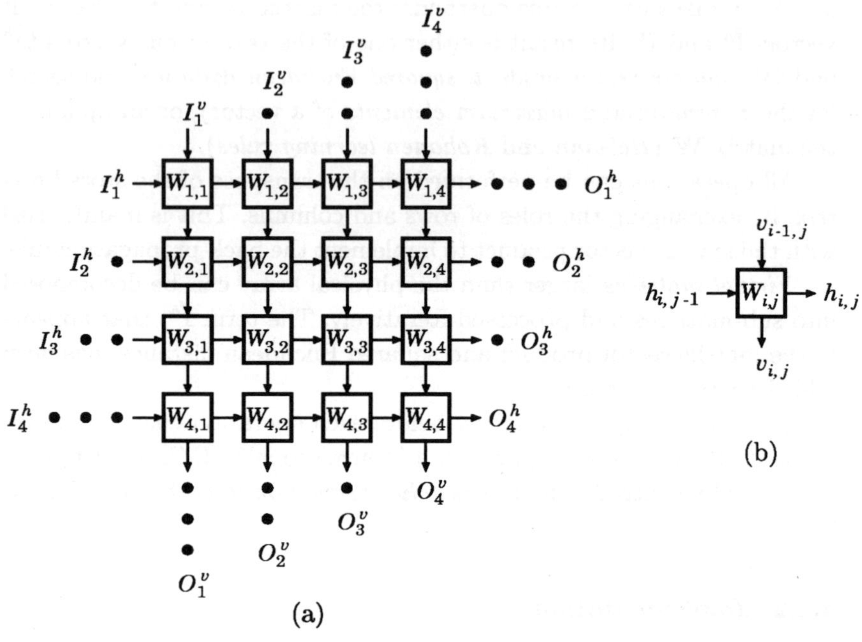

Fig. 4-1. Functional structure of the GENES IV systolic array. (a) Active data streams. (b) Input/output ports of a PE.

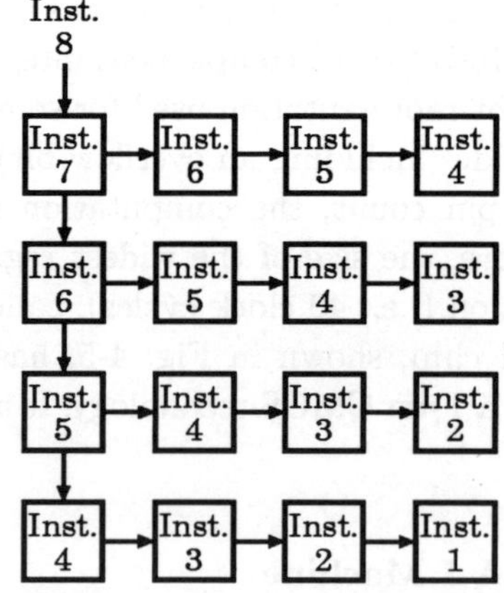

Fig. 4-2. Systolic instruction flow in the GENES IV array.

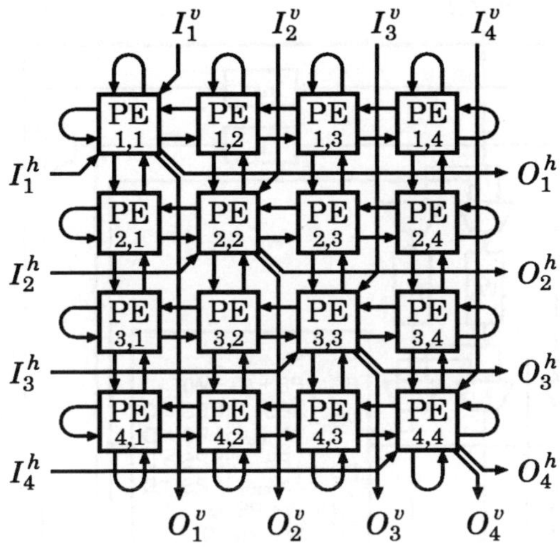

Fig. 4-3. Architecture of the GENES IV systolic array.

The maximum size for the GENES IV array has been chosen as $40 \times 40 = 1600$ PEs. In principle, larger machines can be built by replicating the SIMD module's units.

4.3.1. *The SIMD Module*

As mentioned in Sec. 4.2.1, the GENES IV array is used for $O(m \cdot n)$ or $O(m^2)$ operations, while $O(m)$ operations are performed by dedicated units.

The *delta unit* is a series of dedicated cells computing $(\mathsf{DES} - \mathsf{Y2}) \cdot \mathsf{FY}$. In supervised learning, it is used to generate the error vectors. A custom VLSI chip has been integrated for this purpose (1 μm CMOS, 769 standard cells, 16896 transistors, $3.2 \times 3.2\,\text{mm}^2$).

The main purpose of the *sigma unit* is to apply the 16-bit activation function σ to the 40-bit output of the GENES IV array. To keep the implementation simple and general, a solution based on look-up tables has been chosen. Moreover, to avoid the need for an unrealistic amount of storage (i.e., a 16-bit table with 2^{40} entries represents 2 Tbytes of memory), two tables are used: a *coarse-grain table* mapping the whole input space at reduced precision, and a *fine-grain table* mapping a small

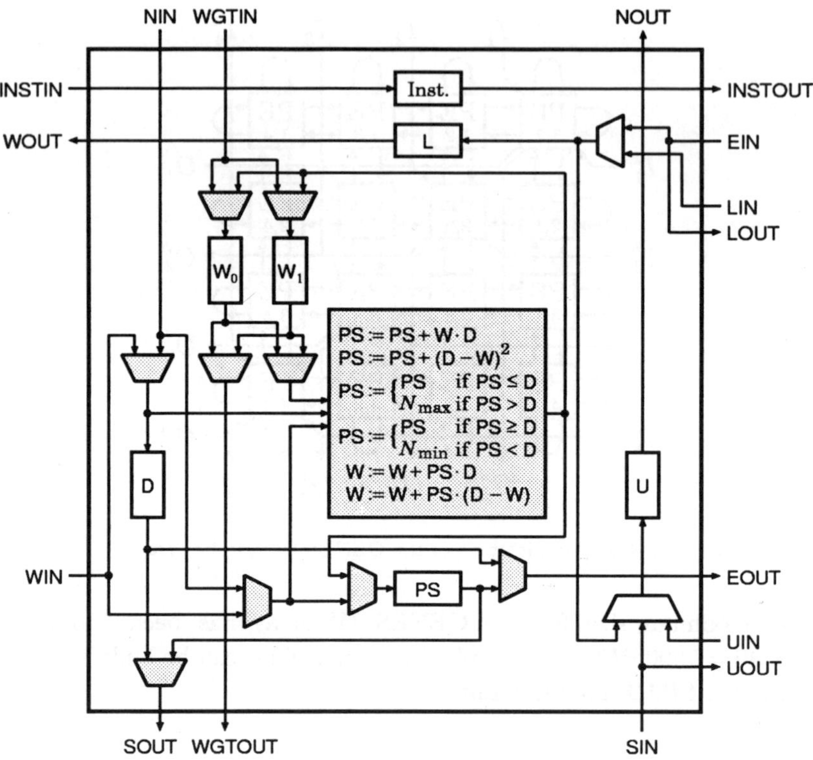

Fig. 4-4. Architecture of the GENES IV diagonal processing element (PE). Shaded units are controlled by the instruction stream, while others are globally controlled by external signals. Non-diagonal PEs are identical, except that the input/output ports UIN, LIN, UOUT, and LOUT are left unconnected, and the control of the corresponding multiplexers is hard-wired.

window at high precision. This fine-grain window can be positioned anywhere in the input space (aligned on multiples of half its size). Its size can be increased by reducing its precision.

Some learning rules require the computation of the first derivative σ' of the activation function. At first sight, a second piece of hardware, identical to the sigma unit, seems to be required. However, the MANTRA I machine takes advantage of the monotonic shape of activation functions to compute their derivative as $\sigma'(p) = \sigma'(\sigma^{-1}(y))$, where $y = \sigma(p)$. This is implemented by the *function-of-Y unit*, a 16-

Fig. 4-5. The GENES IV integrated circuit (3179 standard cells, 71690 transistors, $6.3 \times 6.1\,\mathrm{mm}^2$).

bit look-up table with 2^{16} entries.

To sustain the required data rate, three independent storage units are used for the weights, inputs/outputs, and desired outputs. Several solutions involving static RAMs, FIFOs, dual-port RAMs, and video RAMs have been evaluated[16], before committing to a combination of static RAMs and FIFOs. The chosen implementation uses FIFOs to enter and retrieve data into/from the SIMD module and of static RAMs to store temporary results. It is possible to copy data from a FIFO to memory while loading it into the GENES IV array. It is also possible to write output data to the FIFO and in memory at the same time,

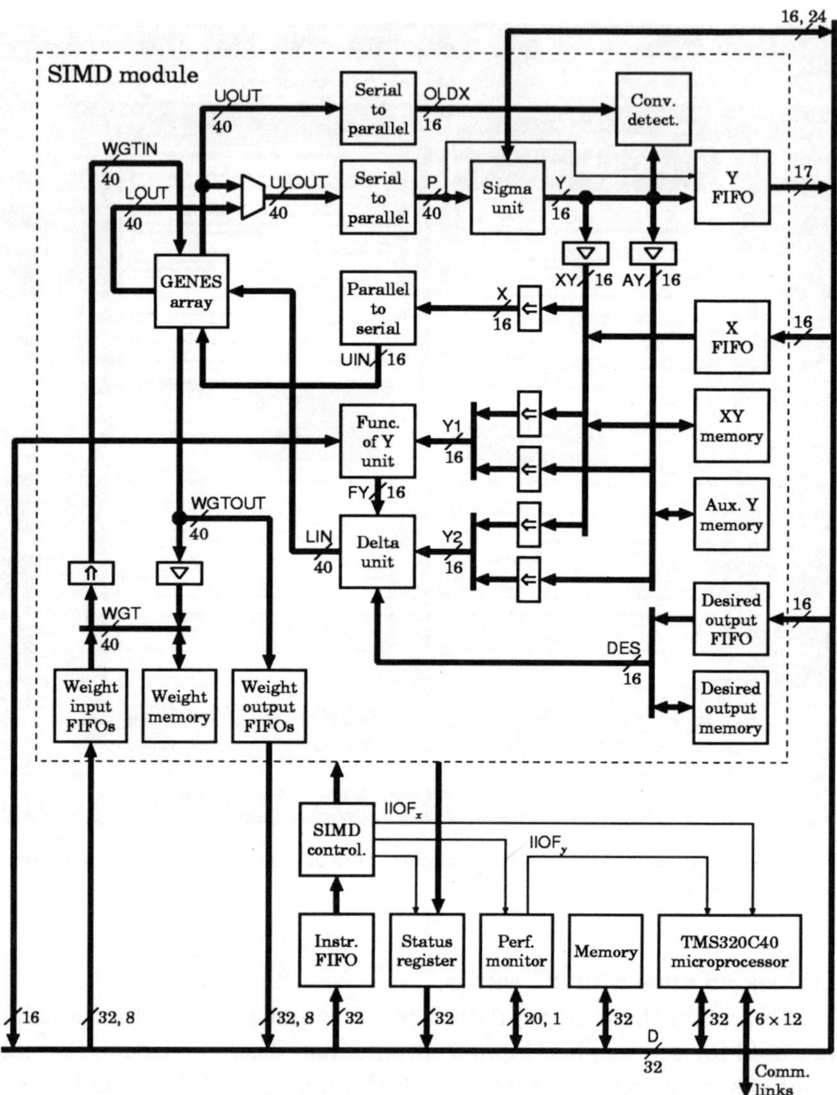

Fig. 4-6. Architecture of the MANTRA I machine. The dashed box represents the SIMD module.

Fig. 4-7. The MANTRA I prototype machine.

allowing to output snapshots of the ANN state.

4.3.2. *The Control Module*

The control module is a complete SISD system, based on the TMS320C40 microprocessor. It controls the SIMD module by dispatching instructions to the *SIMD controller*. The ***instruction FIFO*** serves as an elastic buffer. If this FIFO becomes empty, the SIMD module is frozen. MANTRA I instructions are horizontally coded.

4.3.3. *Hardware*

A prototype of the MANTRA I machine, shown in Fig. 4-7, has been built and tested. This system is based on 20×20 GENES IV PEs (expandable to 40×40 PEs). Four different printed circuit boards are used: (1) a general-purpose processor board (33 chips), (2) the control and storage board (118 chips), (3) the GENES IV input/output board (465 chips), and (4) the GENES IV array board (104 chips). Four such boards are required for configurations larger than 20×20 PEs. The GENES IV clock frequency is 8 MHz, while the microprocessor is running at 16 MHz. Part of the logic is clocked at 32 MHz.

4.4. Programming the MANTRA I Machine

Since resource parallelism is managed in software, programming is a critical aspect of the MANTRA I machine. To hide the complexity of the programming model, the machine is accessed through an *ad hoc* library. The technique used to write libraries consists in decomposing the algorithm in elementary building blocks, each representing the processing of a single block of data independently from concurrent activities. Efficient code is then generated at just-in-time by a fast compaction algorithm[5,7,8].

The Kohonen model (see the present chapter's appendix) for a matrix $\mathbf{w}$ that fits the GENES IV array is presented as an example of mapping ANN algorithms on the MANTRA I machine. The implementation of other models and/or the use of virtual matrices rely on the same principles[16]. Table 4-2 shows the different execution phases. The machine can be viewed as a pipeline, the function-of-Y and delta units being the first two stages, the systolic array the next $2N$ stages (for N^2 PEs), and the sigma unit the last one. For each vector of a batch or *epoch* E, the squared Euclidean distances to all neurons are computed during Phase A. In Phase B, the closest neuron or *winner* is searched. In Phase C, the resulting vector — containing only zeroes, except for the winner(s) — is multiplied by a matrix describing an arbitrary neighborhood. Finally the weight matrix is updated in Phase D. Although the Kohonen model is natively on-line (i.e., non-batch), the algorithm of Table 4-2 is hybrid in that winners are searched by epoch but weights are updated on-line[5]. Theoretical analysis and empirical simulations indicate that this very algorithm implemented on the MANTRA I machine is for all practical purposes equivalent to the original Kohonen model[4].

Table 4-2. Mapping of the Kohonen model on the MANTRA I machine.

	Function-of-Y unit	Delta unit	GENES IV array	Sigma unit
A			$p_i := \lvert\vec{\mathbf{x}} - \mathbf{w}_i^{\mathrm{T}}\rvert^2$	
B			$\mu_i := \begin{cases} p_i & \text{if } p_i = \min(\vec{\mathbf{p}}) \\ N_{\max} & \text{otherwise} \end{cases}$	$v_i := \begin{cases} 1 \text{ if } \mu_i \neq N_{\max} \\ 0 \text{ if } \mu_i = N_{\max} \end{cases}$
C			$\vec{\mathbf{v}} := \alpha \cdot \boldsymbol{\lambda} \cdot \vec{\mathbf{v}}$	$\vec{\mathbf{v}} := \vec{\mathbf{v}}$
D	$\vec{\mathbf{v}} := \vec{\mathbf{v}}$	$\vec{\mathbf{v}} := \vec{\mathbf{v}}$	$\mathbf{w}_i := \mathbf{w}_i + \nu_i \cdot (\vec{\mathbf{x}}^{\mathrm{T}} - \mathbf{w}_i)$	

4.5. Performance Analysis

For the Kohonen algorithm, the ideal performance of the MANTRA I machine is:

$$P_{\text{ideal}} = \frac{\lceil n/N \rceil}{2 \cdot (\lceil n/N \rceil + \lceil m/N \rceil)} \cdot \frac{N^2 \cdot f}{S_{\text{reg}}} \tag{4-1}$$

where m is the number of neurons, n is the number of inputs, N^2 is the number of PEs, f is the clock frequency, and $S_{\text{reg}} = 40$ is the number of clock cycles per macro-cycle. The inverse of the first fraction in Eq. (4-1) is the number of passes through the array to update N^2 weights (e.g., four passes for non-virtual matrices, as indicated in Table 4-2). At a clock rate of 8 MHz and with 20×20 PEs, the peak performance of the current prototype is up to 40 million of connection updates per second (MCUPS), depending on the map size and the number of inputs. In practical cases, the performance degrades from the ideal value and can be expressed as $P = U_{\text{s}} \cdot U_{\text{t}} \cdot U_{\text{d}} \cdot P_{\text{ideal}}$, where:

- U_{s} is the *spatial utilization rate*, expressing the fact that, in dependence with the size of the map, some processors may be left idle during some phases of the computation.
- U_{t} is the *temporal utilization rate*, indicating the proportion of active instructions (i.e., other than NOP).
- U_{d} is the *dynamic utilization rate*, which is the rate of clock cycles when the SIMD module is active over the total. It models situations when the control processor is delaying the parallel module because of unavailability of data or instructions.

Each component of performance reduction is studied separately in the next sections.

4.5.1. *Spatial Utilization Rate*

The spatial utilization rate drops below unity only owing to neuron or input counts not multiple of N. The performance deteriorates because the actual number of connections computed per iteration is reduced by a factor of:

$$U_{\text{s}} = \frac{n \cdot m}{\lceil n/N \rceil \cdot \lceil m/N \rceil \cdot N^2} \tag{4-2}$$

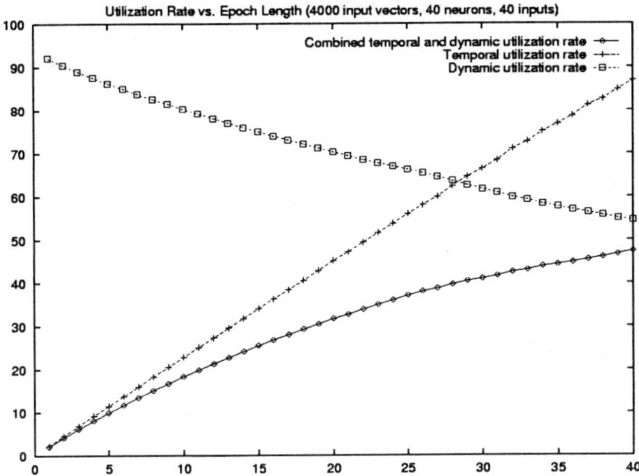

Fig. 4-8. Influence of the epoch length on the utilization rate.

This is essentially the sole effect of neuron or input counts different from multiples of the array size. Hence, it is disregarded in this analysis and only dimensions multiple of N are taken into account. This effect becomes negligible for $n, m \gg N$.

4.5.2. *Temporal Utilization Rate*

The variable that has the greatest influence on the temporal utilization rate is the epoch length $|E|$, whose effect is presented in Fig. 4-8. As typical of pipelined architectures, idle cycles have to be inserted to account for epochs smaller than the pipeline length. Hence, the number of MANTRA I instructions is almost constant for any epoch length smaller than $2\,N$ vectors and, thus, the performance increases linearly. In practice, it grows sub-linearly because more data are sent per iteration, lowering the dynamic utilization rate, as shown in Fig. 4-8. Epochs of $2\,N = 40$ vectors are used in the rest of the analysis.

4.5.3. *Dynamic Utilization Rate*

The most critical component of performance degradation is the communication bottleneck between the control DSP and the SIMD module. The DSP may delay the SIMD module (1) to generate MANTRA I in-

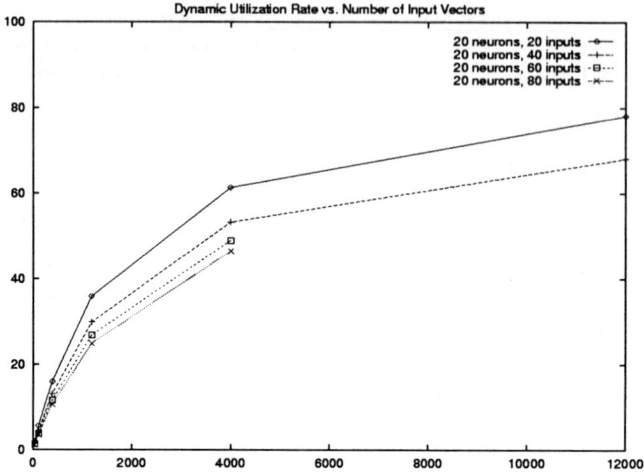

Fig. 4-9. Influence of the number of input vectors on the dynamic utilization rate.

structions, (2) to load input data in the FIFOs, or (3) to retrieve output data from the FIFOs.

The strongest influence on the dynamic utilization rate is that of the number of input vectors, as shown in Fig. 4-9. For small problems, the initial and final overheads make the machine inefficient, due mainly to the preparation required to send the weight matrices to the array. With larger problems, this overhead is averaged over many iterations.

Sending the vectors is the main communication task during the training phase. This communication is responsible for the moderately negative impact of large input counts, presented in Fig. 4-10. On the other hand, the number of neurons has an almost negligible influence on the dynamic performance, as shown in Fig. 4-11. Only communication operations taking place at the beginning and end of the learning process depend on the number of neurons and the corresponding quantities are meanwhile stored in the SIMD module.

4.5.4. *Global Performance*

The performance of the MANTRA I machine measured over a set of Kohonen learning tasks is shown in Fig. 4-12. In the most favorable conditions, a fraction slightly below 70 % of the ideal performance has been measured.

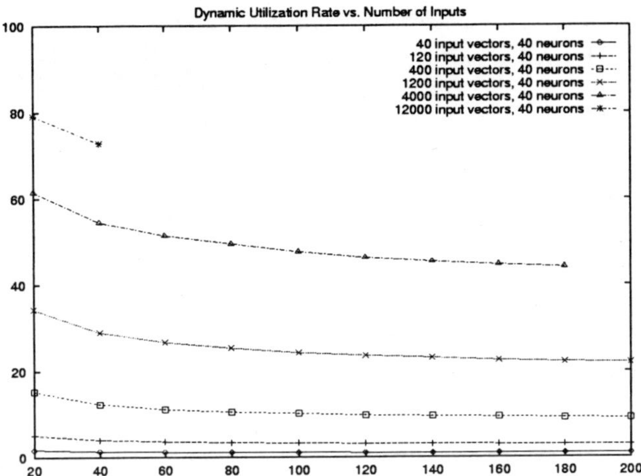

Fig. 4-10. Influence of the number of inputs on the dynamic utilization rate.

4.6. Future Evolution

In this section, the requirements for dedicated massively parallel systems for ANN computation are discussed in the light of the experience acquired with this project. The main drawbacks of the MANTRA I machine are its lack of generality and complex programming model. Future research should mainly concentrate on these two issues. This raises the question of whether dedicated systems can be more cost-effective than general-purpose hardware (e.g., based on RISCs or DSPs). The performance of the MANTRA I machine, presented in Sec. 4.5, is not very impressive compared to today's microprocessors. However, two factors should be taken into account. First, the design of the machine started in early 1992 and the prototype was already working in 1993. Thus, a fair comparison should be based on systems of the same period. More severely, the machine has been built using the technology available to a university and, hence, only few sub-circuits have been integrated and the CMOS technology used offers only a relatively low level of integration (the GENES IV circuit contains 71690 transistors vs. one million transistors for a state-of-the-art microprocessors in the early 90's and several millions today).

Therefore, the MANTRA I machine should be considered as a "test

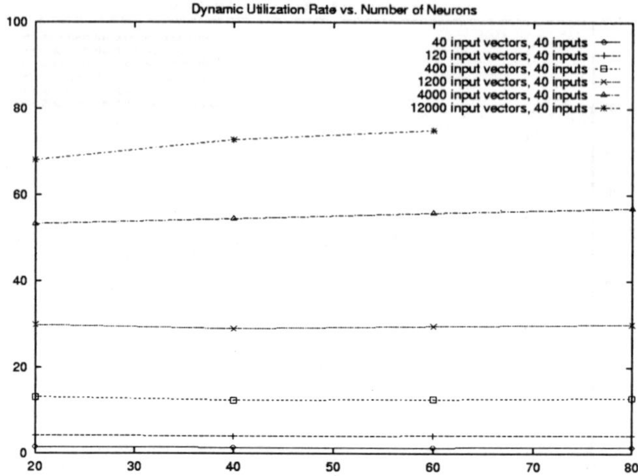

Fig. 4-11. Influence of the number of neurons on the dynamic utilization rate.

prototype" used to study architectural ideas. The performance analysis of Sec. 4.5 shows that it is possible to sustain a good efficiency. At the same time, several weaknesses of the machine, analyzed in this section, have also been outlined. This leads to the conclusion that a carefully engineered system — based on state-of-the-art ASIC technology — preserving the key features of the MANTRA I machine while extending its capabilities toward a greater generality and ease-of-use[8] could be a cost-effective alternative to general-purpose computers.

The natural topology of the MANTRA I machine featuring a bidimensional array for $O(m \cdot n)$ or $O(m^2)$ operations, and a linear collection of dedicated units for $O(m)$ operations could be retained and generalized. A possibility would be to have a bidimensional array of ***synaptic PEs*** (e.g., a generalization of the GENES IV PEs) attached to a linear array of ***neural PEs*** (e.g., resembling today's DSPs). The interconnection topology will depend on the architecture of both types of PEs. In particular, it is debatable whether the input/output ports should be located on the diagonal, as on the GENES IV array. Alternatively, it would also be possible to use bidimensional algorithms to perform the operations that are executed by dedicated units on the MANTRA I machine, at the price of a slight decrease in throughput[8]. Whichever approach is chosen, the system should be directly attached

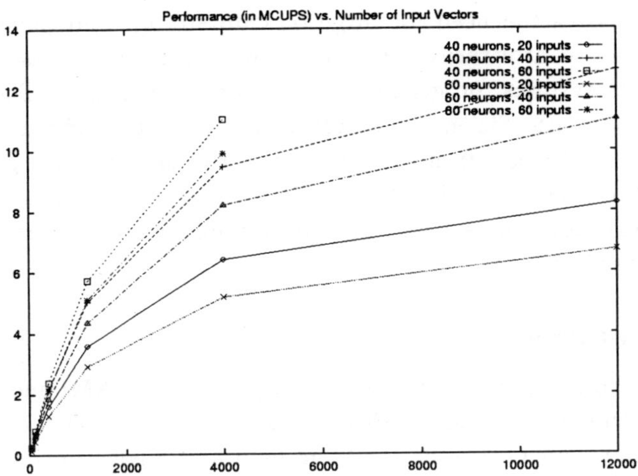

Fig. 4-12. Performance of the system in million of connection updates per second.

to the bulk of memory and not only to some temporary storage.

To keep the programming effort manageable, a necessary feature is that the relevant part of the instruction follows the corresponding data along the whole data path, as in the GENES IV array (but not in the whole MANTRA I machine). However, this is not sufficient, and techniques similar to the implemented just-in-time compaction algorithm should be developed[8] to allow real-time scheduling of the machine.

Finally, parallel or partially parallel (e.g., 4 or 8 bits parallel) communication and computation as well as floating-point numbers could be used for increased performance and ease-of-use.

4.7. Conclusion

A working prototype of a programmable systolic array dedicated to a broad class of ANN models has been presented. The use of a bidimensional systolic mesh as the computational core of the machine differentiates this system from most of those described in literature. Here, PEs correspond to synapses instead of neurons, therefore using efficiently large numbers of PEs or, conversely, running with good utilization rates on smaller problems. Also the data bandwidth necessary for the bidimensional array is much smaller compared to unidimensional systems.

This removes the need for large on-chip storage and makes the PEs adaptable to arbitrarily sized problems. As it is the case with most pipelined systems, programming is a burdensome task. The approach chosen to write efficient code indicates that simple software techniques can relieve the user from a large part of the task. Finally, performance has been discussed to identify the limitations of the approach. Several problems should be solved to produce more general systems, and some have been outlined.

Acknowledgments

The authors wish to thank their colleagues in the LAMI and MANTRA teams and, in particular, Prof. J.-D. Nicoud, Dr. C. Lehmann, Dr. T. Cornu, and G. Caset, for their help and advice during all phases of the project. This work was supported by the Swiss National Fund for Scientific Research through the SPP-IF grant 5003-34353 and by the Swiss Federal Institute of Technology.

Appendix: Kohonen's Self-Organizing Feature Maps

Kohonen's self-organizing feature maps result from an unsupervised method to map a large-dimensionality input space onto an output space usually of much smaller dimensionality. Each neuron is associated with a row $\mathbf{w}_i$ of the weight matrix $\mathbf{w}$, also called *weight vector*. When an input vector $\vec{\mathbf{x}}$ is presented, its Euclidean or Manhattan distance to the weight vector of each neuron is computed. The weights of the closest neuron i_{win} or winner (i.e., $y_{i_{\text{win}}} = \min_i (y_i)$) and its neighbors are moved in the direction of the input vector:

$$\mathbf{w}_i := \mathbf{w}_i + \alpha \cdot \lambda_{i,i_{\text{win}}} \cdot \left(\vec{\mathbf{x}}^{\mathrm{T}} - \mathbf{w}_i\right)$$

The *neighborhood coefficients* $\lambda_{i,j}$ define any topological relationship between neurons.

References

[1] M. A. Glover and W. T. Miller, III. A massively-parallel SIMD processor for neural network and machine vision applications. In J. D. Cowan, G. Tesauro, and J. Alspector, editors, *Advances in Neural Information Processing Systems*, volume 6, pages 843–849. Morgan Kaufmann, San Mateo, CA, 1994.

[2] D. Hammerstrom. A highly parallel digital architecture for neural network emulation. In J. G. Delgado-Frias and W. R. Moore, editors, *VLSI for Artificial Intelligence and Neural Networks*, chapter 5.1, pages 357 – 366. Plenum Press, New York, NY, 1991.

[3] J. Hertz, A. Krogh, and R. G. Palmer. *Introduction to the Theory of Neural Computation*. Addison-Wesley, Redwood City, CA, 1991.

[4] P. Ienne, P. Thiran, and N. Vassilas. Modified self-organizing feature map algorithms for efficient digital hardware implementation. *IEEE Transactions on Neural Networks*, NN-8(2):315 – 330, Mar. 1997.

[5] P. Ienne and M. A. Viredaz. Implementation of Kohonen's self-organizing maps on MANTRA I. In *Proceedings of the Fourth International Conference on Microelectronics for Neural Networks and Fuzzy Systems*, pages 273 – 279, Turin, Sept. 1994.

[6] P. Ienne and M. A. Viredaz. GENES IV: A bit-serial processing element for a multi-model neural-network accelerator. *Journal of VLSI Signal Processing*, 9(3):257 – 273, Apr. 1995.

[7] P. Ienne. Horizontal microcode compaction for programmable systolic accelerators. In *Proceedings of the International Conference on Application Specific Array Processors*, pages 85 – 92, Strasbourg, July 1995.

[8] P. Ienne Lopez. *Programmable VLSI Systolic Processors for Neural Network and Matrix Computations*. Ph.D. thesis no. 1525, EPFL, Lausanne, 1996.

[9] S. R. Jones, K. M. Sammut, and J. Hunter. Learning in linear systolic neural network engines: Analysis and implementation. *IEEE Transactions on Neural Networks*, 5(4):584 – 593, July 1994.

[10] J. C. Lawson, N. Maria, and J. Hérault. SMART: A neurocomputer using sparse matrices. In *Proceedings: Euromicro Workshop on Parallel and Distributed Processing*, pages 59 – 64, Gran Canaria, Jan. 1993.

[11] R. W. Means and L. Lisenbee. Extensible linear floating point SIMD neurocomputer array processor. In *IJCNN International Joint Conference on Neural Networks*, volume I, pages 587 – 592, Seattle, WA, July 1991. IEEE, INNS.

[12] N. Morgan, J. Beck, P. Kohn, J. Bilmes, E. Allman, and J. Beer. The Ring Array Processor: A multiprocessing peripheral for connectionist applications. *Journal of Parallel and Distributed Computing*, 14(3):248 – 259, Mar. 1992.

[13] U. A. Müller, B. Bäumle, P. Kohler, A. Gunzinger, and W. Guggenbühl. Achieving supercomputer performance for neural net simulation with an array of digital signal processors. *IEEE Micro*, 12(5):55 – 65, Oct. 1992.

[14] U. Ramacher. SYNAPSE — a neurocomputer that synthesizes neural algorithms on a parallel systolic engine. *Journal of Parallel and Distributed Computing*, 14(3):306 – 318, Mar. 1992.

[15] M. A. Viredaz. MANTRA I: An SIMD processor array for neural

computation. In P. P. Spies, editor, *Europäischer Informatik Kongreß Architektur von Rechensystemen Euro-ARCH '93*, pages 99 – 110. Springer-Verlag, Berlin Heidelberg, Oct. 1993.

[16] M. Viredaz. *Design and Analysis of a Systolic Array for Neural Computation.* Ph.D. thesis no. 1264, EPFL, Lausanne, 1994.

CHAPTER 5

MIXED-SIGNAL NEURO-FUZZY PROCESSOR IMPLEMENTATIONS: SEQUENTIAL ARCHITECTURES AND CIRCUIT-LEVEL DESCRIPTION

J. Madrenas and E. Alarcón

Department of Electronics Engineering, Universitat Politècnica Catalunya
Gran Capità s/n, Mòdul C4 Campus Nord UPC. E08034 Barcelona, Spain
E-mail: madrenas, ealarcon@eel.upc.es

Mixed-signal VLSI design provides a tradeoff between the fast and compact but fixed analog implementations of neuro/fuzzy feedforward algorithms and the programmable but area and power consuming digital counterparts. In this chapter, a sequentiality study of such systems is performed. Basic blocks for sequential mixed-signal neural and fuzzy computing are proposed, and two sequential example processors are described. Feedback from the designed processors and subcircuits allowed considering the technology constraints for analysis and extension to different sequentiality degrees.

5.1. Introduction

In this chapter, we introduce architecture analysis and guidelines to develop sequential analog processing circuits. By means of switched techniques and analog memory elements, analog sequential processing circuits can be designed. These techniques are applied to the development of mixed-signal neural and fuzzy processors. For this purpose, the characteristics, constraints and properties of analog circuits, that are very different from their digital counterparts, have to be taken into account. At the architectural level, starting from sequential architectures of neural and fuzzy models, a performance analysis in terms of area and speed as a function of the sequentiality –or parallelism- degree is developed.

Efficiency in terms of silicon area occupancy versus computation performance (speed) is fundamental in autonomous intelligent systems. For perception tasks, such systems take advantage of several proposed neural and fuzzy models. In real-time operation, hardware implementation of the required models can be most suitable. However, for a given application, the hardware resources should be adapted to the specific performance requirements in order to reduce silicon and power cost. In this paper we analyze mixed signal sequential architectures that enable this tradeoff.

Analog and mixed-signal hardware implementations of neural networks and fuzzy controllers exhibit area-efficient, high-speed and low-power characteristics compared with digital realizations. This is achieved by fully exploiting the transistor characteristics, at the expense of low resolution and an increased sensitivity to noise, temperature and non-idealities [1-3].

An important weak point in analog implementations is the lack of suitable memory elements. However, in the last years several dynamic and non-volatile analog memory circuits have been developed. This enables the implementation of compact analog architectures.

Although precision requirements for the learning phase of neural and neuro-fuzzy systems can be high, for the recall phase they are less than 8 equivalent bits in most neural [4] and fuzzy [33] applications. Therefore, proper analog circuits can be used to perform this phase. In addition, learning can be done by means of chip-in-the-loop techniques.

Usually, analog implementations are fully parallel one-to-one mappings of the processing network. This results in high complexity of interconnects between processing elements as a consequence of the so-called curse of dimensionality. Despite the compact analog implementations, for real-world applications the number of synapses and processing units usually becomes too high to justify a fully parallel implementation, especially when speed can be reduced and still match the application needs.

In contrast, hybrid parallel/sequential architectures can match the analog processor cost, speed and power consumption to the application constraints. Furthermore, in general, sequentialization exhibits flexibility to emulate a different number of network structures and models (e.g. evolutive networks) with much better efficiency than a fully parallel architecture.

As far as analog sequential processing is concerned, we can take advantage of several developments on switched-current circuits. These techniques perform current-mode, discrete-time processing [5], using the MOS gate capacitor for temporary charge storage.

Since analog dynamic memories are available, sequential processing schemes can be envisaged. A digital block becomes necessary to control the sequential processing, so such systems require mixed-signal design techniques. However, in contrast tomost mixed-signal circuits, a distinct feature of the proposed circuits is that the main processing is performed by the analog part.

Although several mixed-signal sequential realizations have been reported, e.g. [6-9], to our knowledge no study has been done up to now on generalization of the area-delay tradeoff. In this work, we consider two reported sequential realizations that emulate feedforward algorithms (MLP, RBF and Takagi-Sugeno –TSK- fuzzy controllers) for the recall phase. The common points and sequentialization possibilities are analyzed, and a consistent realization style that allows tuning sequentiality is proposed.

In Section 2 we explore the area-delay design space of analog neural and fuzzy processors as a function of sequentiality degree. In Section 3 a testchip containing two sequential analog processors and a digital controller is briefly described. Both case studies, TSK fuzzy- and MLP neural-emulating example processors are described in Sections 5.4 and 5.5. Finally, the conclusions are discussed.

5.2. Sequentiality Analysis

The straightforward solution for an analog implementation of a neural or fuzzy system is one-to-one mapping of each synapse and activation function of processing elements onto physical resources. In this case, internal memory is only needed to store the system parameters.

At the opposite side lies a single processor emulating serially every synapse and activation function. In this case, in addition to parameters, data memory is required to store temporary intermediate values, as it occurs in single-processor digital implementations that sequentially execute a program.

An equivalent mixed-signal elementary processor is shown in Fig. 5-1, for (a) MLP, (b) fuzzy controllers and (c) Radial Basis Functions (RBF). Aside of specific nonlinear functions (sigmoid or distance/t-norm calculations), the multiplier-accumulator is a common element. It employs a temporary storage element (Analog Transfer Cell, ATC). Additionally, necessary elements are a digital controller and parameter storage elements, not shown in the figure.

Between the fully parallel or fully sequential extreme solutions, any combination of processors executing in parallel, each one emulating several processing elements, is possible. In the following subsections two particular reported architectures of this kind are discussed.

5.2.1. *MLP-like Emulation*

During its execution phase, an MLP is composed of layers of neurons, each one performing a nonlinear function of a weighted sum of its inputs. In Fig. 5-2 (a) a neuron is represented, in which all its synapses are being emulated in parallel. A second neuron is represented by shaded lines. It corresponds to the shaded drawing in Fig. 5-2 (b), as will be described later.

In that figure, a synapse-parallel MLP emulating processor for the execution phase (as opposite to learning phase) is shown [8,12,17]. For this kind of systems, either synapse or neuron parallelism can be applied. In the former, synapse aggregation takes place in space, while for the latter it takes place in time.

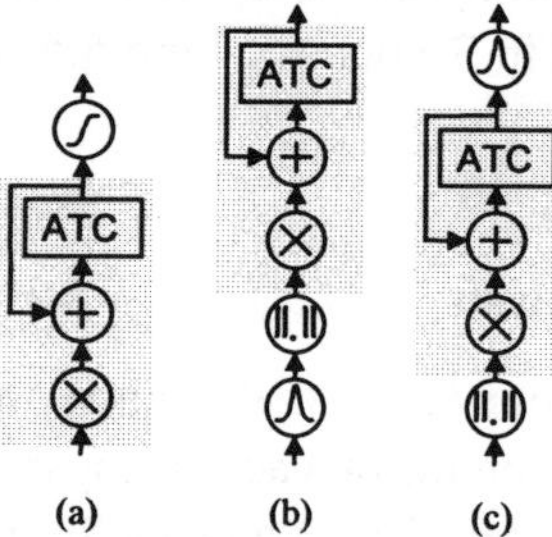

Fig. 5-1. (a) MLP slice. (b) fuzzy slice. (c) RBF slice.

The processor operation principle is based on the sequential emulation of each neuron on a layer-by-layer basis. Let us consider only the black lines. Briefly explained, the input data vector is assumed to be stored in a first row of ATCs. At each clock cycle, all the synapses of one neuron and its activation function are computed. The activation outputs are sequentially stored in the second row of ATCs. After emulating all the neurons on the first neuron layer, the computed outputs are used as inputs to emulate the next layer. The procedure is repeated until last layer is emulated. Two rows of n ATCs, n synapses and one activation function are enough to emulate a network of an arbitrary number of layers, each one consisting of up to n neurons.

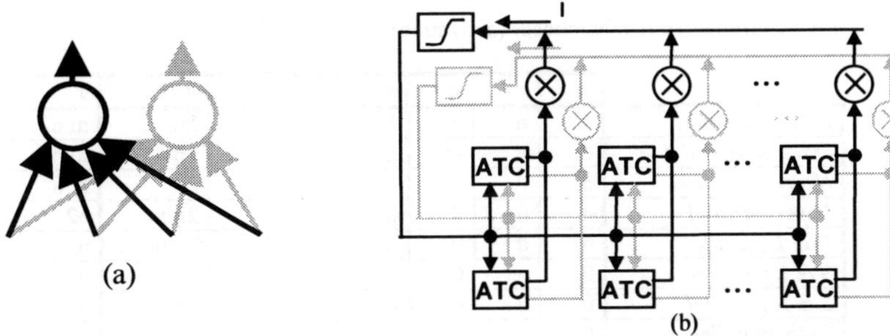

Fig. 5-2. (a) Synapse and neuron parallelism. (b) MLP sequential synapse-parallel realization.

The shadowed drawing in Fig. 5-2 (b) shows the additional elements needed to add one degree of neuron parallelism to the processor. The area overhead is almost duplicated (only ATCs stay the same), but also speed since two activation outputs are calculated at each time.

Comparing synapse and neuron parallelism, the former is more area efficient since the number of required activation functions reduces to one, while for each neuron to emulate in parallel an activation function circuit is needed.

Also, layer parallelism can be considered. As in the case of neuron parallelism, for each parallel layer an extra activation function is required. Furthermore, in this case additional ATCs to store intra-layer data are needed. In Table 5-1, figures for required resources and execution cycles for different degrees of sequentiality schemes are shown. ss, sn and sl stand for synapse, neuron and layer sequentialization indexes, the maximum value representing fully sequential and value 0 meaning fully parallel. For the sake of simplicity, index n is the number of synapses per processing unit and also the number of units per layer. However, extension to different indexes is straightforward. Index m represents the number of layers. The first row corresponds to a fully sequential system, the shadowed row represents the reported example synapse-parallel processor [12], and the last row corresponds to a fully parallel system. Control overhead is not considered.

Table 5-1. Architecture complexity and execution time as a function of sequentiality. # p: Multipliers; # f(·): Nonlinear (activation) functions; T: Clock cycles.

	ss	*sn*	*sl*	*#p*	*#f(·)*	*#ATC*	*T*
Fully sequential	n	n	m	1	1	2n	m·n2
Synapse parallel	0	n	m	n	1	2n	m·n
Neuron parallel	n	0	m	n	n	2n	m·n
Layer parallel	n	n	0	m	m	2n·m	n2
N & L parallel	n	0	0	n·m	n2	2n·m	n
S & L parallel	0	n	0	n·m	m	2n·m	n
S & N parallel	0	0	m	n2	n	2n	m
Fully parallel	0	0	0	m· n2	n2	0	1

The second sequential example architecture was reported in [9]. It is depicted in Fig. 5-3 (a), and, concerning hardware complexity, efficiently maps the Fuzzy Controller for the widely accepted Takagi-Sugeno (TSK) first-order model [10]. This widely-accepted TSK FC model is the best suited for implementation purposes, since, appart from being appealing at theoretical level [34][35][39], it significantly reduces hardware complexity at the output defuzzyfication layer, using input-related output hyperplanes in place of output membership functions. The first-order TSK fuzzy controller is described by the following set of IF-THEN rules:

$$\begin{aligned} &\text{IF x1 is } \Omega_1^l \text{ and ... and xn is } \Omega_n^l \\ &\text{THEN } d^l = c_0^l + c_1^l x_1 + \ldots + c_n^l x_n \end{aligned} \tag{5-1}$$

where Ω_i^l represent input fuzzy sets, c_i^l are constants which define the consequent hyperplanes, and l=1,2,...,M are the number of rules of the fuzzy processor. Given this inference core, for a certain input vector $\vec{x} = (x_1,\ldots,x_n)^T \subset \Re^n$, and when considering the minimum operation as the connective and, the crisp output is computed using the discrete center-of-mass or weighted average method as:

$$o = f(\vec{x}) = \frac{\sum_{l=1}^{M} \omega_l d_l}{\sum_{l=1}^{M} \omega_l} = \frac{\sum_{l=1}^{M} \min_{k=1}^{n}\{F_{l(k),xk}(x_k)\} \cdot \left(\sum_{k=1}^{n} x_k c_k^l\right)}{\sum_{l=1}^{M} \min_{k=1}^{n}\{F_{l(k),x_k}(x_k)\}} \tag{5-2}$$

where F_l, x_k stand for the membership functions associated to the Ω_k^l fuzzy set.

The motivation for the sequential fuzzy controller is again the capability to exchange speed for area and emulation flexibility. Table II summarizes the hardware complexity of the inference block (Fig. 5-3 (b)) in terms of the degree of sequentiality. An n-dimensional input space, with a granularity of m membership functions per input dimension, and the associated mn rules are considered as the general description of the operation of the fuzzy controller.

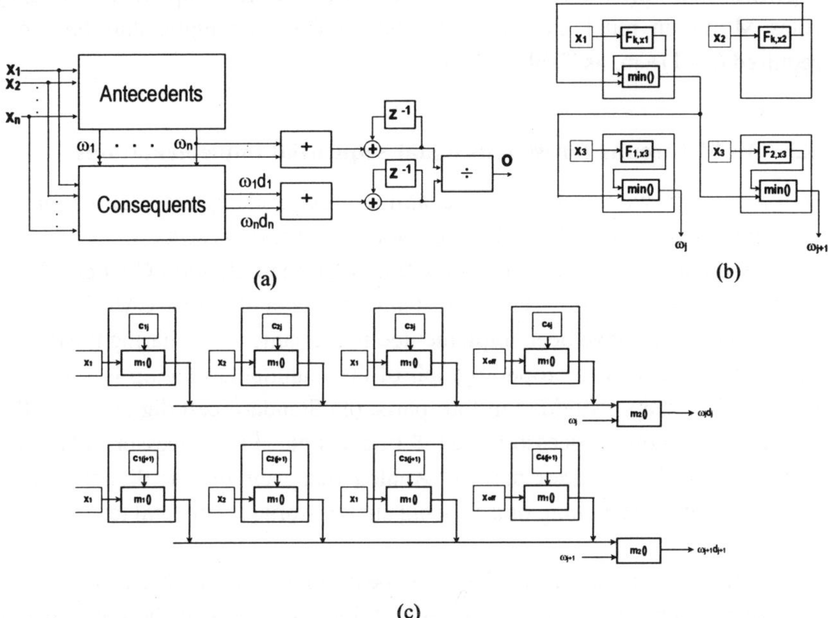

Fig. 5-3. (a) Diagram of the FC analog sequential architecture example. (b) Inference organization. (c) Input-dependent consequents organization.

Considering a designed testchip [11] [39], which is a 3-input and 2-membership-functions per input FC, an architecture exhibiting a certain degree of sequentiality and parallelism is described (shaded row in Table 5-2). In Fig. 5-3 (b), the configuration for the inference block with 3 inputs (x1, x2, x3) and 2 membership functions defined per input is shown. Concerning the consequent block, it consists of a weighted sum of products and thus it can be analyzed with the guidelines given for MLP implementations. The consequent block implements a 1st-order TSK model (di in Fig. 5-3(a)), very suitable for control applications [10]. The blocks labelled Fk,xi in the figure implement the k-th membership function for the i-th input, while the blocks labelled min() provide at their outputs the minimum of their two input signals. The membership functions for the two first inputs are calculated sequentially, while those of the remaining input are calculated in parallel. Thus, at each emulation step two fuzzy rules comprising the 3 inputs are yielded. As a consequence, the 8 rules of the system are calculated in 4 emulation cycles. In a general case, the inference block would provide the complete set of fuzzy rules M=mn in mn-1 cycles, being the number of membership functions required (n –1)+ m, as Table 5-2 shows.

5.3. VLSI Circuits for Mixed-Signal Sequential Implementations

In this section, we introduce two example processor implementations of the sequential architectures [11,12] described in the previous section.

Two sequential processors, emulating MLP (neural) and FC models have been designed in a full-custom testchip, using a 0.8 μm CMOS analog technology. They have been implemented in the same chip, sharing the digital control resources. A microphotograph of the manufactured chip is shown in Fig. 5-4. The chip is divided in four parts: (1) Standard-cell digital controller common for both processors, (2) 8-rule/3-input FC processor, (3) 8-bit 3-synapse MLP processor and (4) Analog delay circuit for digital control signals to reduce switching noise during analog storage. The total testchip area is 10.6 mm2.

A flexible digital controller has been included to check processors performance and their individual cells. A logic diagram of the controller is shown in Fig. 5-5. Its operating principle is as follows. An 8-bit data bus sequentially provides, at a frequency fCK, from a memory or an external host, data bytes that are distributed to control registers. A counter selects at each clock cycle the target control register. The right hand registers (static registers) are written once and store fixed information that remains constant until a reset operation is performed. These registers are used to select processor configurations. On the contrary, the left hand registers (dynamic registers) are precharged and updated synchronously at a rate fS. These registers temporary store changing (dynamic) data during processing as, for instance, weight parameters. fs must be r times slower than fCK, where r stands for the number of dynamic registers.

Table 5-2. Architecture complexity of the inference block as a function of the degree of sequentiality s. # M.F: Number of membership functions. # FR: Number of fuzzy rules delivered in each cycle.

	s	*# M.F.*	*# min.*	*T*	*# FR*
Fully sequential	n	n	n-1	m^n	1
	n-1	(n-1)+m	(n-2)+m	m^{n-1}	m
	...	...	...	...	...
	2	2+m·(n-2)	$1+\sum_{i=1}^{n-2} m^i$	m^2	m^{n-2}
	1	1+m·(n-1)	$\sum_{i=1}^{n-1} m^i$	m^1	m^{n-1}
Fully parallel	0	m·n	$\sum_{i=2}^{n} m^i$	1	m^n

This way, the digital controller can be used as a general control unit of switched sequential processors. In this case it is shared by both implemented processors. This flexible digital controller has been included to check all the processor capabilities. In a final design, the digital controller area could be significantly reduced.

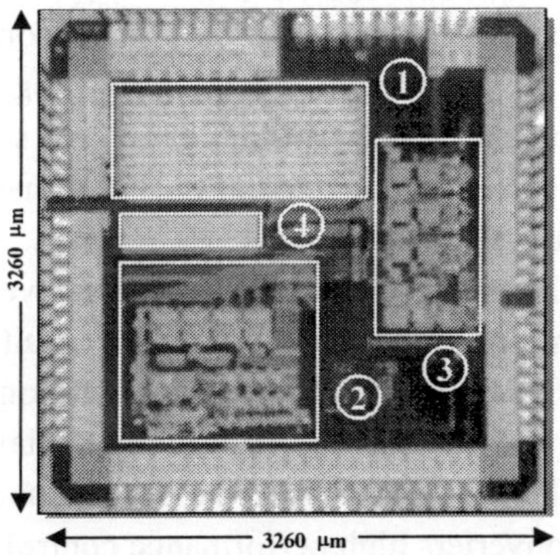

Fig. 5-4. Testchip including MLP and FC sequential processors.

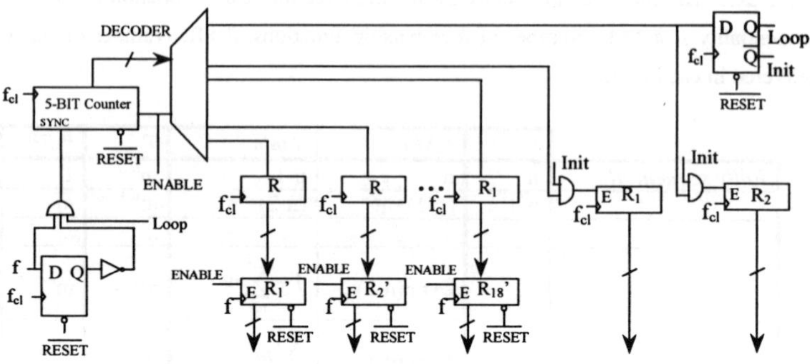

Fig. 5-5. Digital controller for the example processors.

5.4. Case study 1: Mixed-Signal Implementation of a Discrete-time Sequential Takagi-Sugeno Fuzzy Controller

This section presents the design of a mixed-signal VLSI architecture that efficiently maps the processing required to emulate a Fuzzy Controller (FC) of the Takagi-Sugeno (TSK) type [38]. Despite the increasing acceptance of this TSK model due to its flexibility and theoretical advantages, implementations have resorted hitherto to classical Mamdani fuzzy controllers. At circuit level, the proposed FC is constituted by digitally-programmable current-mode analog nonlinear blocks to perform inference, and switched-current memory cells to store in current-mode the partial information resulting from the sequential operation. Details of a novel pulse-width-modulated defuzzyfication scheme are included. Transistor-level post-layout simulation results as well as preliminary experimental results for a 0.8μm CMOS technology are included which validate the operation of the different blocks, their compatibility and the feasibility of the sequential architecture. This FC is intended to provide multidimensional nonlinear-function approximation for switching power converters high performance control [34], [35].

5.4.1. *Motivation*

Fuzzy systems theory proposes a systematic method for mapping human knowledge into a multidimensional input-output nonlinear relation. Several real-world control engineering tasks require this universal approximation characteristic provided by fuzzy inference engines. In the application presented herein, the need for nonlinear control surfaces arises in the area of high-frequency switching power converters, which are inherently nonlinear systems requiring highly nonlinear control laws. In this context, the need for high-frequency operation dictates the development of dedicated hardware implementation.

When the dynamics of the systems are known or capable of being modelled, as is the case for DC-DC switching converters, the analytical control law can be, in general, derived, and thus, there is an a priori knowledge of the nonlinear law to be approximated by the FC. In this situation, the use of neural-like algorithms to adjust or tune both the characteristics of membership functions –position over the universe of discourse, and slope- and the output first-order Sugeno coefficients may be considered [10]. The system being discussed will be trained with the ANFIS neuro-fuzzy learning algorithm proposed in [22][30], by making use of chip-in-the-loop techniques in order to account, at the training phase, for implementation non-idealities.

Concerning the implementation of a FC, the signal processing mode of operation can either be digital [23], analog [24] or mixed analog/digital [25]. The speed constraints demanded by real-time control applications impose limits on digital implementations, and thus analog versions of FC are the natural candidates for these tasks. Apart from the speed issue, compactness, low-area and low consumption also benefit the analog counterparts, as well as the tolerance of trainable neurofuzzy systems to analog implementation imperfections. Nevertheless, the neuro-fuzzy approach imposes the need of fully-adjustable processing blocks, thus precluding the use of implementations with fixed parameters. In this case, mixed-signal circuits consisting of analog processing cores with digitally-programmable characteristics, retain the flexibility and programmability of digital versions while being inherently faster and more compact due to their analog processing mode. The issue of interference coupling must be, however, addressed at the design phase. Lastly, although the approach investigated here is based, as stated, on a mixed analog-digital architecture, promising results have also been obtained when exploiting the analog storing capabilities of floating-gate transistors [31], [32].

5.4.2. *FC Analog Sequential Architecture*

Real-world control tasks require high cardinality of inputs, granularity of input space and number of rules. In this sense, it should be stated that existing FC implementation architectures generally lack sufficient number of fuzzy rules, because of the so-called curse of dimensionality, which in practice results in area-consuming VLSI circuits. The actual implementation of the proposed controller consists of the analog sequential architecture previously presented in section 2 (Fig. 5-3), which, concerning hardware complexity, efficiently maps the FC. There exist some examples of analog sequential architectures in the neural networks implementation field [6][7][8], but not in the area of fuzzy controllers.

In the following section, microelectronic design details are given for the analog and mixed-signal circuits considered to implement the architecture.

5.4.3. *Mixed-Mode VLSI Implementation Details*

Inference Building Blocks

The VLSI implementation of the membership function Fig. 5-6. takes advantage of the large-signal transconductance characteristics of two MOS-transistor differential pairs to obtain the required bell-shaped function, inherently transconducing external voltage signals into current-mode signals. Both position and slope are digitally programmable with 6-bit resolution, being this resolution adequate for this closed-loop controller application [33].

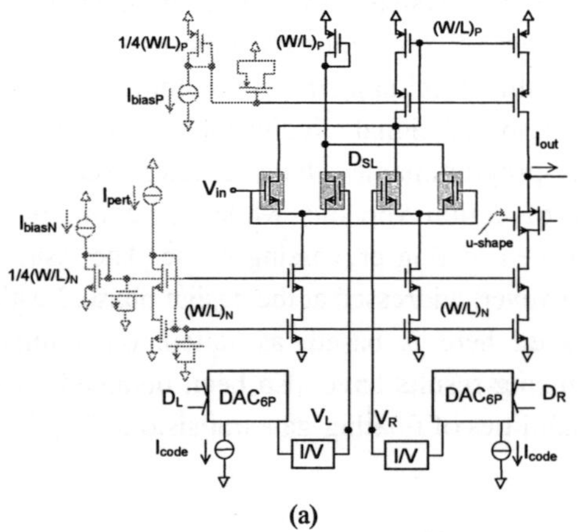

(a)

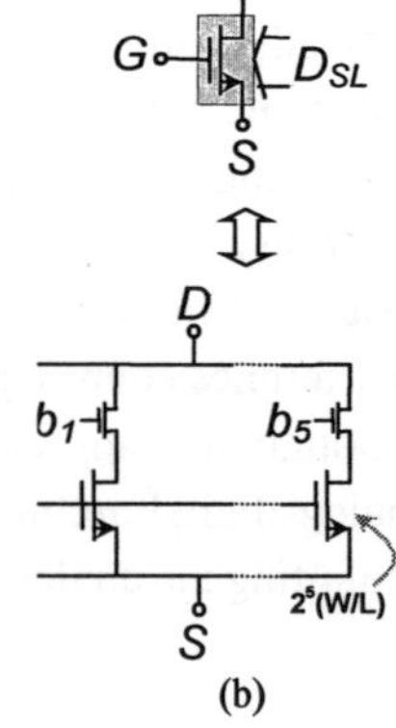

(b)

Fig. 5-6. (cont.)

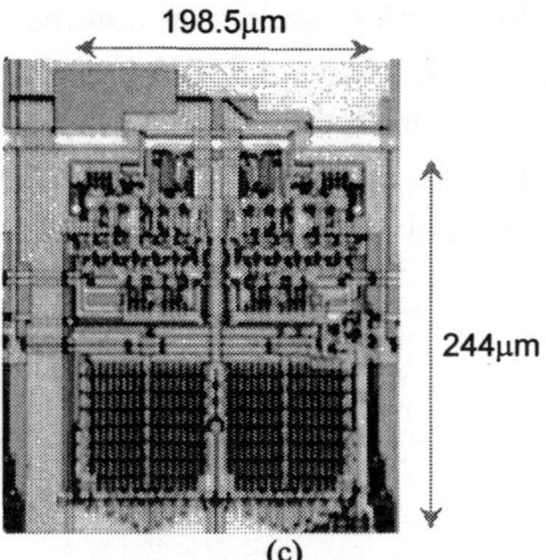

Fig. 5-6. (a) VLSI design for the bell-shaped function. (b) Detail of the transistor array used for digital slope adjustment. (c) layout microphotography.

The use of transistor arrays as input compound transistors in the differential pairs makes their transconductance digitally tuneable to adjust the slope of the characteristic transfer function. For signal steering and bias subtraction purposes, high-compliance one-stage cascode current mirrors are used [26]. These externally voltage-biased mirrors are used throughout the whole ASIC design to attain both high accuracy and copying precision while not reducing the operational margin of the blocks. Voltage offsets are applied at the differential pairs to adjust the position of the membership transition region over the input voltage. These voltage levels are obtained by means of current-steering digital-to-analog converters biased with constant currents, which, in turn, are the main building blocks of the inference consequents part. Active current-voltage conversion is obtained with the I/V converter described in [20], based on a single nonlinearly-biased diode-connected MOS transistor.

For illustration purposes, Fig. 5-6 (c) corresponds to the layout of the digitally-programmable membership function circuit, including (upper half of the figure) two current-steering semialgorithmic D/A converters and continuous-time I/V converters.

Connective minimum operation circuits should be capable of accepting both voltage and current signals at their inputs and outputs for the proper operation of the antecedents block in Fig. 5-3 (b). The natural aggregation property of current-mode operation is needed at the inference output, where the activation degrees ωi are to be added. However, the internal signal transfer should be in voltage-mode. With this end, the MIN circuit used in [15] is adopted (Fig. 5-7). This circuit applies a nonlinear signal compressing (nonlinear function f -1) and subsequent decompressing (f) at input and output ports, similar to the operation basis of a current mirror.

At the internal nodes, the loser-takes all property of a 2-inputs parasitic bipolar transistors differential pair is exploited, thus achieving the MIN operation:

$$I_{out} = f(V_{min}) = f(V_{in1}, V_{in2}) = f\left(min(f^{-1}(I_{in_1}), f^{-1}(I_{in_2})\right) = min(I_{in_1}, I_{in_2} \tag{5-3}$$

Proper biasing for high-speed performance, level-shifting and low-voltage cascoding is included in the MIN circuit.

Consequents Building Blocks

Concerning the consequents block, its operation is based on current-steering digital-to-analog converters using semi-algorithmic techniques [14], which operate as mixed-signal multipliers/weighters and, hence, implement the required output hyperplane coefficients in the first-order TSK model. The use of a two-step algorithmic approach after

$$I_{OUT} = I_{IN}\left(\left(\frac{b_{n-1}}{2} + \ldots + \frac{b_{\frac{n}{2}+1}}{2^{\frac{n}{2}}}\right) + \frac{1}{2^{\frac{n}{2}}}\left(\frac{b_{\frac{n}{2}-1}}{2} + \ldots + \frac{b_0}{2^{\frac{n}{2}}}\right)\right) \tag{5-4}$$

results in a significant reduction in area. Fig. 5-8 depicts the actual 6-bit low-voltage current-steering DAC used in the sequential fuzzy controller.

The blocks labelled as m2() in Fig. 5-3 (c) are in charge of implementing the analog current-mode product between rule activation degrees ωl and the output singletons dl which is needed only in TSK fuzzy controllers. A four-bipolar-transistor translinear cell has been designed for the hardware implementation of this current multiplier, for its compact current-mode operation. This cell [5], naturally obtains current-domain products (6) by virtue of voltage summations around the so-called translinear loop and silicon-junction exponential nonlinear relationships.

$$I_{OUT} = \frac{I_{in1} \cdot I_{in2}}{I_{in3}} \tag{5-5}$$

The shaded pnp bipolar transistors in Fig. 5-9 stand for parasitic lateral bipolar transistors available in standard CMOS technology. Although both current gain and Early voltages are low, the dynamic response is sufficient for this application (ft~200MHz). MOS-based regulating cascoding techniques are applied at the multiplier output transistor to lower output conductance to a negligible level.

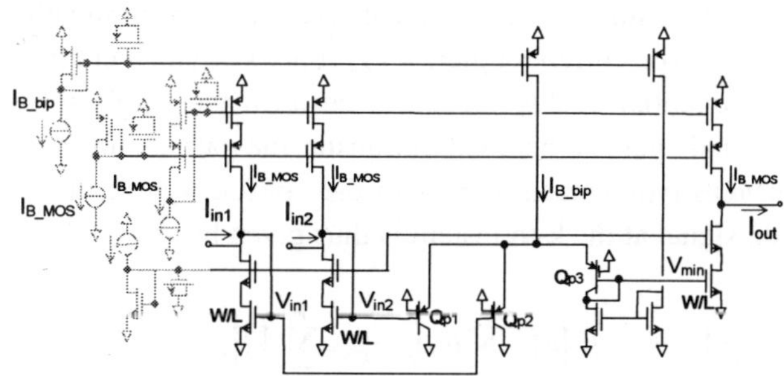

Fig. 5-7. Current/voltage-mode minimum circuit.

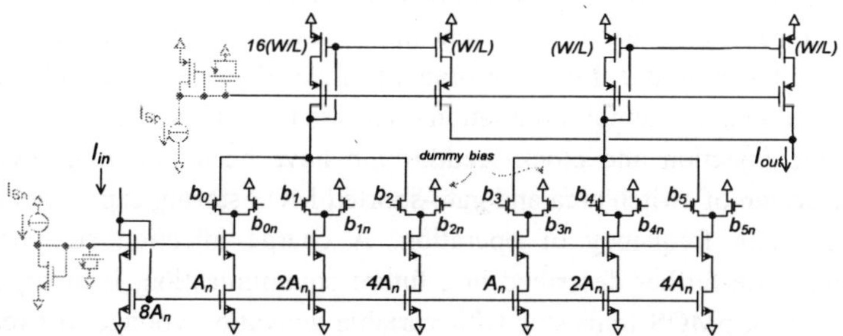

Fig. 5-8. Mixed-signal multiplier design based on high-speed current-mode digital-to-analog converter.

Switched-Current Storage Section and PWM time-domain division Blocks

Since the defuzzyfication method computes signal addition from all fuzzy rules (2), the values of those rules, ω, and the rule-weighted singletons - $d^l\omega^l$

- are sequentially delivered in current-mode by both inference and consequent branches at the rate imposed by the system clock. Therefore, the controller features accumulation or discrete-time integration instead of concurrent aggregation. At circuit level, the analog storing-integrating stage required by the sequential operation has been implemented using switched-current (SI) methods. The storing core of this part of the circuit is the current copier [13], which is capable of loading and delivering a current sampler by storing its nonlinearly-related voltage sampler in a low-quality capacitor, available in digital single-poly technologies. Fig. 5-10 (a) shows a detailed schematic of the section comprising switched-current processing.

For each branch –both antecedent and consequent currents- the designed circuit features storing and discrete-time accumulating, by using complementary-type current copiers and proper switching control signals. For each mn-1 sequential cycles needed to perform the whole set of IF-THEN rules, an internal cycle is included to transfer the partial current sum stored at the main nMOS type current-copier to the pMOS type current copier. The accumulated signal at the k-th instant is thus given by:

$$i[k]_{NMOS} = i_{in}[k] + \sum_{j=1}^{k-1} i[j]_{PMOS} = \sum_{j=1}^{k} i[j]_{NMOS} \tag{5-6}$$

When all the emulation steps to obtain each rule are done, the final values of stored currents are steered to current-mode sample-and-hold circuits, which will drive the subsequent division stage. Regulated cascode stages have been added to alleviate channel-modulation effects in the current copier. The effects of charge injection and clock feedthrough have been minimised with the proper design of switch area and gate-shorted MOS storing capacitors for the corresponding frequency of operation. A charge injection compensation scheme, to be further described in a future communication, is incorporated, which uses a pMOS transistor with variable activation voltage to tweak the charge injected by the nMOS diode-closing transistor. In order to properly control this SI integrator, a non-overlapping three-phase control signals generator is included in the design. This clock generator makes use of one-side starved digital inverters to obtain slow diode-opening waveforms with adjustable decay time, as shown in Fig. 5-10 (b).

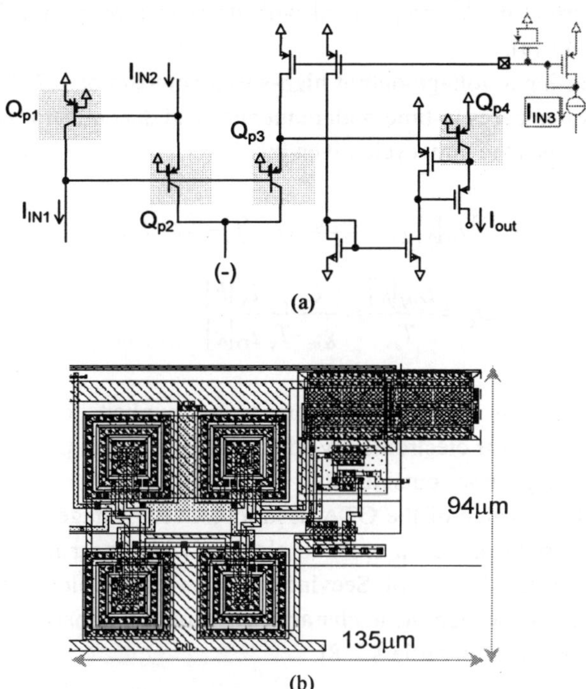

Fig. 5-9. (a) Diagram of the parasitic-bipolar translinear current-multiplier with regulated-cascode current output. (b) Sample layout.

As long as the division stage is concerned, it should be stressed that, due to the sequential operation of the FC, concurrent normalization is not allowed, and thus the division operation is required. Its operation is embedded in a PWM stage which drives the output signal.

Finally, the last stage features the capability of implementing the division operation required by the center-of-area defuzzyfication method, by means of the intrinsic division obtained by a PWM modulator, which is performed by virtue of its signal-to-time conversion. The resulting output signal for the FC is, thus, in voltage-mode and PWM modulated. This novel solution is hardware-efficient and appropriate for the fully-integrated fuzzy control of switched power converters, or, in general, any power plant requiring a PWM drive. The circuit is based on the periodic integration of the current associated to the denominator (aggregation of consequents-weigthed rule values), thus creating a current-mode ramp signal with its peak level proportional to the signal Fig. 5-11 (a).

A current-input voltage-output high-speed comparator [27] Fig. 5-11 (b) continuously compares in time both numerator and denominator accumulated signals, being thus the duty cycle given by:

$$i_N[k]=\frac{g_m}{C_{\text{int}}}\int_{t_k}^{t_k+t_{ON}} i_D[k]\cdot dt \xrightarrow{t_k\le t\le t_{k+1}} i_N[k]=\frac{g_m}{C_{\text{int}}}\cdot i_D[k]\cdot t_{ON}$$

$$D_k=\frac{t_{ON}[k]}{T_s}=\frac{C_{int}}{g_m\cdot T_s}\frac{i_N[k]}{i_D[k]} \tag{5-7}$$

where the division operation appears in an explicit form. The current integration constant is electronically adjustable by varying Cint or gm. Cint is an external capacitor outside the integrated circuit, and gm is the transconductance value of the OTA used to convert voltage-mode integrated signals into current-mode. In order to obtain a wide linear margin, the OTA used in the design is that of Seevinck [29], which exploits the class AB compensation of the nonlinear characteristics of the transistors, as in high slew-rate operational amplifiers [28].

5.4.4. *Post-Layout Simulation and Experimental Results*

The presented FC has been designed and full-custom laid-out using a 0.8µm CMOS analog technology and simulations using extracted circuits show proper operation. As an example, Fig. 5-12 shows two performance validation results. Fig. 5-12 (a) shows a post-layout simulation of the inference block including four membership functions and three MIN operators, programmed sequentially in position and slope but with saw-tooth input signal to show the membership characteristics over the whole input range. Fig. 5-15 (b) validates the operation of the switched section, both SI accumulators and PWM stage, for continuous-time inputs. On the other hand, Fig. 5-12 (c) illustrates, with a fully-transistorised sequential-mode simulation the operation of the whole FC, showing the multidimensional membership function current-sample accumulation and PWM controller output signal.

sw_dioP
vdd_tweak
I_{reg}
$I_{in}[1]+ \ldots + I_{in}[n-1]$
sw_T&H
vdd_tweak
I_{reg}
I_{out}
$I_{in}[n]$
sw_in
sw_out
$I_{in}[1]+ \ldots + I_{in}[n-1]+I_{in}[n]$
sw_dioN
M_C
M_A
M_M
vss_tweak
(a)
CLK
sw_N
sw_in
M/T&H
sw_mem
M_p
V_{in}
V_{out}
M_n
V_{biasN}
M_{Bn}
V_{biasP}
M_{Bp}
(b)

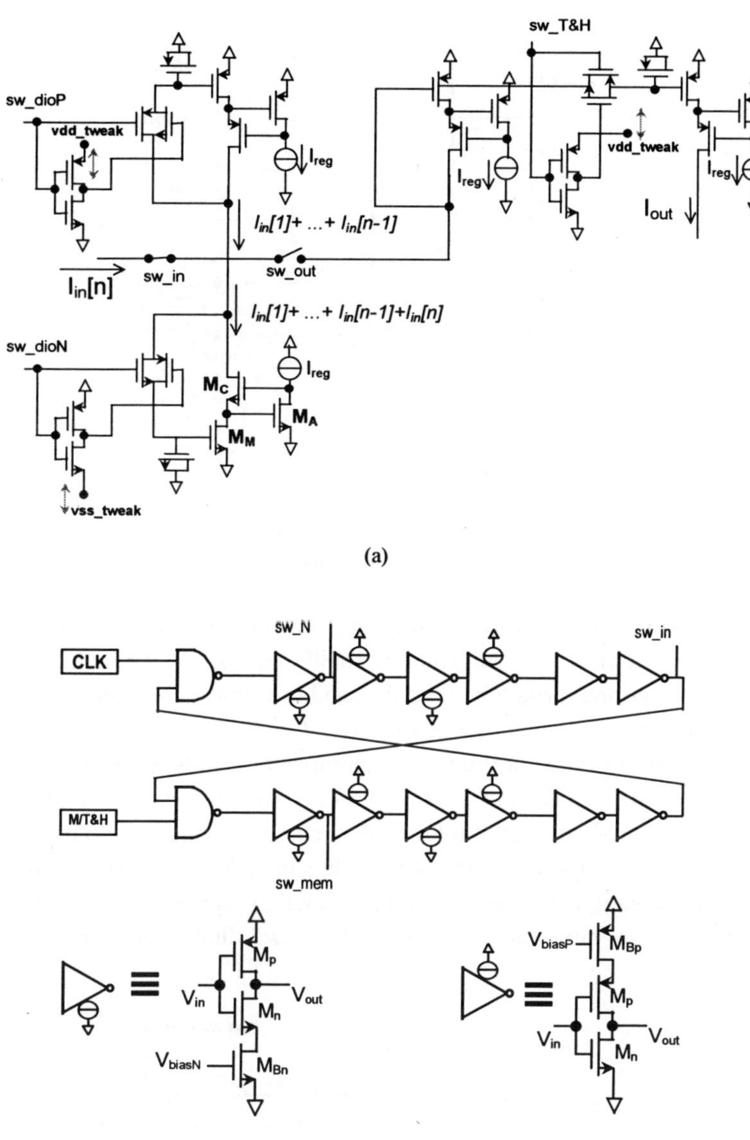

Fig. 5-10. (cont.)

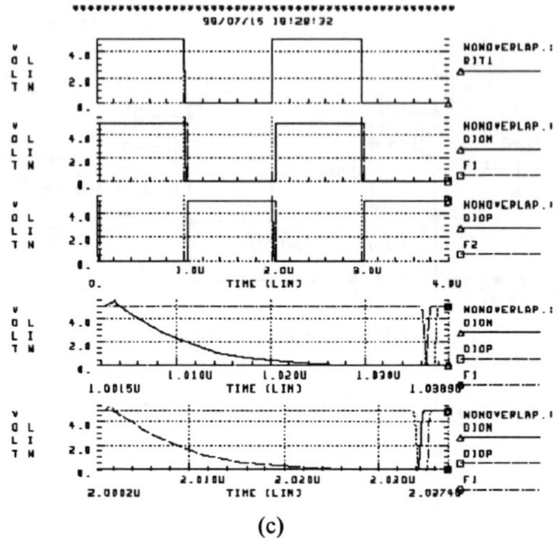

(c)

Fig. 5-10. (a) Diagram of the switched-current storing/integrating stage (b) Non-overlapping adjustable-slope clock generator (c) Obtained clock waveforms.

Fig. 5-13 shows a view of the final layout. The analog part of this application specific controller features small size –about 0.65mm2 for this eight-rule/three-input prototype testchip-, and low power consumption.

So as to conclude with the description of the fuzzy controller prototype, Fig. 5-14 depicts sample preliminary experimental results of the operation of some continuous-time subblocks. Fig. 5-14 (a) validates the programmability of the mixed-signal membership function circuit, by digitally varying both the position and the slope of a membership function. Fig. 5-14 (b) shows and indirect observation of the minimum function applied to membership functions in two dimensions, resulting in a two-dimensional membership function.

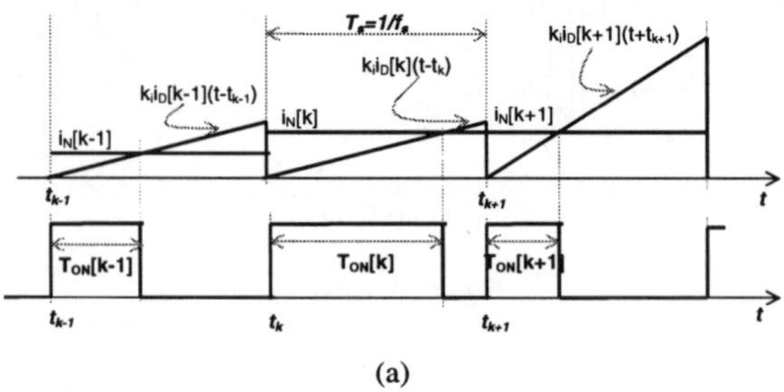

(a)

Fig. 5-11. (cont.)

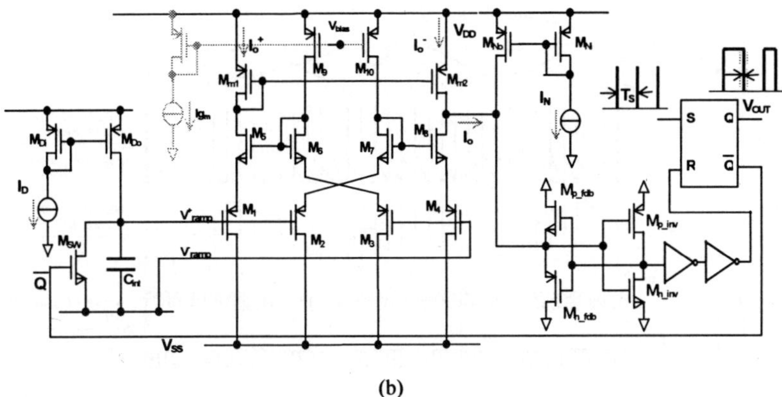

(b)

Fig. 5-11. (a) Time diagram of the signals involved in the time-domain division operation (b) Circuit-level diagram of the current-mode PWM division stage.

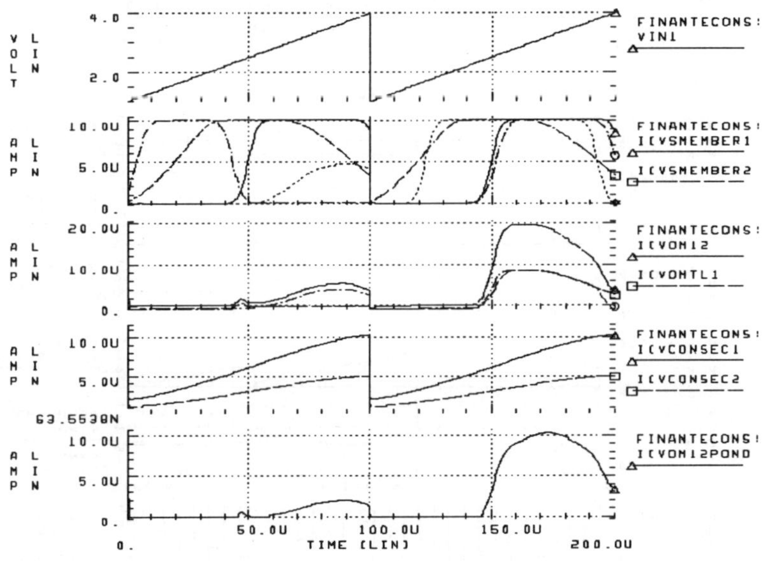

(a)

Fig. 5-12. (cont.)

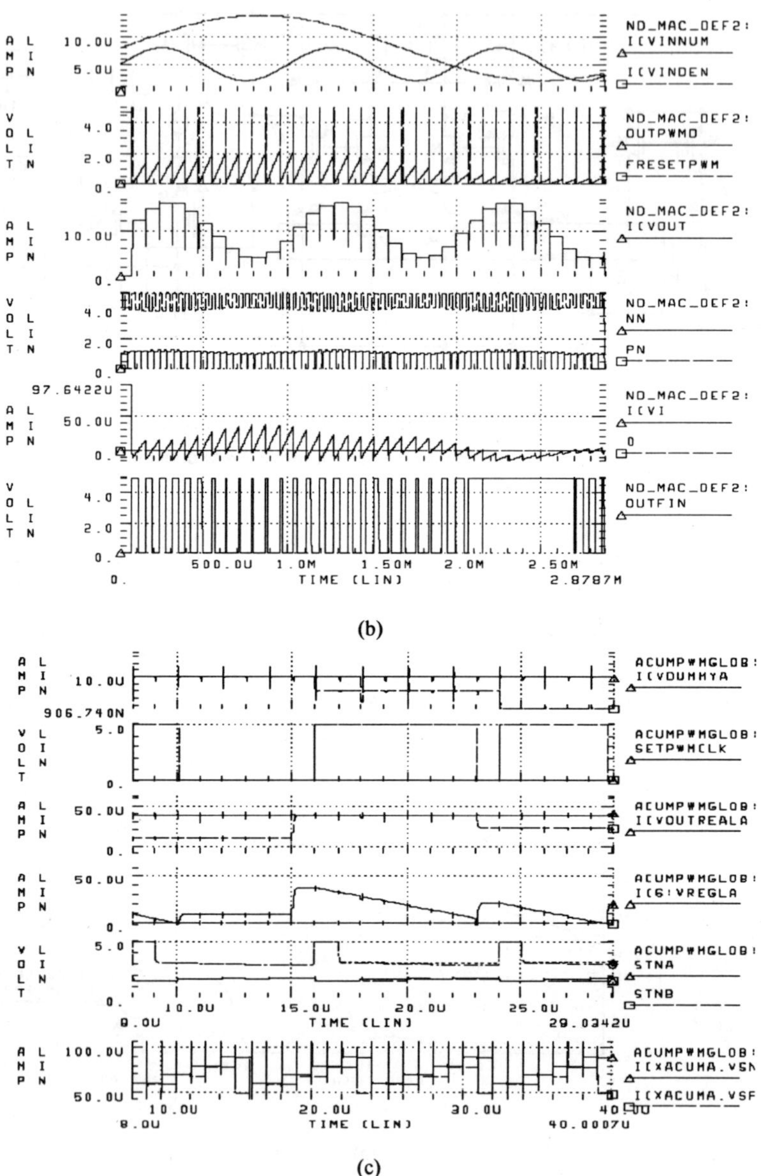

(b)

(c)

Fig. 5-12. (a) Inference time-domain simulation (b) Switched-mode simulation of SI accumulator and PWM stage (c) Discrete-time switched-current integration and PWM waveforms.

5.5. Case Study 2: VLSI Design of a Flexible-Structure Sequential Mixed-Signal Neural Processor

5.5.1. *General Facts*

In the second case study we introduce a neural processor. The good performance of artificial neural networks in solving complex tasks such as classification, prediction, interpolation, information compression or recognition is well-known [40]. Although the most popular model is clearly the Multi-Layer Perceptron (MLP), a large amount of different neural models have been extensively proposed [41].

In order to avoid the interconnect explosion of the fully parallel implementation, a systolic-ring analog architecture was reported in [42]. This initial architecture was essentially the one explained in Section 2, with an important modification: in a systolic approach, data circulates across a number of registers. This is not acceptable for analog storage, since the stored value would suffer from progressive degradation. Therefore an architectural modification was performed to limit each analog value that is computed to a single storage operation [43], resulting in the final architecture that has been described in Section 2. A proof-of-concept implementation has been done for a 3-synapse processor.

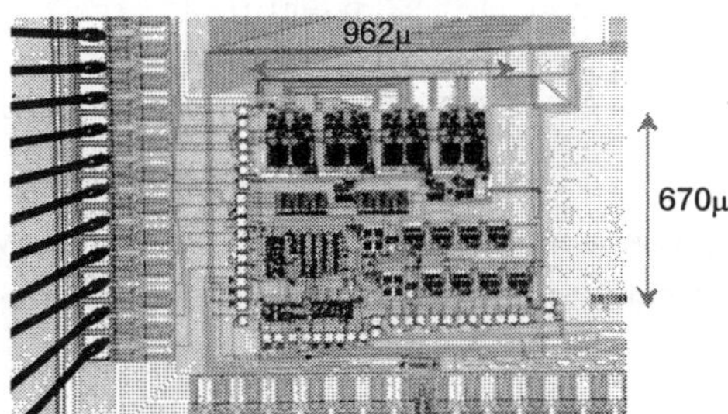

Fig. 5-13. Designed layout for the sequential FC. Dimensions shown in μm.

As explained in Section 2, the benefits of this architecture are: reduced routing overhead, compact circuitry, unlimited number of layer emulation capability, direct scalability and very flexible network structure. The maximum number of neurons in a single layer is only limited by the implemented number of synapses. Also, it has been shown that this architecture can emulate a broad number of neural models [44].

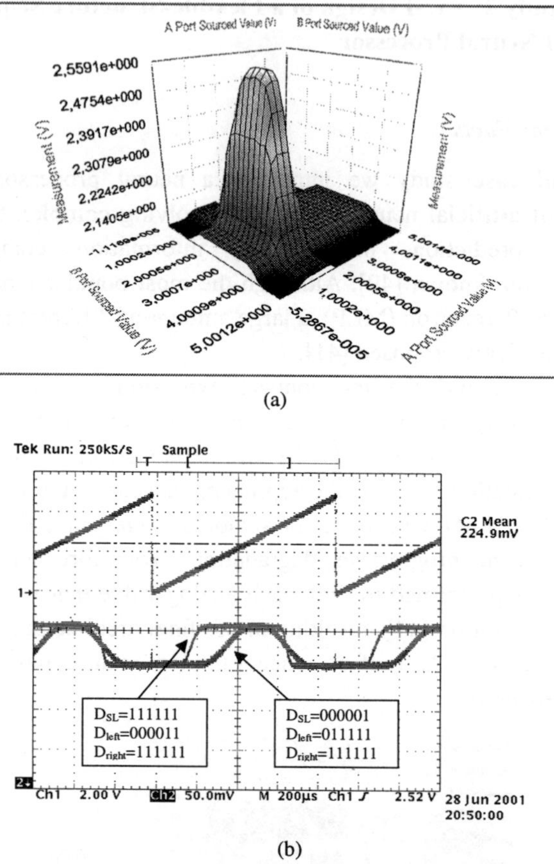

Fig. 5-14. Preliminary experimental results from the FC testchip. (a) Membership function programmability. (b) Bidimensional membership function by means of minimum function.

5.5.2. *Processor Description*

In an MLP model, synapses consist of the product between the synapse weight and the input value, while a sigmoid is used as the activation function. Since the current-mode approach has been adopted, synapse addition is obtained simply by connecting synapse outputs. Concerning network parameters, they are stored in a conventional digital memory, and retrieved as they are required.

The processor implementation is divided into the analog datapath, where processing takes place, and the digital controller, that was described previously, which is in charge of delivering the routing data and the overall synchronization.

In Fig. 5-15, the analog datapath of the circuit is shown at a block level for a 3-synapse processor. Extension to an arbitrary number of synapses is straightforward. The electrical information carrier is current.

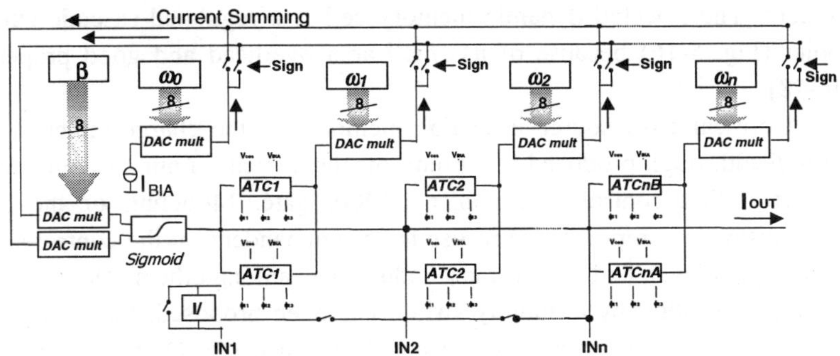

Fig. 5-15. Analog datapath.

Voltage or current input data can be loaded either serially or in parallel from the bottom analog inputs and are stored in a row of ATCs. The two rows of ATCs store input and output activations, exchanging their role at each new layer emulation (since outputs of one layer become inputs of the next one). The row of ATCs storing the analog input data are multiplied with the digital synapse weights by means of digital-to-analog converters (DACs). Depending on the weight sign, the output current is driven to a positive or negative current aggregation line. After a programmable scaling are used to apply the (slope) parameter, both currents enter a sigmoid circuit. The sigmoid output is stored into one of the output row ATCs.

Other important features included in the processor are: digitally programmable offset and gain compensations to reduce process tolerance, and a delay control in atc switches to guarantee a correct operation.

The parallel registers that control datapath routing and weight values are updated at each clock period. The clock cycle has been established in 1 MHz, and the weight resolution 8 bits plus a sign bit for the first test circuit. This resolution has been shown to be enough for the execution phase of many real applications [45].

In the following sections, the key issues in the design of the main building blocks are discussed.

5.5.3. *Analog Transfer Cell*

Systems that use analog memory require a careful design of this critical element. The selected dynamic memory cell is a regulated-cascode current copier (Fig. 5-16) because of its small area overhead and good properties [46-48].

A critical error source in analog memories is the charge injection and clock feedthrough produced by turning off the switch, in our case an NMOS transistor, that controls the copier NMOS gate. Modeling of the error mechanisms is complicated, because is very dependent on the control signal slope, signal level, and copier transconductance, among others [48].

Compensation by a dummy switch has been proposed, but it has been shown that its compensating effect is not clear. Since a PMOS transistor has an opposite sign charge injection, we propose the use of a complementary switch. By controlling the voltage swing of the PMOS gate during switching, charge injection is adjusted so the error can be virtually eliminated.

Fig. 5-17 depicts the voltage error for a single NMOS switch (solid line) and for 2 different CMOS switches. The dashed lines correspond to a PMOS width of 1.4 times the NMOS switch width. The dotted lines to a PMOS width 2 times larger than that of the NMOS switch. The two lines represent the error change with a 3 V swing of the control voltage Vinj (Fig. 5-16). For PMOS switch dimensions within 1.4 and 2 times the NMOS switch, the switch-induced voltage (and thus current) error can be compensated by varying the VSS value. The compensation can be done either externally or by means of an internal feedback loop [49].

5.5.4. *Current-Steering Mixed-Signal Multiplier*

Synaptic-weight product can be implemented with continuous-time circuits –i.e. with multiplier blocks, as open-loop transconductors [50] or bipolar Gilbert cells- or with mixed-signal multipliers based on digital-to-analog converters. In general, the aggregation operation needed at the output end of the synaptic cells dictates the use of current-domain signal representation. In addition, for the architecture being considered, input signals to the synaptic multipliers are the currents delivered by the current-copier-based ATCs. In this situation, a current-steering switched current mirror digital-analogue converter is used.

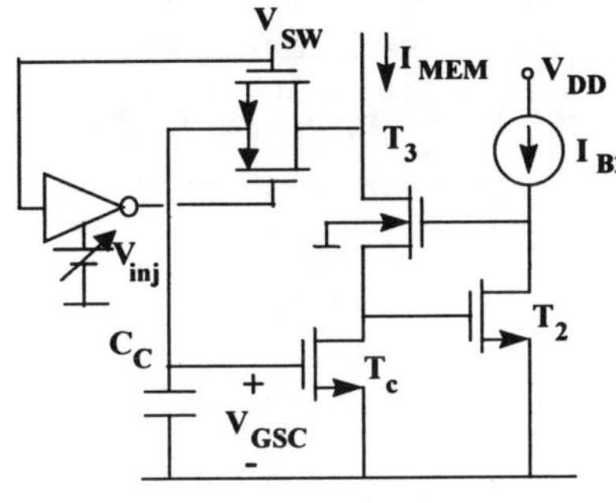

Fig. 5-16. Regulated-cascode ATC with charge compensation compensation.

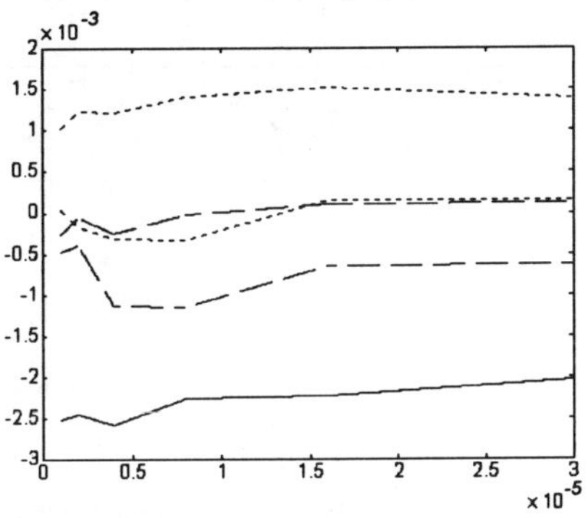

Fig. 5-17. Simulation of the switch error X-axis: current(A). Y-axis: Voltage(V).

The signal-scaling action in current-steering DACs is obtained by means of active-device scaling and, thus, no need for passive components appears, avoiding the use of linear capacitors (which, in turn, require one extra polysilicon layer process step). The current-steering DAC, being in fact a switched current mirror, achieves high-frequency operation because of the inherent reduction of voltage swings, thus minimising sensitivities to parasitic capacitances [51].

The designed DAC (Fig. 5-18) is composed of a matrix-arranged array of unit cascoded current sources. The use of high-compliance low-voltage cascode stages within the converter assures the reduction of channel length modulation effect and consequently improves integral linearity and monotonicity, while keeping supply-voltage requirements low.

The use of semialgorithmic techniques [52] reduces matching requirements (imposed by fully algorithmic DACs) while maintaining the linear relation between active chip area and resolution. In order to limit current mismatch to ½ LSB, the maximum allowable mismatch [53] between current sources is given by:

$$\sigma_N^2\left(\Delta i/_i\right) < 1/_{2^{n+2}} \xrightarrow{n=8\,bits} \sigma_8^2\left(\Delta i/_i\right) \leq 0.19\% \tag{5-8}$$

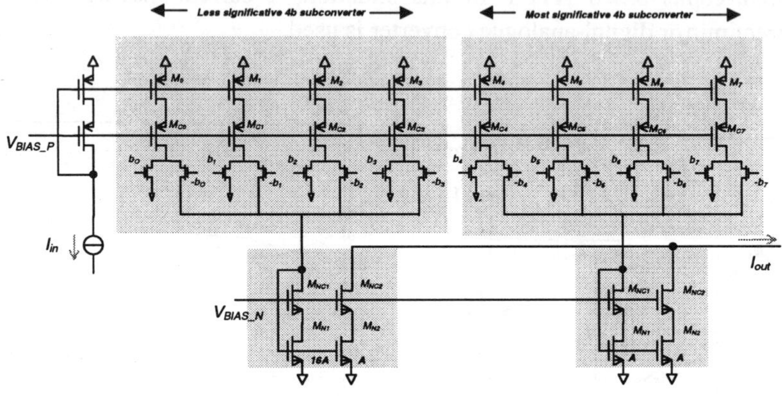

Fig. 5-18. Mixed-signal semialgorithmic D/A multiplier.

For an n-bit semialgorithmic DAC, the transfer function given in (2) for Iin input current and DW digital code illustrates the use of fine and coarse n/2-bit subconversion blocks.

$$I_{OUT} = D^w \cdot I_{IN} = I_{IN}\left(\left(\frac{b_{n-1}}{2} + \ldots + \frac{b_{\frac{n}{2}+1}}{2^{\frac{n}{2}}}\right) + \frac{1}{2^{\frac{n}{2}}}\left(\frac{b_{\frac{n}{2}-1}}{2} + \ldots + \frac{b_0}{2^{\frac{n}{2}}}\right)\right) \tag{5-9}$$

In each four-bit subconverter in Fig. 5-18, digital codes are applied to differential current switches (consisting of p-type fully saturated differential pairs) which steer the current sourced by the current generator matrix to either the output summing node or a dummy node.

The use of differential switches ensures current flowing for each cell in the matrix, hence long recovery times are avoided and settling time requirements are fulfilled for low-level currents.

The validation of this mixed-signal current-mode multiplier is inferred from the post-layout HSPICE one-run Monte Carlo simulation shown in Fig. 5-19. This simulation is carried out under a constant current input of Iin=10μA and cyclic digital stimuli. Device size information is included within statistical simulation parameters to ensure the correct mismatch effect. Only a part of the whole conversion margin is depicted. Some performance evaluation parameters are an active area occupation of about 0.01 mm^2, and settling times bounded by ≅ 200 ns.

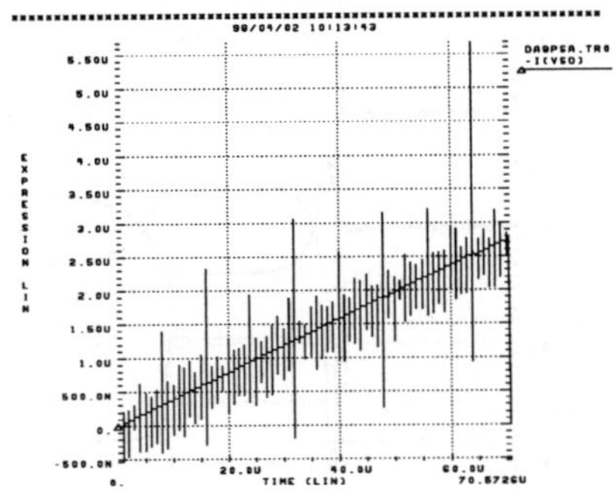

Fig. 5-19. Transient waveform for the central codes of the semialgorithmic D/A multiplier.

5.5.5. *WTA-Based Sigmoid*

As was shown in Fig. 5-15, the positive and negative synaptic currents are accumulated in different lines. These currents should be subtracted, but to avoid the mismatch error introduced by extra current mirror stages, both lines are applied to a differential-input sigmoid circuit.

In Fig. 5-20 (a), the sigmoid circuit generator is shown. It is based on a current-mode Winner-Take-All (WTA) circuit [54] and consists of a differential pair with two transistors providing a common-mode feedback.

The sigmoid function is the natural V-I characteristic in a differential pair of exponential devices. For MOS transistors operating in strong inversion, the characteristic reasonably approaches the sigmoid form. The output current Isigm is connected to a common line that can be routed to any ATC or to the circuit output.

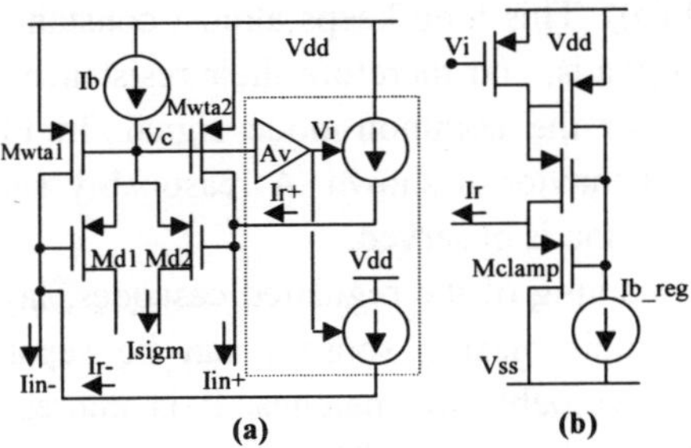

Fig. 5-20. (a) WTA-based sigmoid circuit. (b) Rightmost current sources implementation detail.

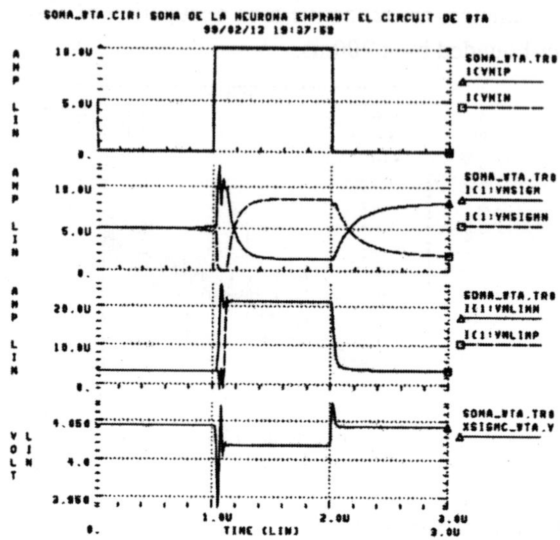

Fig. 5-21. Simulation of the WTA sigmoid transient response to 10 uA common-mode and 100 nA differential steps.

However, a simple differential pair is not enough since inputs are in current form, so a previous I-V conversion is necessary. Furthermore, a high resistance value is needed to obtain a significant voltage drop at sub- A levels. For this purpose, the common-mode feedback transistors (Mwta1 and Mwta2) working in saturation region are used as active loads.

Since the MOS transistor small-signal output resistance depends on the bias current, to keep it constant, the common-mode current has to be suppressed. An additional feedback loop performs this task. vc is amplified and it controls two regulated cascode current sources connected to the WTA inputs (Fig. 5-20 (b)). This loop keeps almost constant the currents flowing through the active loads, and therefore their resistance, making the sigmoid slope independent of the common-mode signal. In Fig. 5-21, a transient simulation of this behavior is shown. A reasonably short settling time and common-mode rejection is observed.

Concerning matching of the regulated cascodes, layout techniques have been used. Also, their biasing sources can be separately fine tuned to compensate the unavoidable mismatches. It is interesting to highlight the function of the Mclamp transistors. They are normally off, but when one input voltage rises too much, the corresponding Mclamp turns on, fixes that voltage, thus preventing the Mwta to operate in the linear region. Moreover, it prevents the feedback source to become out of regulation by sinking the extra current. Finally, the always saturated operation speeds up the processing.

5.5.6. *The Complete Neural Processor*

Fig. 5-22 shows an enlarged photograph of the MLP processor. Its area occupancy is 0.79 mm^2, and it includes three synapse slices (label 1 in the picture), a sigmoid function (2) and an 8-bit scaling circuit (3). Each synapse slice contains two ATCs (4) and an 8-bit mixed digital-analog multiplier (5), as well as control logic. The synapse-slice scheme allows a straightforward change of the number of synapses. Offset is digitally adjusted by means of an additional constant-input DAC multiplier.

The analog cells have been designed to be compatible and consistent in terms of voltage levels, input and output resistances and to keep error under 0.5% in order to guarantee 8-bit precision. A conservative 1 MHz clock is used in the first version

The complete system has been functionally simulated and validated using macromodels.

The processor has been designed in such a form that each cell can be separately tested. Furthermore, by programming the appropriate control digital words, the data flow can be modified with a high freedom degree. For instance, data transfer between ATCs, and weighted current addition.

5.5.7. *Cost-Performance Tradeoff*

Having described the neural processor, now we explore the design space for the MLP circuit as a design example that can be adapted to the application requirements of cost and performance. As a cost parameter we use an area occupancy index, whereas speed is used as the performance index, measured as the latency in clock periods needed by the processor to generate the output vector from an input vector.

Table 5-3 displays the measured area of the three key elements of the MLP processor. Let us assume a real application requirement of three layers of neurons, each layer including 10 neurons, and 10 synapses per neuron. This represents for Table I, nn = ns = 10 and nl = 3. Using the expressions of that table with these figures, and calculating the area from Table III, we can plot each table row in an area-delay plot (Fig. 5-23). Of course, the number of possible combinations is much larger, however an interesting set of designs is already obtained.

Table 5-3. MLP key elements area.

8-bit A-D multiplier	sigmoid	ATC	
25	9	13.5	mm2 (×10-3)

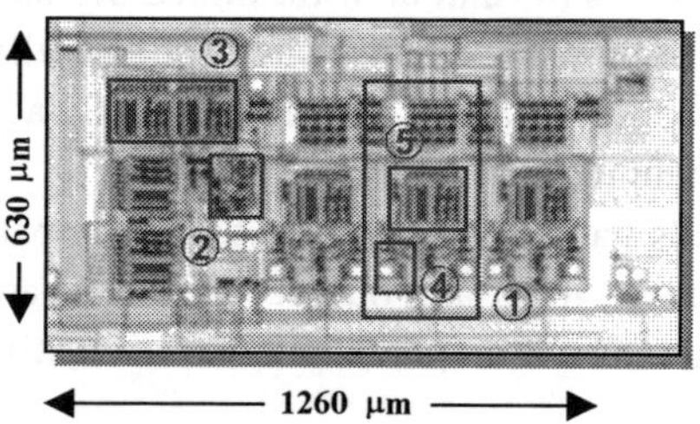

Fig. 5-22. MLP mixed-signal sequential test processor.

Although the plotted area figures are not directly the processor area, because glue logic and routing area is not accounted, the global area should be, with some deviation, proportional with the key element area. It is very interesting to see that, although some solutions are sub-optimal, there is a set of designs that trade off cost for performance. A hyperbolic curve of fairly constant area-delay product can be deduced as a straight line in the log-log plot of Fig. 5-23. Hence, the most suitable option can be selected for a given application constraints.

5.6. Conclusions

Design alternatives to digital or fully parallel analog implementations of neural or fuzzy systems that can fit speed applications requirements while minimizing area resources by means of mixed-signal VLSI have been demonstrated and analyzed.

Issues on sequential design have been analyzed and applied to neural and fuzzy processing. Mixed-signal sequential implementations exhibit high flexibility to emulate different system structures. Proper selection of sequentiality allows an application-specific efficient selection of the point in cost-performance (speed and area) design space.

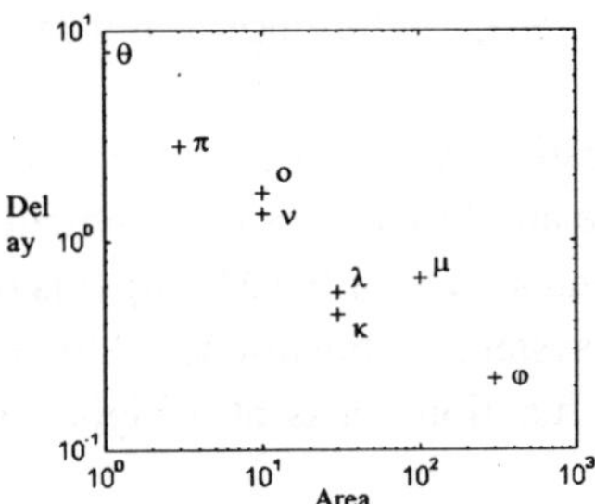

Fig. 5-23. MLP area-delay tradeoff plot.

An important point is that sequentialization considerations are not technology independent, but take into account technology and subcircuit constraints. A consistent design style has been applied.

Two sequential processors that demonstrate feasibility of the approach have been implemented and validated. The example processors are integrated in a single testchip, sharing a common digital controller. This controller reads from an external bus control data and parameters, to configure and provide data (slopes, weights) to the processors.

In the fuzzy processor, both antecedents and consequents are sequentially calculated, thus reducing the area requirements. The section devoted to the fuzzy controller reported on the design and implementation of the required mixed analogue/digital circuits. Among others cells used in this architecture, the description of the fully adjustable membership function circuit, the voltage/current-mode minimum circuit, current-steering DACs, switched-current analogue memories and PWM division circuits has been presented. Layout-extracted fully-transistorized simulations as well as preliminary results validated the operation of the FC.

The neural processor has the property to allow emulation of the recall phase of any feedforward network structure provided that the largest layer has no more units than the processor number of synapses. Building blocks suitable for the analog implementation have been presented and discussed.

In this testchip, static CMOS standard logic is used. To minimize digital noise, a careful design of guard rings, and separate power lines and pins has been done., a fully differential style design for the analog blocks is to be employed to increase digital noise immunity. Although not differential, the designed logic blocks are equally valid to estimate the alternatives in this sequentiality study. Furthermore, a low-noise logic, as FSCL (Folded Source-Coupled Logic) [55], can be employed for the digital circuits to reduce digital noise source.

Competitive applications envisaged for the described processors are those driven by cost-sensitive and low-power requirements. Especially, when input data is in analog form and analog or PWM output is of interest (for instance, in control and embedded systems applications). Although digital processors can perform the described functions, it is at a higher area overhead and power consumption cost.

Feedback obtained from designed processors and subcircuits allows considering the technology constraints for analysis and extension to different sequentiality degrees, and a quantitative experimental comparison in terms of speed, area and power consumption between the several solutions has been presented.

Acknowledgement

The authors acknowledge Dr. J. Manuel Moreno for fruitful discussions, Dr. A. Poveda for valuable advices at circuit-level design, Mr. Jordi Cosp for support in the digital design, and Mr. Laurent Imbernon for experimental measurements.

References

[1] M. Glessner and W. Pöchmüller. *Neurocomputers.* Chapman and Hall, 1994.

[2] D. Del Corso. Hardware Implementations of Artificial Neural Networks. J. Mira, F. Sandoval (Eds.), *Lecture Notes in Computer Science,* pages. 405-419, Springer-Verlag, 1995.

[3] J. Madrenas, et al. Analog and Mixed-Signal Neural VLSI Processors: Taxonomy, Comparison and Performance Evaluation, *Proc. MIXDES'96*, Lodz (Poland), 1996, pages 389-400.

[4] R. J. Schalkoff. *Artificial Neural Networks.* McGraw-Hill, 1997.

[5] C. Toumazou, F. J. Lidgey and D. Haigh, (Eds.) *Analog IC design: The current-mode approach,* IEE Peter Peregrinus, London, 1991.

[6] N. Yazdi, M. Ahmadi, G. A. Jullien and M. Shridhar. Pipelined Analog multiplier feedforward neural networks. *Proceedings of the IEEE International Symposium on Circuits and Systems (ISCAS'93),* page 2768.

[7] J. C. Lee, B. J. Sheu, J. Choi and R. Chellappa. A mixed-signal VLSI neuroprocessor for image restoration. *IEEE transactions on circuits and systems for video technology*, 2(3), pages 319-324, September 1992.

[8] J. M. Moreno, F. Castillo, J. Cabestany, J. Madrenas and A. Napieralski, An Analog Systolic Neural Processing Architecture, *IEEE Micro*, 14(3), pages 51-59, June 1994.

[9] J. M. Moreno, J. Madrenas, E. Alarcón and J. Cabestany. Analog Sequential Architecture for Neuro-Fuzzy Models VLSI Implementation. *proceedings of the 7th International Conference on Artificial Neural Networks (ICANN97)*, Laussane, Switzerland, pages 8-10 October 1997.

[10] T. Takagi and M. Sugeno. Fuzzy Identification of Systems and its Applications to Modelling and Control. *IEEE Trans. on Systems, Man and Cybernetics*, 15(1). Jan. 1985.

[11] E. Alarcón, M. Iannazzo, J. Madrenas, J.M. Moreno, S. Gomáriz, F. Guinjoan and A. Poveda. Implementation of an Application-Specific Fuzzy Controller by means of a Mixed-Signal Sequential Architecture. *proceedings of the 41st IEEE Midwest Symposium on Circuits and Systems (MWSCAS98)*, Notre-Dame, Indiana, Aug. 1998.

[12] J. Madrenas, E. Alarcón, J. Cosp and J.M. Moreno. VLSI Design of a flexible-structure Sequential Mixed-signal Neural processor, *proceedings of the 6th International Conference Mixed-Signal Design of Integrated Circuits and Systems (MIXDES'99)*, Kraków (Poland), June 1999.

[13] C. Toumazou, J. B. Hughes and N. C. Battersby (Eds.). *Switched-Currents: An analog technique for digital technology.* IEE Peter Peregrinus, London, 1994.

[14] N. Paulino and J. E. Franca. A CMOS digitally programmable current multiplier. *proceedings of the IEEE International Symposium on Circuits and Systems (ISCAS'96),* pages 254-257, May 1996.

[15] J. Ramírez-Angulo, K. Treece and P. Andrews. Current-Mode and Voltage-Mode VLSI Fuzzy Processor architecture, *proceedings of the IEEE International Symposium on Circuits and Systems (ISCAS'95)*, pages 1156.

[16] S. Espejo, R. Domínguez-Castro, F. Medeiroy and A. Rodríguez-Vázquez. Tunable feedthrough cancellation in switched-current circuits. *IEE Electronics Letters*, 30(23), pages 1912, 10th November 1994.

[17] J. M. Moreno, J. Madrenas and J. Cabestany. Systolic Modular VLSI Architecture for Multi-Model Neural Network Implementation. *Proceedingsof the 4th International Conference on Microelectronics for Neural Networks and Fuzzy Syst.*, pages 118-124, 1994.

[18] K. L. Fong and A. T. Salama. A 10 bit semi-algorithmic current mode DAC. *proceedings of the IEEE International Symposium on Circuits and Systems (ISCAS'93)*, page 978.

[19] C. A. A. Bastiaansen, D. W. J. Groeneveld, H. J. Schouwenaars and H. A. H. Termeer. A 10-b 40-MHz 0.8μm CMOS current-output D/A converter. *IEEE Journal of Solid-State Circuits*, 26(7), page 917, July 1991.

[20] K. Bult and H. Wallinga, A class of analog CMOS circuits based on the square-law characteristic of an MOS transistor in saturation. *IEEE Journal of Solid-State Circuits*, 22(3), pages 357, June 1987.

[21] E. Alarcón, A. Poveda, J. Madrenas, E. Vidal, S. Gomáriz and F. Guinjoan, Novel Pulse-width-modulated Current-mode Defuzzifier for the fuzzy control o switching DC-DC converters, *proceedings of the 42nd IEEE Midwest Symposium on Circuits and Systems (MWSCAS99),* Las Cruces, NMSU, New Mexico, USA, Aug. 1999.

[22] J. R. Jang. ANFIS: Adapted Network based Fuzzy Inference System. *IEEE Transactions on Systems, Man and Cybernetics*. 23(3), pages 665-685. May/June 1993.

[23] M. J. Patyra, J. L. Grantner, K. Koster. Digital Fuzzy Controller: Design and Implementation. *IEEE Transactions on Fuzzy Systems*, 4(4), pages 439-459, Nov. 1996.

[24] S. Guo, L. Peters and H. Surmann. Design and application of an analog fuzzy logic controller. *IEEE Transactions on Fuzzy Systems*, 4(4), page 429, Nov. 1996.

[25] F. Vidal and A. Rodríguez Vázquez. Using building blocks to design analog neurofuzzy controllers. *IEEE Micro*, pages 49, Aug. 1995.

[26] P. J. Crawley and G.W. Roberts. Designing Operational transconductance amplifiers for low-voltage operation. *proceedings of the IEEE International Symposium on Circuits and Systems (ISCAS'93),* pages 1455.

[27] A. Rodríguez Vázquez, R. Domínguez Castro, F. Medeiro and M. Delgado Restituto. High resolution CMOS current comparators: Design and applications to current-mode function generation. *Analog Integrated Circuits and Signal Processing*, 7, pages 149-165, 1995.

[28] T. Fiez, H. Yang, C. Yu and D. Allstot. A Family of High-Swing CMOS Operational Amplifiers. *IEEE Journal of Solid-State Circuits*, 24(6), Dec 1989.

[29] E. Seevinck and R. Wassenaar. A Versatile CMOS linear Transconductor/ Square-Law Function Circuit. *IEEE Journal of Solid-State Circuits*,.SC22(3), June 1987.

[30] S. Gomáriz, E. Alarcón, J. A. Martínez, A. Poveda, J. Madrenas and F. Guinjoan. Minimum-time Control of a Buck Converter by means of fuzzy logic approximation. *proceedings of The 24th Annual Conference of the IEEE Industrial Electronics Society (IECON'98),* Aachen, Aug. 1998.

[31] P. Hasler, C. Diorio, B. A. Minch and C. Mead. Single Transistor Learning Synapse with Long-Term Storage. *proceedings of the IEEE International Symposium on Circuits and Systems (ISCAS'95)*.

[32] J. Madrenas, Ivorra, E. Alarcón and J. M. Moreno. Injector Design for Optimized Tunelling in Standard CMOS Floating-Gate Analog Memories. *proceedings of The 41st IEEE Midwest Symposium on Circuits and Systems (MWSCAS98)*, Southbend, Indiana, Aug. 1998.

[33] I. Del Campo, J. M. Tarela, Consequences of the digitization on the performance of a fuzzy logic controller. *Fuzzy Systems, IEEE Transactions on*, 7(1), pages 85 –92, Feb. 1999

[34] S. Gomáriz, E. Alarcón, F. Guinjoan and A. Poveda. Analytical Considerations in the Design of a Nonlinear State-Dependent Takagi-Sugeno Fuzzy Controller for a Boost Switching Power Regulator, *proceedings of the 9th IEEE Mediterranean Conference on Control and Automation - MCCA'01*, Dubrovnik, Croatia, June 2001.

[35] S. Gomáriz, E. Alarcón, F. Guinjoan, E. Vidal-Iriarte and L. Martínez-Salamero. PWM-sliding global control of a boost switching regulator by means of first-order Takagi-Sugeno fuzzy control. *Proceedings of the 2001 IEEE International Symposium on Circuits and Systems - ISCAS'01*, Sydney, Australia, May 2001.

[36] J. Madrenas, E. Alarcón, J. Cosp, J. M. Moreno, A. Poveda and J. Cabestany. Mixed-Signal VLSI for Neural and Fuzzy Sequential Processors. *Proceedings of the 2000 IEEE International Symposium on Circuits and Systems - ISCAS'00*, Geneve, Switzerland, June 2000.

[37] J. Madrenas, E. Alarcón, J. Cosp, J. M. Moreno, A. Poveda and J. Cabestany. Design Space tradeoff in VLSI implementations of Mixed-Signal Neuro-Fuzzy Processors. *Proceedings of the Fifth International Symposium on Artificial Life and Robotics - AROB'00*, Oita, Japan, Jan. 2000.

[38] E. Alarcón, J. Madrenas, J.M. Moreno, J. Cosp, S. Gomáriz, F. Guinjoan and A. Poveda. Mixed-signal implementation of a discrete-time sequential Takagi-Sugeno Neurofuzzy Controller, *Proceedings of the 6th International Conference Mixed-Signal Design of Integrated Circuits and Systems (MIXDES'99)*, Kraków (Poland), June1999.

[39] E. Alarcón, *Microelectronic design of controllers for switching power converters.* PhD dissertation, Univ. Politecnica Catalunya, Barcelona, 1999.

[40] R. Lippmann. An Introduction to Computing with Neural Nets. *IEEE Acoustics, Speech and Signal Processing Magazine*, 4(2). 4-22, April 1987.

[41] J. Madrenas, J. M. Moreno and J. Cabestany. Analog and Mixed-Signal Neural VLSI Processors: Taxonomy, Comparison and Performance Evaluation. *MIXDES'96, Proceedings*, pages 389-400, Lodz (Poland).

[42] J. M. Moreno et al.. An Analog Systolic Neural Processor Architecture, *IEEE MICRO*, 14(3), pages 51-59, Jun. 1994.

[43] J. Madrenas et al.. A Current-Mode Sequential CMOS A/D Variable-Structure Processor for Neural Networks Emulation. *Proc. of the Design of Circuits and Integrated Systems Conf.*, pages 536-541. Madrid (Spain), 1998.

[44] J. M. Moreno, J. Madrenas and J. Cabestany. Systolic Modular VLSI Architecture for Multi-Model Neural Network Implementation. *Proc. of the 4th Int. Conf. on Microelectronics for Neural Networks and Fuzzy Systems*, pages 118-124, 1994.

[45] Blayo et al., R1-C and R2-C: Hardware Implementations. *ELENA ESPRIT BRA 6891: Enhanced Learning for Evolutive Neural Architectures*. 1994.

[46] J. Madrenas, J. M. Moreno and J. Cabestany, CMOS Current-mode Analog Basic Blocks for Neural Processing. *Parts 1 and 2. MIXDES'95 Proc.*, pages 109-120, Kraków (Poland), 1995.

[47] J. Daubert, D. Vallancourt and Y. P. Tsividis. Current Copier Cells. *Electr. Letters*, pages 1560-1562, Dec. 1988.

[48] D. Macq and P. Jespers. Charge Injection in Current Copier Cells. *Electronics Letters*, 29(9), pages 780-781, April 1993.

[49] S. Espejo, R. Domínguez-Castro, F. Medeiro and A. Rodríguez-Vázquez. Tunable feedthrough cancellation in switched-current circuits. *Electronics Letters*, 30(23), pages 1912-1914, 10th Nov. 1994.

[50] S. Satyanarayana, Y. Tsividis and H. P. Graf. A Reconfigurable VLSI Neural Network. *IEEE Journal of Solid-State Circuits*, 27(1), pages 67, Jan. 1992

[51] C. A. A. Bastiaansen, D. W. J. Groeneveld, H. J. Schouwenaars and H. A. H. Termeer. A 10-b 40-Mhz 0.8μm CMOS current-output D/A converter, *IEEE Journal of Solid-State Circuits*, 26(7), pages 917, July 1991.

[52] K. L. Fong and A. T. Salama. A 10 bit semi-algorithmic current mode DAC, *proceedings of the IEEE International Symposium on Circuits and Systems (ISCAS'93)*, pages 978.

[53] C. Abel et al.. Characterization of Transistor Mismatch for Statiscal CAD Submicron CMOS Analog Circuits. *Proc. ISCAS'93*, page 1401, Chicago.

[54] J. Lazzaro et al.. Winner-Take-All Networks of O(n) Complexity. *Advances in Neural Information Processing Systems*, 1, pages 703-711, Touretzky D.S., ed., Morgan Kaufmann Publishers, 1989.

[55] D. J. Allstot, San-Hwa Chee, S. Kiaei and M. Shrivastawa. Folded source-coupled logic vs. CMOS static logic for low-noise mixed-signal Ics. *IEEE Trans. Circuits Syst. I: Fundamental Theory and Applications*, 40, pages 553–563, Sept. 1993.

CHAPTER 6

CMAC NEURAL NETWORKS AND SYSTOLIC IMPLEMENTATION

B.D. Liu
Department of Electrical Engineering
National Cheng Kung University
1, University Rd., Tainan, Taiwan 70101
E-mail: bdliu@cad.ee.ncku.edu.tw

Y.H. Kuo
Department of Computer Science and Information Engineering
National Cheng Kung University
1, University Rd., Tainan, Taiwan 70101
E-mail: yhkuo@cad.csie.ncku.edu.tw

J.S. Ker
Department of Electronic Engineering
Kun Shan University of Technology
949, Da-Wan Rd., Yung-Kang, Tainan County, Taiwan 71003
E-mail: kerjs@cad.csie.ncku.edu.tw

Neural networks operate in a massively parallel structure and with complex computations. The issues of cost, performance, precision, and reliability all must be considered carefully when one implements a neural network with hardware. In this chapter, we will introduce the Cerebellar Model Articulation Controller (CMAC) neural network and show how it operates. Secondly, we will introduce an enhanced model, named the higher-order CMAC, and illustrate its excellent approximation capability function. Next, we present a cost-efficient systolic array architecture to implement the higher-order CMAC with digital hardware. Finally, a case study, used to solve the color calibration problem, is presented to demonstrate how to realize higher-order CMAC with the presented systolic array architecture.

6.1. Introduction

In this chapter, we will introduce the Cerebellar Model Articulation Controller (CMAC) neural network, which is an alternative to backprogated multilayer networks, and present a cost-efficient systolic array architecture to realize the CMAC neural network. CMAC has the excellent features of local generalization and fast learning convergence. It is very suited for solving complex non-linear function approximation application problems. There are many successful applications in the fields of robot control, pattern recognition, signal processing, and color calibration.

Theoretically, backpropagated multilayer neural networks are capable of representing arbitrary mappings, provided that a sufficient number of nodes are included in the hidden layers. Because all of the weights are updated with each training pattern, these networks can construct global approximations of the interested functions. However, the global nature of weight-updating tends to blur the details of local characteristics, slows the rate of learning and recall, and most importantly, makes the accuracy of the resulting function approximation sensitive to the presentation order of the training patterns. Backpropagated multilayer neural networks are not suitable for producing applications requiring a real-time response.

Albus's CMAC model [1-4] is one of the famous models that exhibit the local weight updating feature, instead of inheriting the global weight updating feature of the backpropagated multilayer network. Many biological researches revealed that biological sensorimotor control structures in the cerebellum are organized using neurons that possess locally-tuned overlapping receptive fields. Because only a small subset of the neurons/weights is active at each point in the input space, individual weights account for a significant fraction of the local output error. The main benefits of using local approximation techniques to represent complex nonlinear functions are low computation complexity, fast learning and the ability to train the network in one part of the input space without corrupting what was already learned in more distant regions.

In this chapter, Section 2 introduces the standard CMAC-based neural networks, Section 3 introduces the higher-order CMAC model, Section 4 presents an efficient weight cell addressing formula for the access of active weight cells, Section 5 presents a cost-effective systolic array implementation of the higher-order CMAC based on the addressing formula introduced in Section 4, and Section 6 gives a design example which is used to solve the color calibration problem of color reproduction systems.

6.2. Standard CMAC-Based Neural Networks

The Cerebellar Model Articulation Controller (CMAC) was proposed by Albus to formulate the processing characteristics of the cerebellum [1-4]. It is essentially an extension of the single layer neural network. This model can learn a broad category of nonlinear functions. Unlike the global weight updating scheme used in the back-propagation styled neural networks, CMAC is characterized by a local weight updating scheme that has the advantages of low computation complexity and fast learning capability. This system avoids the drawback in which the accuracy of the function approximation is sensitive to the presentation order of the training data. This sensitivity drawback severely limits the applicability of neural networks with global weight updating to problems that require on-line learning.

6.2.1. *Detail Operations of CMAC*

Fig. 6-1 illustrates the detailed operation flow of a standard CMAC neural network. The scalar output y can be obtained by a sequence of operations, including input quantization, address segmentation, address segment concatenation, weight cell addressing, and output computation.

6.2.1.1. *Input Quantization ($F: X \rightarrow Q$)*

The primary input vector x is quantized into a quantization vector $q=(q_1,\cdots,q_i,\cdots,q_M)$, where q_i represents the quantization interval to which the component x_i belongs, that is,

$$q_i = F_i(x_i) = \left\lfloor \frac{x_i - X_{\min,i}}{D_i} \right\rfloor \tag{6-1}$$

where $x_i \in [X_{\min,i}, X_{\max,i}]$ and D_i are the width of each quantization interval on the i-th input dimension space. Therefore, the problem space is quantized into some discrete states in CMAC, and the resolution of the i-th input dimension is equal to Q_i+1, where Q_i is the maximum quantization value over the i-th input dimension space. The resolution determines the precision of the CMAC. There is a tradeoff among the CMAC output accuracy, the required weight memory size and the learning efforts. Finer resolution results in better accuracy but requires larger memory size.

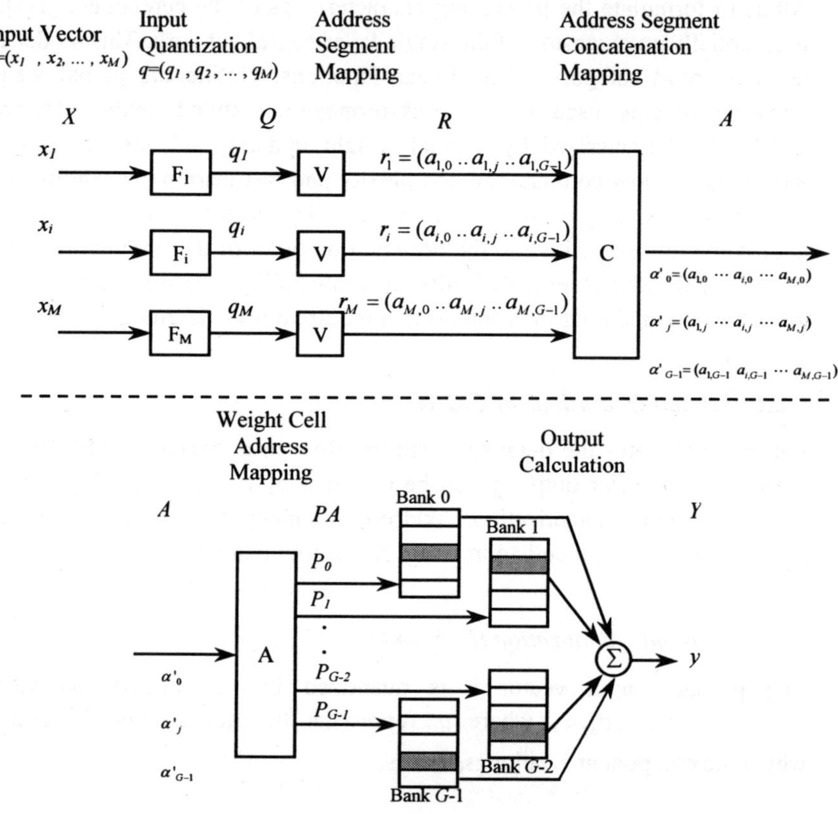

Fig. 6-1. Generic operation flow of CMAC.

6.2.1.2. *Weight Cell Address Generation (S: Q → PA)*

The main spirit of CMAC lies in determining the activation locations or addresses of a small sub-set of weight cells for a given input vector. Once the addresses are determined, the corresponding response of the CMAC can be easily calculaed.

In Albus's orginal proposal [3], weight address generation can be achieved by performing a series of mappings. The required mappings, as illustrated in Fig. 6-1, are address segmentation (Q → R), address segment concatenation (R → A), and weight cell addressing (A → PA). Regardless the kind of implemented mapping formula, there is one basic criteria to follow. The weight cell activation mechanism must present the local generalization feature. For two input vectors X_1 and X_2, two individual sets of weight cells V_1 and V_2 will be activated. The number of elements in the intersection of V_1 and V_2 is inverse proportional to the distance of X_1 and X_2. The smaller the distance is, the greater the number of overlapped elements. If the distance between X_1 and X_2 is larger then the generalization scope, the intersection of V_1 and V_2 is an empty set.

6.2.1.3. *Output Function and Learning Rule*

The output function just accumulates the weights retrieved by the corresponding active weight addresses:

$$y = \sum_{j=0}^{G-1} w_j \tag{6-2}$$

The most famous CMAC learning rule is the LMS weight updating algorithm [3]. The following equation is used to update the activated weight cells:

$$w_j = w_j + \beta \frac{(y_d - y)}{G} \tag{6-3}$$

where y_d is the desired output, β is the learning factor and G is the generalization factor.

By applying the LMS weight updating algorithm, the dynamics of the target function will be properly distributed over the weight cells, w_j's. During the training phase, activated weights are retrieved and adaptively updated according to CMAC's generalization rule so that the corresponding output of the training sequence can not only approximate the desired degree of accuracy, but also respond to any other input data that can properly generalize the appropriate outputs.

6.2.2. *Generic Architecture of CMAC*

Fig. 6-2 illustrates a generic CMAC architecture. For a certain input vector, only a small subset of weight cells, for example, w_0, w_3, w_6 and w_{N_A-1}, will contribute to the evaluation of an output value. Formally, CMAC approximates a nonlinear function $y = f(x)$ by using two primary mappings [5]:

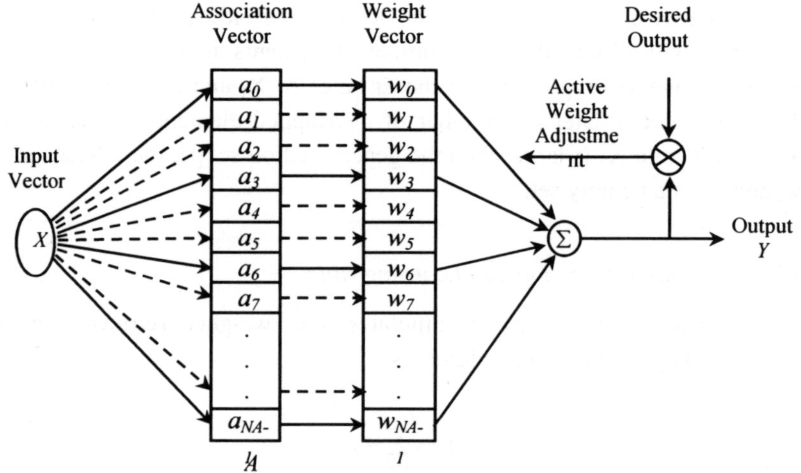

Fig. 6-2. Standard CMAC-based neural network model.

$$S : X \Rightarrow A$$
$$P : A \Rightarrow Y$$

where X is a continuous M-dimensional input space, A is an N_A-dimensional association space and Y is a one-dimensional output space. For the systems with multiple output dimensions, this CMAC model can be easily extended. The function S(x) maps each point x in the input space onto an association vector α that has N nonzero elements. For a standard CMAC model, the association vector contains only binary elements, either 0 or 1. The second mapping function P(α) computes a scalar output value y by projecting the associated vector onto a vector w of adjustable weights so that the scalar output can be easily obtained by evaluating the inner product of the two vectors α and w.

$$y_i = P(\alpha) = \alpha^T w \tag{6-4}$$

In comparison to the generic operation flow introduced in the previous sub-section, the function S(x) actually executes the input quantization mapping, address segmentation, address seqgment concatenation and physical address calculation. In this way, we can directly generate all of the required weight addresses for any input vector through the S(x) function. The other function P(α) is equivalent to the output function.

From the hardware implementation perspective, CMAC can be viewed as a special look-up table structure. For each input vector, there must be an efficient mechanism to determine a set of fixed-number weights to generate the corresponding response based on CMAC's generalization characteristics. A CMAC hardware module consists mainly of a weight storage unit, a weight address generation unit, and an output computation unit. A learning unit would be necessary if on-chip learning is required. The weight storage unit stores the weights to evaluate the output value. The weight address generation unit is responsible for implementing the mapping S(x), that is, deciding a set of weight addresses corresponding to those active (non-zero) association cells. The output computation unit executes the mapping P(α).

Two dominant factors in the CMAC hardware implementation should be managed carefully. First, traditional CMAC implementation requires a huge address space for the weight memory, but the weight memory usually has a very low utilization ratio. Hence, reducing the size of the weight memory is an important design issue. Second, the speed of weight address generation is the key factor that influences the overall CMAC performance, so a well-constructed architecture with a good performance/cost ratio should be developed for the weight address generation. In this chapter, these issues are solved by employing the direct weight address mapping approach. Based on this approach, a multiple weight memory bank structure and a pipelined weight address generator are adopted, which can drop out the redundant weight cells and result in fast address mapping.

6.3. Higher-Order CMAC (BCMAC)

As stated in the previous section, the representation of a non-linear function, $y_i = f(x_1, x_2, \cdots, x_M)$, by a CMAC is accomplished using two primary mappings, *S*: $X \Rightarrow A$ and $P_i : A \overset{W_i}{\Rightarrow} Y_i$, where *X* is a continuous *M*-dimensional input space, *A* is an N_A-dimensional association cell space, Y_i is the *i*-th output space, and W_i is the corresponding weight cells in which the mapping information of Y_i is stored.

The mapping S(x) can be further characterized using the following three sub-mappings [5]:

$$R: X \Rightarrow T$$
$$Q: T \Rightarrow L$$
$$E: L, T \Rightarrow A$$

where *R* is a receptive field function, *Q* is a quantization function, *E* is an embedding function, *T* is a matrix of the receptive field activation values and *L* is an array containing column vectors for identifying the locations of the maximally activated receptive fields along each input dimension.

Lane [5] tried to improve the overall function approximation precisions in CMAC and he found that incorporating the receptive field functions in terms of high-order polynomials, for example, cubic B-splines, can improve the CMAC-based function approximation precision. Fig. 6-3 illustrates the B-Spline shaped of the order n = 1, 2, and 4. B-spline receptive representation is preferable to the other spline formulations because of its superior numerical and computational properties [6].

Standard CMAC uses a rectangular function as the receptive field function. This is just a special case of the higher-order CMAC with order n = 1. Because the standard CMAC produces discontinuous staircased function approximation without inherent analytical derivatives, the output precision for the input vectors within each quantization region is worse than that of the trained patterns. By using B-spline receptive field functions, Lane [5] developed the higher-order CMAC model, called B-spline CMAC (BCMAC), to learn both functions and function derivatives. With the same quantization function, BCMAC can produce continuous function approximation. It can improve the overall output precision.

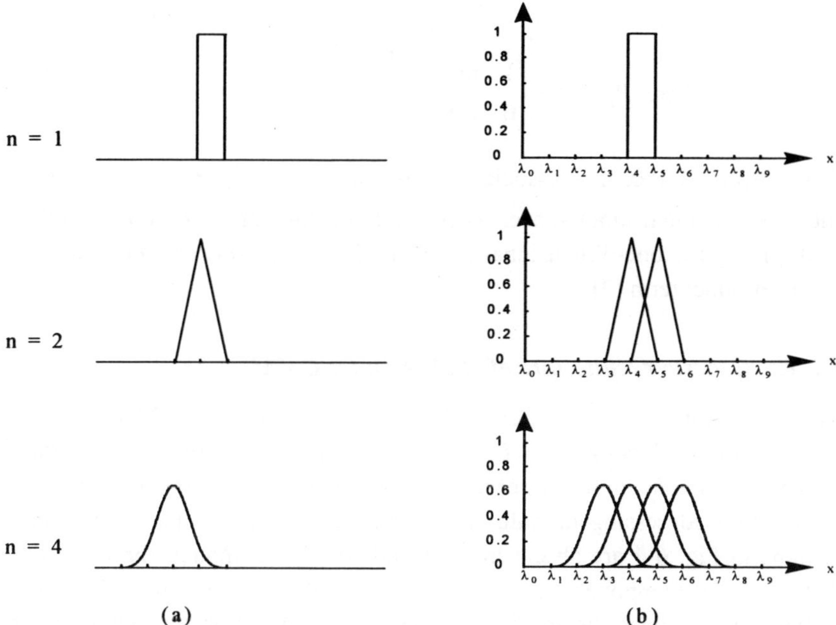

Fig. 6-3. B-Splines of order n = 1, 2, and 4 when x ∈ [λ4, λ5]. (a) B-Spline shape. (b) nonzero B-Spline location.

Given Q partitions over the interval x∈[a,b), an n-th order spline function $\xi(x)$ of the form:

$$f(x) \cong \xi(x) = \sum_{j=1}^{Q+n-1} w_j \cdot B_{n,j}(x) \tag{6-5}$$

can be constructed to approximate f(x) using a linear combination of B-splines Bn,j(x) weighted by the wj's coefficients. The sequence of normalized B-splines, Bn,1(x), Bn,2(x), ..., and Bn,Q+n-1(x), constructed on a knot set {λ0, λ1, ..., λQ}, forms a basis set for all polynomial splines of the order n on Q partitions of the interval x∈[a,b). Each Bn,j(x) can be obtained from the following recurrent relation,

$$B_{n,j}(x) = \frac{x-\lambda_{j-n}}{\lambda_{j-1}-\lambda_{j-n}} \cdot B_{n-1,j-1}(x) + \frac{\lambda_j - x}{\lambda_j - \lambda_{j-n+1}} \cdot B_{n-1,j}(x)$$
$$\text{where } B_{1,j}(x) = \begin{cases} 1 & \text{for } x \in [\lambda_{j-1}, \lambda_j) \\ 0 & \text{otherwise.} \end{cases} \tag{6-6}$$

The B-spline index j, is associated with the region $\lambda_{j-1} \le x < \lambda_j$. For the multi-dimensional input space, receptive field functions can be obtained by multiplying the one-dimensional receptive field functions contained in each tensor product term [7].

6.3.1. *Generic Architecture of Higher-Order CMAC*

From the implementation point of view, a higher-order CMAC generic architecture is illustrated in Fig. 6-4. Basically, it is composed of three processing units: multi-dimensional receptive field evaluation module, weight cell addressing module and output generation module. The shaded regions in Fig. 6-4 are absent in the standard CMAC model, shown in Fig. 6-2. The key component is still the weight cell addressing module, which utilizes and realizes the local generalization feature. For any input vector, efficiently determining a proper set of weight cells is most important. Based on the simple addressing methodolgy, we will present a cost-effective systolic array approach to implement the CMAC-based models.

6.3.2. *B-spline Computation Issue*

The feature incorporating B-spline receptive field functions with the CMAC model contributes to the excellent approximation capability of the higher-order CMAC on continuous nonlinear control functions. However, $B_{n,j}(x)$ is derived from a recurrent equation, whose value can be evaluated when the values of $B_{n-1,j-1}(x)$ and $B_{n-1,j}(x)$ are available.

Therefore, the computation flow of $B_{n,j}(x)$ is equivalent to traversing a binary tree where $B_{n,j}(x)$ is the root node.

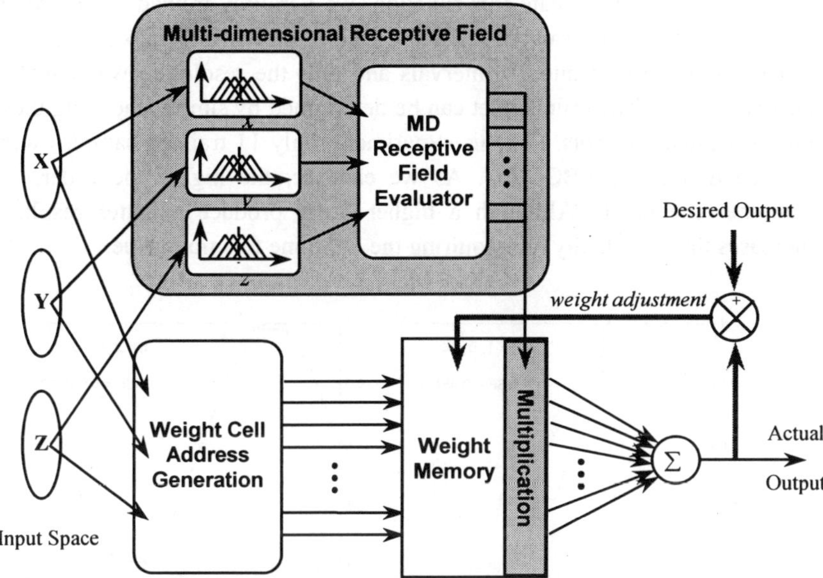

Fig. 6-4. Architecture of higher-order CMAC model.

Because B-spline receptive field computation is beyond the scope of this chapter, its implementation details are not included in this material. A more efficient computation hierarchy is derived in [8,9] to simplify the operations and speedup the computational performance.

6.3.3. *Simulation Results of BCMAC*

Some experiments are performed to verify the function approximation capability and performance of the BCMAC model. The order of the BCMAC model determines the continuity of the approximated function. There are some important parameters that might affect the quality of the approximation results, such as the quantization levels and the input space resolution, the span of the training patterns, and the continuity and smoothness of the control functions.

6.3.3.1. *Effect of the Order of B-Spline Receptive Fields*

Fig. 6-5 shows the simulation result for approximating $f(x) = sin(x*\pi/180)$ using different B-Spline function orders, where n is denoted as the order parameter. To approximate this function, we iterativly trained the BCMAC with a set of training samples. For the results shown in Fig. 6-5, the input space was quantized into 10 intervals and thus the resolution was equal to 360/10, i.e., 36. The training set can be determined by simply choosing those points located on knots. For this experiment, only 11 training samples were applied to train the BCMAC. As we can see, the higher the order, the smoother the result. Although a higher order produce a better result, it increases the complexity for acquiring the B-Spline function value.

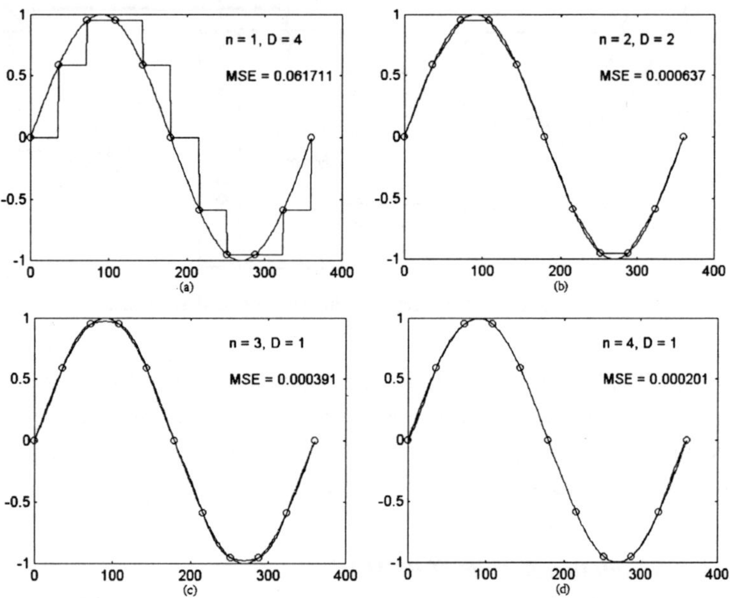

Fig. 6-5. Simulation results of f(x) = sin(x*π/180) by different B-Spline orders.

There are four approximation results illustrated for the BCMAC's with different order n. Fig. 6-5 (a) shows that the outside testing feature for a 1st-order BCMAC, i.e., the standard CMAC model, is a stair-case. The corresponding root-mean-square error (RMSE) is equal to 0.0617 which is too large to be satisfied. For such a 1st-order BCMAC to be adopted in a real application, the quantization level must be large enough so that the errors due to the stair-case approximation phenomenon become less serious and implicit. Fig. 6-5 (b) shows that the approximation capability between training samples is linear for a 2nd-order BCMAC. It can be referred to as linear interpolation. Fig. 6-5 (c) and 6-5 (d) show that BCMAC's with orders larger than 2 exhibit good continuous function approximation capability. In comparison with the lower-order BCMAC's in Fig. 6-5 (a) and 6-5 (b), we can see that the BCMAC exhibits excellent capability in approximating unknown nonlinear continuous functions with fewer knots on each input dimension. Such a feature requires fewer weight cells to be maintained.

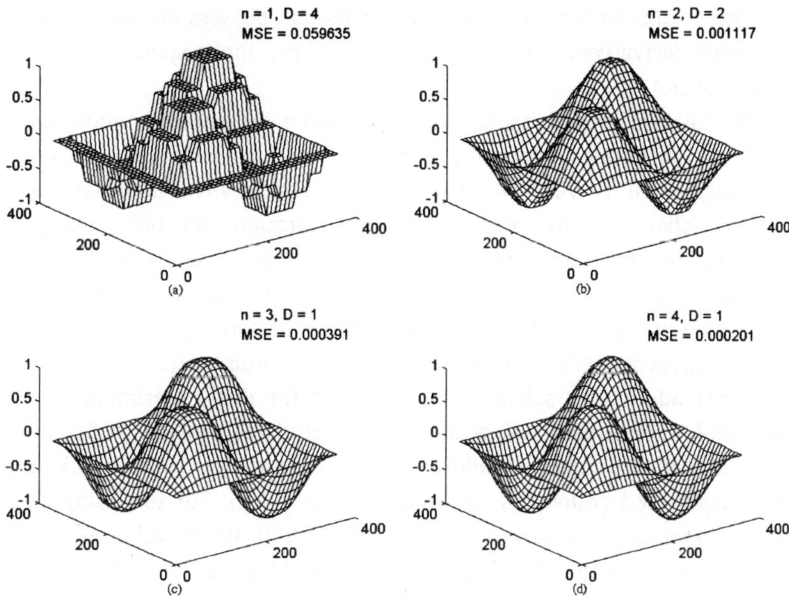

Fig. 6-6. Simulation results of $f(x,y) = \sin(x\pi/180)\times\sin(y\pi/180)$ by different order B-Splines.

Comparing the simulation results shown in Fig. 6-5, we can see that the higher-order CMAC model can exhibit excellent capability in approximating unknown nonlinear continuous functions with fewer knots on each input dimension. The same effect still remains when we apply the same experiment to the two-dimensional input spaces, where the simulation result is shown in Fig. 6-6. Clearly, the greater BCMAC the order, the greater approximation continuity acquired.

6.3.3.2. *Approximating with Various Addressing Schemes*

Because the non-zero elements (*N*) for the fully-connected addressing scheme will grow exponentially large with the number of input dimensions, i.e., $N = G^M$, for a model size problem, such as the case when the number of inputs $M = 6$, $G = 4$, and $n = 4$, the number of elements that must be computed in the associated tensor products, $N = 4^6 = 4096$, is quite large. As discussed in [5], using fewer than G^M basis functions is a potential solution, but this results in function approximations that possess discontinuities in the function derivatives across some faces of the input space receptive field hypercubes.

In the following experiment, we will explore the approximation results by applying different addressing schemes. For the proposed systolic array architecture, it is very flexible to apply any desired addressing scheme without changing the architectural configuration. By only changing the contents of the addressing matrix J and the value of N, any desired addressing scheme can be attached to the same physical computation unit.

As shown in Fig. 6-7, the approximation results form applying four different addressing schemes on a 4th-order BCMAC are illustrated. They are the main diagonal addressing scheme (N = G = 4 for this experiment), the main diagonal & anti-diagonal addressing scheme ($N = 2\times G = 8$), the main diagonal & sub-diagonal addressing scheme ($N = 3\times G = 12$), and the fully-connected addressing scheme ($N = G^2 = 16$). For the computation of multi-dimensional non-zero receptive fields with these addressing schemes, the results indicate that the tensor product operation fails to result in desirable convergent approximation results for those addressing schemes excluding the fully-connected addressing scheme. Therefore, instead of applying the tensor product operation, we employed the geometrical average operation to evaluate the multi-dimensional receptive fields. We achieved better convergent results. In Fig. 6-7 (a), the approximated surface is distorted because it does not converge to a desire state. For the case using the main diagonal &anti-diagonal addressing scheme, as shown in Fig. 6-7 (b), there are some slight fallen regions in the approximated surface. For this experiment, we can conclude that the approximation result for the main diagonal & sub-diagonal schemes, as shown in Fig. 6-7 (c), is good enough in comparison with the result for the fully-connected addressing scheme, as shown in Fig. 6-7 (d).

6.4. Weight Cell Address Generation

To enhance the weight cell storage utilization, the entire weight memory for one output dimension is divided into N memory banks (for example, $N = G = 4$ in Fig. 6-1). Using this memory bank structure, all weights related to one computation for the output value can be retrieved in parallel. Moreover, we introduced an address mapping scheme that integrates the three mapping stages (address segment generation, address segment concatenation, and physical weight address mapping) into a single mapping stage so that we can take redundant weight cells off and speed up the weight address computation.

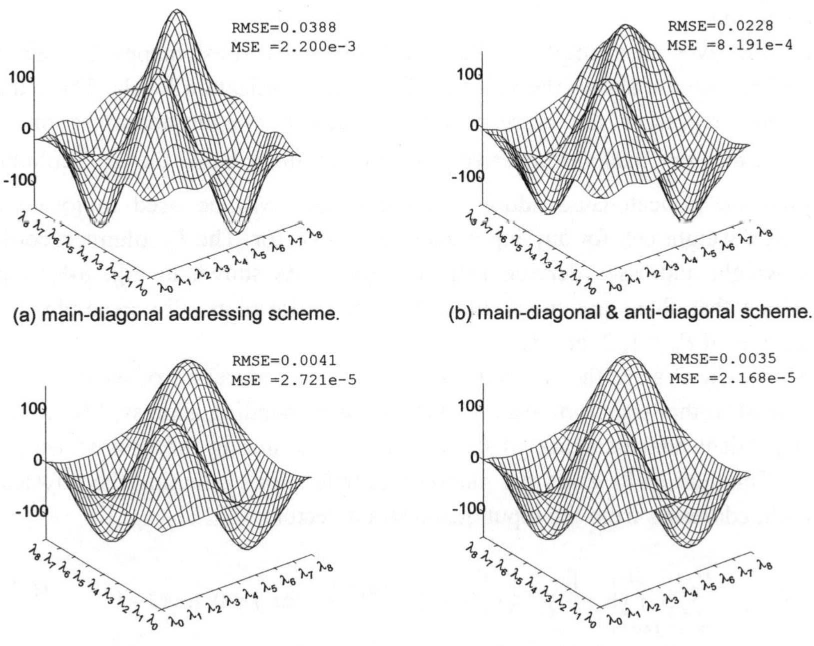

(a) main-diagonal addressing scheme. (b) main-diagonal & anti-diagonal scheme.

(c) main diagonal & sub-diagonal scheme. (d) fully-connected addressing scheme.

Fig. 6-7. Approximation results for different addressing schemes.

6.4.1. *Direct Weight Address Mapping*

To explain this direct mapping scheme, the weight addresses P_j's are generated first by assuming that the weight memory is a continuous one-dimensional array. The corresponding evaluation formula is given in the following:

$$\begin{aligned} P_j &= \sum_{m=1}^{M-1}\left(\prod_{n=m+1}^{M}(Q_n + G + 1)\times a_{m,j}\right) + a_{M,j} \\ &= (Q_2 + G + 1)\cdots(Q_M + G + 1)\cdot a_{1,j} + \\ &\quad (Q_3 + G + 1)\cdots(Q_M + G + 1)\cdot a_{2,j} + \cdots + a_{M,j} \quad \text{for j = 0 to G-1} \end{aligned} \tag{6-7}$$

where G is the generalization factor and M denotes the number of input dimensions.

Fig. 6-8 gives an example to illustrate the direct address mapping scheme. The first column shows the value of the input quantization levels. The other columns represent the corresponding generated weight addresses on G weight memory banks, respectively. In each memory bank, the $\alpha_j^{'}$ column shows the concatenated address segments that will be used to locate a specific weight cell for any input quantization vector. The P_j column records the weight addresses derived from Eq. (6-7). As shown in Fig. 6-8, it is obvious that there are many unused memory locations, for example, the locations of $P_0 = 1, 2,$ and 3.

Because the P_j's are the one-dimensional memory addresses of weight cells, we must further transform each P_j into the corresponding address, PN_j, in the independent address space of the memory bank storing the desired weight cell. Equation (6-8) gives the mapping function for evaluating the physical weight addresses from the input quantization vector.

$$PN_j = \sum_{m=1}^{M-1}\left(\prod_{i=m+1}^{M}\left(1+\left\lceil\frac{Q_i - j}{G}\right\rceil\right)\times\left\lceil\frac{q_m - j}{G}\right\rceil\right) + \left\lceil\frac{q_M - j}{G}\right\rceil \quad \text{for } j = 0 \text{ to } G-1 \tag{6-8}$$

Using Eq. (6-8), all of the weight cells are distributed in G independent memory banks. The weights are continuously allocated in each memory bank. The unused weight spaces have been dropped so that the total weight memory size can be maintained as small as possible for the physical hardware implementation.

q_1	q_2	q_3	Bank 0						Bank 1						Bank 2						Bank 3					
			α'_0				P_0	PN_0	α'_1				P_1	PN_1	α'_2				P_2	PN_2	α'_3				P_3	PN_3
			$a_{1,0}$	$a_{2,0}$	$a_{3,0}$				$a_{1,1}$	$a_{2,1}$	$a_{3,1}$				$a_{1,2}$	$a_{2,2}$	$a_{3,2}$				$a_{1,3}$	$a_{2,3}$	$a_{3,3}$			
0	0	0	0	0	0	=	0	0	1	1	1	=	133	0	2	2	2	=	266	0	3	3	3	=	399	0
0	0	1	0	0	4	=	4		1	1	1	=	133		2	2	2	=	266		3	3	3	=	399	
0	0	2	0	0	4	=	4		1	1	5	=	137		2	2	2	=	266		3	3	3	=	399	
0	0	3	0	0	4	=	4		1	1	5	=	137		2	2	6	=	270		3	3	3	=	399	
0	0	4	0	0	4	=	4	1	1	1	5	=	137	1	2	2	6	=	270	1	3	3	7	=	403	1
0	0	5	0	0	8	=	8		1	1	5	=	137		2	2	6	=	270		3	3	7	=	403	
0	0	6	0	0	8	=	8		1	1	9	=	141		2	2	6	=	270		3	3	7	=	403	
0	0	7	0	0	8	=	8	2	1	1	9	=	141	2	2	2	10	=	274	2	3	3	7	=	403	
0	1	0	0	4	0	=	44	3	1	1	1	=	133	0	2	2	2	=	266	0	3	3	3	=	399	0
0	1	1	0	4	4	=	48	4	1	1	1	=	133		2	2	2	=	266		3	3	3	=	399	
.	.	.	.	.	.		.		.	.	.		.		.	.	.		.		.	.	.		.	
.	.	.	.	.	.		.		.	.	.		.		.	.	.		.		.	.	.		.	
7	7	7	8	8	8	=	1064	26	9	9	9	=	1197	26	10	10	10	=	1330	26	7	7	7	=	931	7

Fig. 6-8. Illustration of direct weight address mapping scheme. (M=3, G=4, Q1=Q2=Q3=7).

6.4.2. *Extended Direct Weight Mapping*

As stated in [5], a variety of weight addressing schemes can be developed to activate a set of weights for the computation of one output value. Based upon the standard CMAC addressing scheme, we derived a direct weight cell address mapping equation in Eq. (6-8).

Eq. (6-8) gives the mapping function for evaluating the physical weight cell addresses from the input quantization vector. Although it can efficiently generate the set of activation addresses and reduce the size requirement for the weight cell memory, only the standard CMAC addressing scheme, i.e., the main diagonal addressing scheme, is considered. It becomes inappropriate for the weight cell address generation of the higher-order CMAC model. In order to increase the flexibility, we extended the direct address mapping equation to produce any desired addressing scheme. The extended equation is shown in the following form:

$$PN_j = \sum_{m=1}^{M-1}\left(\prod_{i=m+1}^{M}\left(1+\left\lceil\frac{Q_i-J_{i,j}}{G}\right\rceil\right)\times\left\lceil\frac{q_m-J_{m,j}}{G}\right\rceil\right)+\left\lceil\frac{q_M-J_{M,j}}{G}\right\rceil \quad \text{for } j=1 \text{ to } N \tag{6-9}$$

where J is an $M\times N$ addressing matrix used to specify the desired addressing scheme. M denotes the number of input dimensions and N denotes the number of specific non-zero terms in the hypercube determined by the location of the input vector and the generalization parameter G. Each entry

$J_{i,j}$ in J refers to the relative coordinate on dimension i of the j-th non-zero term in the corresponding hypercube.

By changing the contents in the addressing matrix, any desired addressing scheme can be realized without modifying the hardware. Suppose that the number of activated non-zero B-splines in each input dimension, G, is 4, and the number of M is 2. The addressing matrices $J_{2\times4}$ and $J_{2\times8}$ for the main diagonal addressing scheme and for the main diagonal & anti-diagonal addressing scheme respectively are shown in the following:

$$J_{2\times4} = \begin{bmatrix} 0 & 1 & 2 & 3 \\ 0 & 1 & 2 & 3 \end{bmatrix}, \text{and} \quad J_{2\times8} = \begin{bmatrix} 0 & 1 & 2 & 3 & 0 & 1 & 2 & 3 \\ 0 & 1 & 2 & 3 & 3 & 2 & 1 & 0 \end{bmatrix}.$$

We can explain how the above equation is derived from another viewpoint. In Fig. 6-9, the CMAC model is configured with M = 2, G = 4, and the main-diagonal addressing scheme is adopted, which implies that N = G = 4. Each circle in Fig. 6-9 represents a specific weight cell where meaningful information is stored. For a given quantized input pattern (q_1, q_2), a unique set of weights enclosed within a hypercube will be addressed to compute the corresponding output value. The location and size of the hypercube is determined according to the following basic guidelines: first, the hypercube originates at the grid where the quantized input pattern is located; second, the width of the hypercube along the direction of each input dimension is equal to G quantization intervals.

Because the main diagonal addressing scheme is used, the distribution of employable weight cells is very regular over the entire logical weight space. All of the active weight cells are located on the main diagonal line of the hypercube that originates from the grid (q_1, q_2), where the q_i value must be divisible by G, i.e., (q_i mod G) = 0. As illustrated in Fig. 6-9, the main diagonal addressing scheme results in a sparse distribution of actually used weight cells over the entire weight memory space. In other words, only a small subset of memory cells is really used to store the weights that will contribute to the evaluation of the outputs.

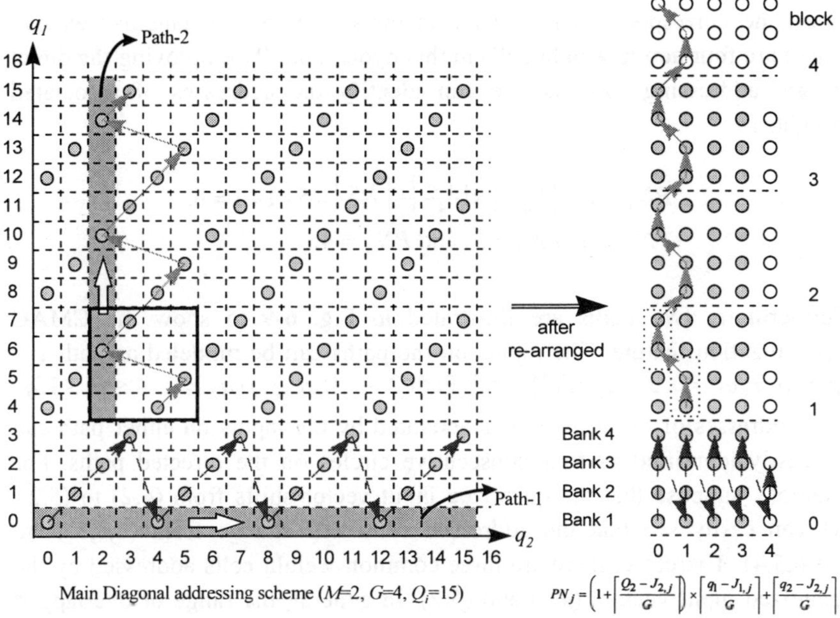

Fig. 6-9. Illustration of the extended weight cell addressing formula.

The extended direct weight addressing equation (Eq. (6-9)) can drop redundant memory cells (the grids without circles on them) so that the memory size and hardware cost can be reduced. The reason is described as follows. First, the weight cells on the same row in the j-th memory bank are shifted towards the vertical axis q_1 with $J_{2,j}$ grids, so that each shifted weight cell is aligned to the same column of the nearest weight cell in Bank 1. Second, the weight space is squeezed into a compact area so that only those circled weight cells exist. As a consequence, the weight memory space can be implemented in N memory banks, where each memory bank is composed of a sequence of blocks. The block size of each memory bank is equal to the number of weights on the same row $1+\left\lceil \frac{Q_2 - J_{2,j}}{G} \right\rceil$, as illustrated in Fig. 6-9. The number of blocks in the j-th bank is $\prod_{i=1}^{M-1}\left(1+\left\lceil \frac{Q_i - J_{i,j}}{G} \right\rceil\right)$, for j = 1 to N. All of the weights in any memory bank are continuously allocated and can be easily and efficiently addressed.

For example, given the input pattern located in the quantized region $(q_1,q_2) = (4,2)$, the corresponding hypercube is marked with a rectangular widow. There are four active weight cells in this hypercube. By employing the direct weight addressing formula, we can identify the addresses of associated weights as

$$PN_1 = \left(1+\left\lceil \tfrac{15-0}{4} \right\rceil\right)\times\left\lceil \tfrac{4-0}{4} \right\rceil+\left\lceil \tfrac{2-0}{4} \right\rceil = 5\times 1+1=6,$$
$$PN_2 = 6,\ PN_3 = 5,\ \text{and}\ PN_4 = 4.$$

Furthermore, two paths are illustrated in Fig. 6-9 to show the CMAC generalization feature. The inputs on the paths can be modeled as Path-1 = $\{(q_1,q_2)\mid q_1 = 0, 0\le q_2 \le 15\}$ and Path-2 = $\{(q_1,q_2)\mid q_2 = 2,\ 4\le q_1 \le 15\}$. The addresses of the weight cells activated by the inputs on either path can be easily identified as four consecutive circles on the directed paths. For example, suppose that the quantized input vector shifts from (4,2) to (5,2), we can easily see that the addressed cells will change from (6,6,5,4) to (11,6,5,4). Therefore, there are three common weight cells addressed by the quantized input vectors (4,2) and (5,2). In general, the range of overlapped weight cells is inversely proportional to the distance between two input vectors and there will be no common weight cell when the distance is not less than the generalization parameter G. For example, the distance between (4,2) and (8,2) is 4, so there is no common weight cell between the activated weight cells (6,6,5,4) and (11,11,10,8).

6.5. Systolic Architecture for Addressing Weight Cells

To efficiently implement the weight cell address generation unit, the operations performed in Eq. (6-9) are re-formulated into the following equivlent formula:

$$R_j(0) = 0$$
$$R_j(i) = \left(1+\left\lceil \tfrac{Q_i - J_{i,j}}{G} \right\rceil\right)\times R_j(i-1)+\left\lceil \tfrac{q_i - J_{i,j}}{G} \right\rceil \quad \text{for } i=1 \text{ to } M, \text{and}$$
$$PN_j = R_j(M)$$

Referring to Eq. (6-10), we developed a linear systolic array architecture to implement the extended direct addressing mechanism. The corresponding systolic schedule is illustrated in Fig. 6-10.

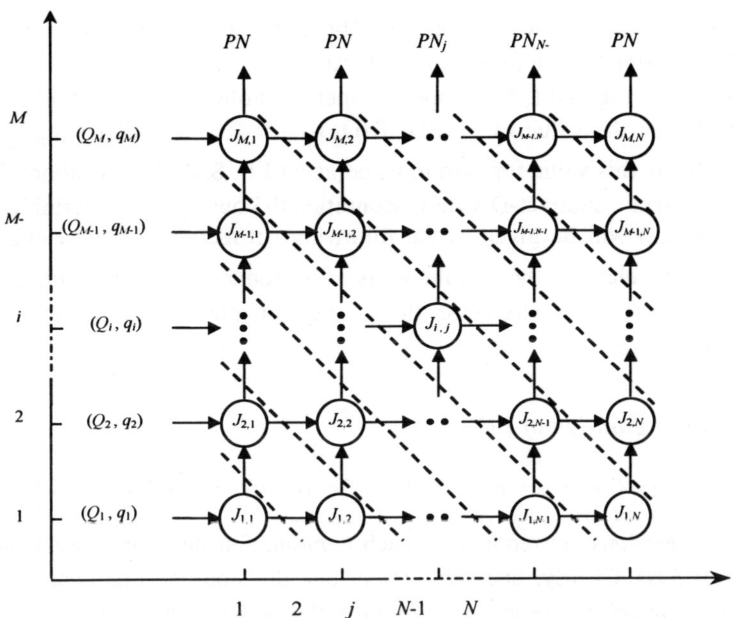

Fig. 6-10. Systolic schedule for the extended addressing mechanism.

In Fig. 6-10, the processing elements (PEs) in the same row i, evaluate the corresponding component $R_j(i)$ in column j, for j = 1 to N, according to the values of Q_i, q_i, $J_{i,j}$, and the component $R_j(i-1)$ generated by its precedent PE. The PEs in the same row can be operated concurrently or in any sequential order. However, for the PEs in the same column j, there exists data dependency between adjacent PEs, so that the PEs must be operated sequentially in the ascendant order of i.

6.5.1. *Architecture Mapping*

The systolic schedule is obtained by scheduling along the direction of the schedule vector $\vec{s} = [1 \quad 1]^T$. The PEs in the same partition (in Fig. 6.10, each partition identified by adjacent dotted lines) can operate simultaneously. To reduce the overall hardware cost, we projected the systolic schedule along the projection vector $\vec{d} = [0 \quad 1]^T$ to obtain a linear systolic array architecture, as shown in Fig. 6-11. Alternatively, if we project the systolic schedule along the vector $\vec{d} = [1 \quad 0]^T$, an alternative parallel computational architecture can also be obtained [9]. The detailed configuration of each PE will be described in the following subsection.

As shown in Fig. 6-11, the proposed linear systolic array has M PEs. Each PE evaluates the value of $(1+SDUR(Q_i,J_{i,j})) \times R(i-1) + SDUR(q_i,J_{i,j})$ and then delivers this value forward to its adjacent PE. SDUR is the abbreviation of the term Subtract-Divide-Unconditional-Round. The definition of $SDUR(q_i,J_{i,j})$ will be given in the following subsection. After M cycles of latency have elapsed, the systolic array will produce one weight address per cycle at the last PE. Therefore, the throughput rate is equal to one weight address per cycle.

6.5.2. *Configuration of PE*

By examining the recurrent formula in Eq. (6-10), we find that two $\left\lceil \frac{q_i - J_{i,j}}{G} \right\rceil$ SDUR operations are required in each computation iteration to evaluate the value of $R_j(i)$. Clearly, one subtraction, one division and one unconditional round operation are needed to accomplish the corresponding SDUR function. From the viewpoints of speed and area, it is too expansive to directly implement the corresponding operations into the physical hardware circuits. Therefore, we looked further for an optimized computation scheme to implement the SDUR operation with an efficient digital hardware implementation. We found that the result of $SDUR(q_i, J_{i,j})$, for $J_{i,j}$ = 0, 1, ..., G-1, equals either $\left\lfloor \frac{q_i}{G} \right\rfloor$ if the remainder R of $\frac{q_i}{G}$ is not greater than $J_{i,j}$ or $\left\lfloor \frac{q_i}{G} \right\rfloor + 1$ if the remainder R is greater than $J_{i,j}$ [9].

Accordingly, the SDUR operation can be carried out using just one integer division $\frac{q_i}{G}$ and one increment operators. After analyzing the data sequence fed to a PE, we found that the value of Q_i and q_i are constants during the evaluation session of PN_j's for a certain input vector X, and only the value of $J_{i,j}$, for j = 1 to N, is varied. Therefore, the SDUR operation can be further accelerated and simplified by extracting the division operation from the SDUR operator to shorten the computation time.

accelerated and simplified by extracting the division operation from the SDUR operator to shorten the computation time.

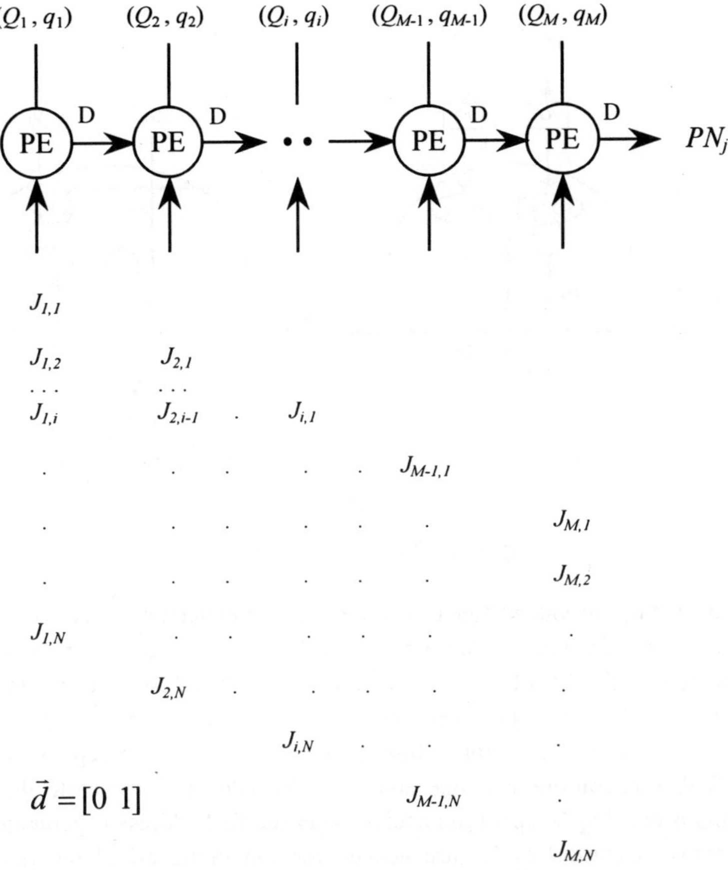

Fig. 6-11. Architecture mapping for the linear systolic array.

Fig. 6-12 illustrates the detailed configuration of a processing elements correspond to the quotient and remainder part of $\frac{Q_i}{G}$ and $\frac{q_i}{G}$, respectively, and latched in the PE. Therefore, the operation performed in an SDUR operator not only becomes rather simple, but also takes less computation time. As shown in Fig. 6-12 (b), the value of SDUR(q_i,$J_{i,j}$) can be easily determined, which equals to u_i, or u_i+1, by comparing the values of v_i and

$J_{i,j}$: the value of SDUR is equal to either u_i when $v_i \le J_{i,j}$, or u_i+1 when $v_i > J_{i,j}$.

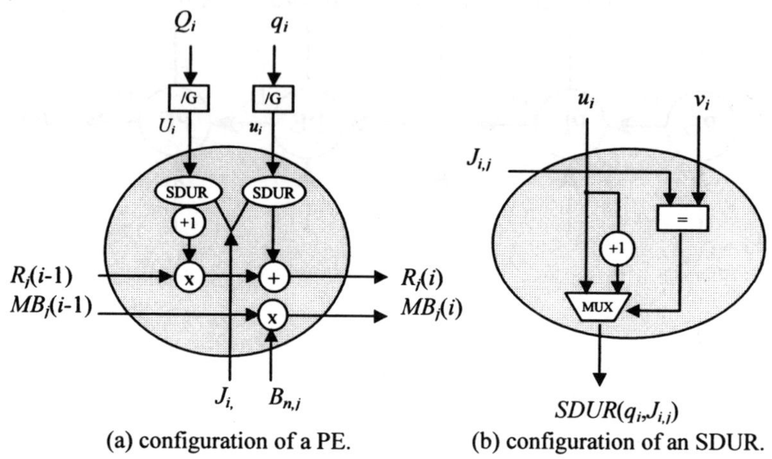

(a) configuration of a PE. (b) configuration of an SDUR.

Fig. 6-12. Detailed configuration of PE.

Based on the introduced linear systolic array architecture, a cost-effective hardware implementation for a higher-order CMAC model can be easily developed. Only M PEs are needed to compute all of the weight cell addresses. The PE operations are rather simple: two multiplications (including one for the computation of multi-dimensional receptive fields), two SDUR operations and one addition. The latency to generate the first address is equal to M operation cycles. After the first address is generated, N successive operation cycles are needed for generating all of the required weight cell addresses. In other words, the latency time for the first output vector is equal to M+N operation cycles. The overall throughput of the linear systolic array is equal to $1/(N\times T_{cyc})$ for generating successive output vectors.

6.6. VHDL Modeling of the Systolic Implementation

To implement the presented architecture for a higher-order CMAC (BCMAC) model with digital hardware, we can employ high-level synthesis methodology to design the circuits. All of the modules are modeled in

VHDL. In this section, the VHDL models are demonstrated with a practical design for the color calibration application. The specifications for implemented BCMAC chip are: the number of input and output dimensions, M and K, are equal to 3. The maximum quantization levels, which are programmable by an external host controller, for each input dimension set to the same value. The order of the B-spline receptive field function n is 4. The generalization parameter G is 4 and the weight cell addressing scheme is programmable, that is, N is programmable. The valid values for the color input dimensions and weights are represented by 8-bit binary values, that is, D = 8.

6.6.1. *Block Diagram of BCMAC*

The block diagram of the BCMAC prototyped circuit for the color calibration is shown in Fig. 6-13. It is partitioned into three sub-modules for creating the VHDL behavioral models. The module ADR_GEN is designed to evaluate the weight cell addresses PN_js and the multi-dimensional B-spline reveptive fields. The OPT_GEN module is designed to calculate the output response with the retrieved weight values and multi-dimensional B-spline receptive fields. Interactions between these two modules are managed by a third module ACC_GEN, which generates all of the timing and control signals to control the operations of the systolic array based BCMAC. The ACC_GEN also interacts with the external host controller or processor.

6.6.2. *Systolic Weight Cell Addressing Module*

As introduced in Section 6.5, the linear systolic array consists of M PE's, where $M = 3$ for the color calibration application. Each PE evaluates the value of $(1{+}SDUR(Q_i, J_{i,j})) \times R(i\text{-}1) + SDUR(q_i, J_{i,j})$ and the value of $MB_{i-1} \times B_{(n,j)_i}$, then delivers these values forward to its adjacent PE. After M cycles of latency have elapsed, the systolic array will produce one weight cell address per cycle at the last PE. Therefore, the throughput rate equals to one weight cell address per cycle.

Because the structure of each PE is very regular, as illustrated in Fig. 6-13, it is easily to write the corresponding VHDL codes for each PE. The following VHDL codes constitute the description of the linear systolic array.

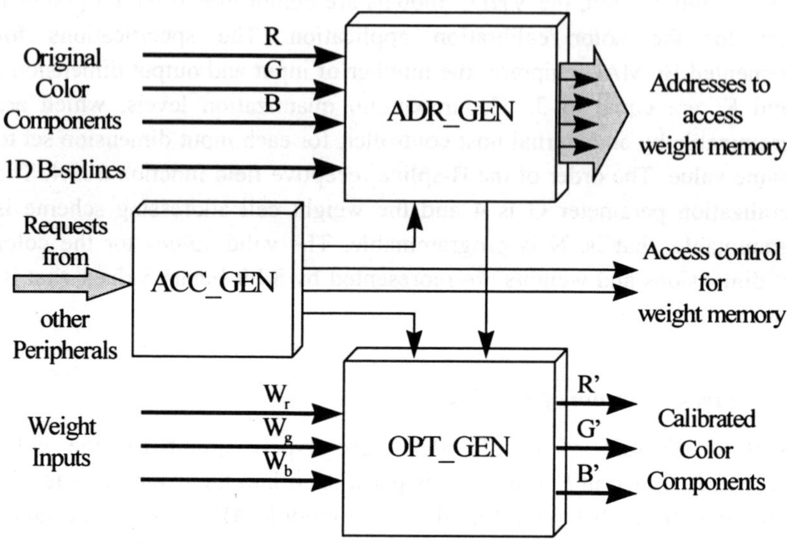

Fig. 6-13. Block diagram of the higher-order CMAC chip.

```
pe1: block
begin
   So12 <= SDUR630(US1, VS1, J1);

   process(CLK)
   begin
      if CLK'event and CLK = '0' then
          R1 <= So12; -- output of PE1
      end if;
   end process;

   process(CLK2)
   begin
      if CLK2'event and CLK2 = '1' then
          if CLK = '1' then
            MB1 <= Bnj1(5 downto 0); -- output of PE1
          end if;
      end if;
   end process;
```

```
end block pe1;

pe2: block
begin
  So21 <= SDUR631(UB,  VB,  J2);
  So22 <= SDUR630(US2, VS2, J2);
  MUL21: MUL6_6  Port Map ( -- 6-bit * 6-bit multiplier
         X(5 downto 0) => R1(5 downto 0),
         Y(5 downto 0) => So21(5 downto 0),
         Z(11 downto 0) => M21(11 downto 0) );

  process(CLK)
  begin
    if CLK'event and CLK = '0' then
       R2 <= M21(11:0) + So22; -- output of PE2
    end if;
  end process;

  MUL22: MUL6_6 Port Map (
         X(5 downto 0) => MB1(5 downto 0),
         Y(5 downto 0) => Bnj2(5 downto 0),
         Z(11 downto 0) => iMB2(11 downto 0) );

  process(CLK2)
  begin
    if CLK2'event and CLK2 = '1' then
        if CLK = '1' then
           MB2 <= iMB2(11:6); -- output of PE2
        end if;
    end if;
  end process;
end block pe2;

pe3: block
begin
  So31 <= SDUR631(UB,  VB,  J3);
  So32 <= SDUR630(US3, VS3, J3);
  MUL31: MUL12_6  Port Map ( -- 12-bit * 6-bit multiplier
         X(11 downto 0) => R2(11 downto 0),
         Y(5 downto 0) => So31(5 downto 0),
         Z(17 downto 0) => M31(17 downto 0) );

  process(CLK)
  begin
    if CLK'event and CLK = '0' then
        R3 <= M31(17:0) + So32;-- output of PE3, ie.,
           -- weight address output of the systolic array
    end if;
```

```
   end process;

   MUL32: MUL6_6 Port Map (
          X(5 downto 0) => MB2(5 downto 0),
          Y(5 downto 0) => Bnj3(5 downto 0),
          Z(11 downto 0) => iMB3(11 downto 0) );

   process(CLK2)
   begin
     if CLK2'event and CLK2 = '1' then
          if CLK = '1' then
             MB3 <= iMB3(11:6); -- output of PE3, ie.,
                 -- the multi-dimensional receptive fields
          end if;
     end if;
   end process;
end block pe3;
```

6.6.3. *Output Generation Module*

The output generation module is designed to calculate the BCMAC output in response to a certain input vector. It performs the following operation:

$$Y = \sum_{j-1}^{N} w_j \times MB_j \tag{6-11}$$

where j = 1 to N, w_j is the j-th non-zero weight retrieved from the weight cell memory, and MB_j is the corresponding j-th multi-dimensional B-spline receptive field value.

For the implementation used in color calibration, there are three output dimensions. Therefore, three copies of the above equation are implemented. In the following, the VHDL descriptions of an output dimension, $R_{in} + \Delta R$, is given. $G_{in} + \Delta G$ and $B_{in} + \Delta B$ can be easily realized with another two similar block constructs.

```
evaluate_r: block  begin
    MUL_WR: MUL8_6 Port Map (
              X(7 downto 0) => WR(7 downto 0),
              Y(5 downto 0) => MB(5 downto 0),
              Z(13 downto 0) => iMBWR(13 downto 0) );
    Opndr2 <= iMBWR(13) & iMBWR(13 downto 6);

    process(START, OP_CLK)
```

```
    begin
        if START = '0' then
            Opndr1 <= "000000000";
        elsif OP_CLK'event and OP_CLK = '1' then
            if vld2int(P_STATE) = 4 then
                Opndr1 <= '0'&Rin; -- initial value
            Else
                Opndr1 <= Rcor; -- result of Multiply-an
            d-Add
            end if;
        end if;
    end process;

    Rcor  <= Opndr1 + Opndr2;
    Ov_rflag <= Rcor(8) and (Opndr1(8) nor Opndr2(8));
    Un_rflag <= (not Rcor(8)) and
                (Opndr1(8) and Opndr2(8));

    -- process to handle the overflow/underflow problem
    -- of a pixel.
    process(Un_rflag, DONE)
    begin
        if Un_rflag = '1' then
            Ro <= "00000000";
        elsif DONE'event and DONE = '1' then
            if Ov_rflag = '1' then
                Ro <= "11111111";
            else
                Ro <= Rcor(7:0); -- Output of the BCMAC
            end if;
        end if;
    end process;
end block evaluate_r;
```

6.6.4. *Access Control Module*

The access control module ACC_GEN is designed to control the operations of the entire BCMAC circuit. In the previous two sub-sections, we described the simplicity of the ADR_GEN and OPT_GEN module operations. Their VHDL descriptions model just the corresponding data flow operations. The information for the control path design is gathered in the access control module ACC_GEN.

There are several basic building blocks in the ACC_GEN module, including Kernel Control, Quantization, Register Access, and Weight Access. Because the details of the ACC_GEN module are out of the scope of this book, descriptions of these blocks are not included in this material. If you are interested in how they are can be realized, you can refer to [8].

6.7. Conclusions

Neural network hardware implementation is still an important issue in research. Because neural networks operate in a massively parallel structure and with complex computations, the issues of cost, performance, precision and reliability must all be considered carefully when implementing a neural network with hardware. Design productivity is also an important issue. In this chapter, we developed a cost-effective hardware architecture to implement the standard CMAC model and the higher-order CMAC model. Some conclusions about the hardware implementation of the CMAC-based models will now be presented.

For the standard CMAC and BCMAC (higher-order CMAC) models, as described in Sections 6.2 and 6.3, the size of the weight memory can be reduced using the introduced direct weight cell addressing approach. In addition, we presented a cost-effective hardware architecture to realize the operation of any CMAC-based model. A linear systolic array architecture was derived so that we can efficiently generate all of the weight cell addresses for the corresponding non-zero terms in a relative hyper-cube window. Only M PEs are needed for the N weight cell address computations. The throughput of such a systolic array architecture is equal to $1/(N \times T_{cyc})$, i.e., N computation cycles are required to generate two adjacent output vectors.

In Section 6.6, we presented a case study example that employs the presented BCMAC systolic architecture to solve the color correction problem for a color reproduction environment.

Most literature about CMAC neural network models is concentrated on the application to simplify robot control [10-13]. Among others, some papers concentrated on a discussion of learning convergence [14-15] and new learning methodologies [16-17] for the CMAC model. However, few relevant publications have discussed how to efficiently implement the CMAC model from the aspect of hardware design. In this chapter, we focused on the implementation issue for CMAC-based neural networks.

References

[1] J. S. Albus. A Theory of Cerebellar Function. *Mathematical Biosciences*, 10, page 25, 1971.

[2] J. S. Albus, *Theoretical and Experimental Aspects of Cerebellar Model*, Ph. D. Dissertation, Dep. Biomed. Eng., Univ. Maryland, 1972.

[3] J. S. Albus. A New Approach to Manipulator Control: The Cerebellar Model Articulation Controller (CMAC). *Journal of Dynamic Systems, Measurement, and Control*, 97, page 220, 1975.

[4] J. S. Albus. Data Storage in the CMAC. *Journal of Dynamic Systems, Measurement, and Control*, 97, page 228, 1975.

[5] S. H. Lane, D. A. Handelman, and J. J. Gelfand. Theory and development of higher-order CMAC neural networks. *IEEE Control Systems*, 22, page 23, 1992.

[6] M. G. Cox. Practical Spline Approximation. In *Lecture Notes in Mathematics 965: Topics in Numerical Analysis*, Springer-Verlag, New York, page 79, 1984.

[7] P. J. Hartley. Tensor product approximations to data defined on rectangular meshes in N-Space. *The Computer J.* 19, page 348, 1975.

[8] J. S. Ker. *Hardware Design of the CMAC-based Neural Network and Its Application on Color Reproduction Systems*. Ph.D. Dissertation, National Cheng Kung University, 1996.

[9] J. S. Ker, Y. H. Kuo, and B. D. Liu. Systolic Implementation of a Higher-order CMAC and Its Application in Color Calibration. *IEE Proceedings - Circuits, Devices, and Systems*, 144, page 129, 1997.

[10] W. T. Miller, III, R. P. Hewes, F. H. Glanz, and L. G. Kraft, III. Real-time Dynamic Control of an Industrial Manipulator Using a Neural-network-based Learning Controller, *IEEE Trans. on Robot. Automat.* 6, page 1, 1990.

[11] W. T. Miller. A Nonlinear Learning Controller for Robotic Manipulators. *Proc. SPIE, Intelligent Robots and Computer Vision*, 726, page 416, 1986.

[12] F. H. Glanz and W. T. Miller. Shape Recognition Using a CMAC Based Learning System. *Proc. SPIE, Intelligent Robots and Computer Vision*, 848, page 294, 1987.

[13] F. H. Glanz and W. T. Miller. Deconvolution Using a CMAC Neural Network. *In Proc. 1st Annual Conf. of the Int. Neural Network Society, IEEE*, New York, page 440, 1988.

[14] D. Ellison. On the Convergence of the Multidimensional Albus Perceptron. *The Int. J. of Robotic Research*, 10, page 338, 1991.

[15] Y. F. Wong and A. Sideris. Learning Convergence in the Cerebellar Model Articulation Controller. *IEEE Trans. on Neural Networks*, 3, page115, 1992.

[16] D. E. Thompson and S. Kwon. Neighborhood Sequential and Random Training Techniques for CMAC. *IEEE Trans. on Neural Networks*, 6, page196, 1995.

[17] C. S. Lin and H. Kim. Selection of Learning Parameters for CMAC-Based Adaptive Critic Learning. *IEEE Trans. on Neural Networks*, 6, page 642, 1995.

CHAPTER 7

QUADRANT INTERLOCKING FACTORIZATION ON SYSTOLIC AND WAVEFRONT ARRAY PROCESSORS

M.P. Bekakos

Department of Electrical & Computer Engineering,
School of Engineering, Democritus University of Thrace,
67 100 Xanthi, Greece
E-mail: mbp@aueb.gr

O.B. Efremides

Department of Informatics,
Athens University of Economics and Business,
104 34 Athens, Greece
E-mail: obe@aueb.gr

D.J. Evans

Department of Computing, Faculty of Engineering & Computing,
The Nottingham Trent University,
Nottingham, U.K.
E-mail: D.J.Evans@doc.ntu.ac.uk

In this Chapter, the Quadrant Interlocking Factorization (QIF) procedure is implemented on Systolic and Wavefront Array Processors. At first, the fast direct hardware implementation of the QIF procedure, for the solution of a set of linear equations on a special-purpose pipelined + shared memory architecture with multilayered meshes, is investigated. The proposed Multilayered DEWavefront Array Processor complex (M-DEWAP) consists of three square meshes interconnected via shared memory circuits, while pipelined techniques are adopted. The potential of a block generalized WZ factorization procedure for direct hardware implementation on a special-purpose network of Wavefront Array Processor Modules (WAPms) is then investigated. The definition of a wavefront module is given and the

development of an Eigenvalue Evaluator Engine (E^3) based on this module is discussed. Finally, a new systolic array is introduced for the solution of linear system equations occurring in scientific engineering calculations using the parallel Quadrant Interlocking Elimination method of Evans *et al.* [8].

7.1. Introduction

The QIF or ***butterfly*** algorithm, established by Evans ***at al.*** [7], can be expressed in a recursively regular manner for VLSI implementation to obtain the factorization and forward/backward processes.

The efficient utilization of VLSI advanced technology in large processor array complexes confronts some fundamental problems imposed by this technology. Such problems, for example, arise from the need for a restricted local communication and self-timed schemes to avoid the clock skew incurring in signal distribution when the computing structure is considerably large; also, from the need for innovative descriptive tools to support the visualization, description and verification of the implemented parallel software, as well as the need for minimizing the design complexity, and hence the cost, through the utilization of iterative modular structures and programmable processor modules, instead of dedicated modules.

The advanced characteristic of the M-DEWAP architecture, investigated herein, is its computational notation introduced through the adopted notion of advancing waves of data and computational activity in a continuous mode. This activity propagation resembles a physical wave propagation phenomenon and eliminates the need for global synchronization requirements.

The generalized WZ factorization procedure for the eigenvalue problem was proposed in Benaini ***et al.*** [4] and Benaini ***et al.*** [5] as an extension of the WZ factorization introduced by Evans [10]. This new approach seeks the factorization of a real matrix A in a form $A = WZW^{-1}$, where matrices W and Z are of butterfly form. So, the eigenvalues of A are those of Z and they can be computed in $O(n)$ sequential time steps and in $O(1)$ parallel time steps. An area efficient combined utilization of both pipelined and wavefront conceptual computational tools is attempted for a mixed pipelined/parallel solution of general linear systems.

A special-purpose network of modules, each module consisting of a WAP (Kung *et al.* [15] and Evans et al. [12]), is used to implement the factorization procedure. The important feature of this WAP is its computational notation which removes the need for global synchronization and also proves to be useful for programming the engine and describing the algorithm.

The development of systolic arrays for problems in linear algebra has to date been largely restricted to existing sequential algorithms, like the LU factorization, gaussian elimination and QR decomposition. However, recently a class of parallel methods based on Quadrant Interlocking transformations have been introduced by Evans [11], which show improvement in a parallel environment. There is no fundamental reason other than simplicity which indicates that a sequential to systolic algorithm conversion gives the best systolic array. Indeed, it is interesting to question this implicit assumption of array designers.

A solution to the following question *"are the Quadrant Interlocking (QI) methods better suited to systolic computation than basic sequential method?"* is presented. Intuitively the QI methods have a number of advantages. Firstly, they are aimed at a parallel machine from the outset, and secondly, they have improved performance over other parallel implementations for solving linear systems. The QI elimination algorithm has been presented previously by Evans *et al.* [8]. The challenging task here is to transfer these improvements to systolic array implementations.

7.2. The Butterfly Procedure for Solving Linear Systems

The problem considered here is given by 7-1, where A is a non-singular compact dense $(n \times n)$ matrix, x is an unknown $(n \times 1)$ column matrix and b is a given $(n \times 1)$ column matrix.

$$A \cdot x = b, \tag{7-1}$$

The matrix A, according to the method exploited herein, is factorized into two matrices W and Z of the form:

$$W = \begin{bmatrix} 1 & & & \mathbf{0} & & & 0 \\ w_{21} & 1 & & & & 0 & w_{2n} \\ w_{31} & w_{32} & 1 & & 0 & w_{3,n-1} & w_{3n} \\ \cdot & \cdot & \cdot & \cdot & \cdot & \cdot & \cdot \\ \cdot & \cdot & \cdot & \cdot & \cdot & \cdot & \cdot \\ \cdot & \cdot & \cdot & \cdot & \cdot & \cdot & \cdot \\ w_{n-2,1} & w_{n-2,2} & 0 & & 1 & w_{n-2,n-1} & w_{n-2,n} \\ w_{n-1,1} & 0 & & & & 1 & w_{n-1,n} \\ 0 & & & \mathbf{0} & & & 1 \end{bmatrix} \tag{7-2}$$

and

$$Z = \begin{bmatrix} z_{11} & z_{12} & z_{13} & \cdots & z_{1,n-2} & z_{1,n-1} & z_{1n} \\ & z_{22} & z_{23} & \cdots & z_{2,n-2} & z_{2,n-1} & \\ & & z_{33} & \cdots & z_{3,n-2} & & \\ & & & \cdot\ \ \cdot & & & \\ & \mathbf{0} & & \cdot & & \mathbf{0} & \\ & & & \cdot\ \ \cdot & & & \\ & & z_{n-2,3} & \cdots & z_{n-2,n-2} & & \\ & z_{n-12} & z_{n-1,3} & \cdots & z_{n-1,n-2} & z_{n-1,n-1} & \\ z_{n1} & z_{n2} & z_{n3} & \cdots & z_{n,n-2} & z_{n,n-1} & z_{nn} \end{bmatrix} \tag{7-3}$$

respectively, and the following relationship holds:

$$A = W \cdot Z. \tag{7-4}$$

The matrix structures are similar to the capital letters W and Z, which they can be seen to possess an interlocking quadrant appearance. In a more compact form, the matrices W and Z can be written as

$$W = [W_1, W_2, \cdots, W_n] \quad and \quad Z^T = [Z_1, Z_2, \cdots, Z_n], \tag{7-5}$$

where W_i and Z_i, $i = 1(1)n$ are the column vectors of the matrices W and Z^T, which can be expressed by the following general forms:

(i) For n *odd*,

$$W_i \equiv \begin{cases} [\underbrace{0,\cdots,0,1}_{i}, w_{i+1,i},\cdots,w_{n-i,i},0,\cdots,0]^T, & i = 1(1)\tfrac{1}{2}(n-1), \\ [\underbrace{0,\cdots,0,1}_{i}, 0\cdots,0]^T, & i = \tfrac{1}{2}(n+1), \\ [\underbrace{0,\cdots,0}_{n-i+1}, w_{n-i+2,i},\cdots,w_{i-1,i},1,0,\cdots,0]^T, & i = \tfrac{1}{2}(n+3)(1)n \end{cases} \tag{7-6}$$

and

$$Z_i \equiv \begin{cases} [\underbrace{0,0,\cdots,0}_{i-1}, z_{ii},\cdots,z_{i,n-i+1},0,\cdots,0]^T, & i = 1(1)\tfrac{1}{2}(n+1), \\ [\underbrace{0,0,\cdots,0}_{n-i}, z_{i,n-i+1,i},\cdots,z_{ii},0,\cdots,0]^T, & i = \tfrac{1}{2}(n+3)(1)n \end{cases} \tag{7-7}$$

(ii) For n *even*,

$$W_i \equiv \begin{cases} [\underbrace{0,\cdots,0,1}_{i-1}, w_{i+1,i},\cdots,w_{n-i,i},0,\cdots,0]^T, & i = 1(1)(\tfrac{1}{2}n-1), \\ [\underbrace{0,\cdots,0}_{i-1},0\cdots,0]^T, & i = \tfrac{1}{2}n, \tfrac{1}{2}n+1 \\ [\underbrace{0,\cdots,0}_{n-i+1}, w_{n-i+2,i},\cdots,w_{i-1,i},1,0,\cdots,0]^T, & i = (\tfrac{1}{2}n+2)(1)n \end{cases} \tag{7-8}$$

and

$$Z_i \equiv \begin{cases} [\underbrace{0,0,\cdots,0}_{i-1}, z_{ii},\cdots,z_{i,n-i+1},0,\cdots,0]^T, & i = 1(1)\tfrac{1}{2}n, \\ [\underbrace{0,0,\cdots,0}_{n-i}, z_{i,n-i+1,i},\cdots,z_{ii},0,\cdots,0]^T, & i = (\tfrac{1}{2}n+1)(1)n \end{cases} \tag{7-9}$$

In order to solve the system given by Eq. (7-1) using the butterfly method, as it can be seen from Eq. (7-4), the system

$$(WZ)\cdot x = b, \tag{7-10}$$

can be solved instead of that given by Eq. (7-1). Hence, we need to solve two related and simpler linear systems of the form

$$W \cdot y = b, \tag{7-11}$$

and

$$Z \cdot x = y \tag{7-12}$$

Each computational stage of the factorization and solution parts of the QIF method can be generally distinguished into two sub-stages performing, respectively,

(i) the solution of (2×2) linear system(s), and
(ii) the modification of the appropriate matrix element(s).

More analytically, for a VLSI implementation, the butterfly algorithm can be expressed in a recursively (single stage) regular manner to obtain the following homogenized steps:

(i) Factorization Process
The elements of the matrices W and Z can be evaluated in $\lfloor \frac{1}{2}(n-1) \rfloor$ distinct stages. At each computational stage i we have to solve $(n-2i)$ (2×2) linear systems to evaluate the corresponding w_{ij}'s (moving, at each step, one column inwards from each side of the matrix A), and to perform the modification of the $(n+2i)^2$ elements a_{ij} of the inner square (at every stage) of the coefficient matrix A, ending up with a central peak matrix element if n is odd or a (2×2) central submatrix if n is even. It can be easily observed, from the forms of matrices W and Z given in Eq. (7-6), (7-9) and the equality given by Eq. (7-4), that the values of the elements of the first and last rows of the matrix Z are as follows:

$$z_{1i} = a_{1i} \quad and \quad z_{ni} = a_{ni}, \quad for\ i = 1(1)n. \tag{7-13}$$

The elements of the first and last columns of the matrix W are then evaluated by solving $(n-2)$ sets of (2×2) linear systems given by

$$z_{11}w_{i1} + z_{n1}w_{in} = a_{i1},\, z_{in}w_{i1} + z_{nn}w_{in} = a_{in},\, for\ i = 2(1)(n-1). \tag{7-14}$$

This then completes the first stage of the factorization process, whereas in preparation for the next stage the elements of the matrix A are modified according to the following formula:

$$a_{ij} = a_{ij} - w_{i1}z_{1j} - w_{in}z_{nj}, \quad for\ i,j = 2(1)(n-1). \tag{7-15}$$

In general, at the ith stage of the factorization process the relationships that hold are

$$z_{ij} = a_{ij} \quad and \quad z_{n-i+1,j} = a_{n-i+1,j}, \quad for\ j = i(1)(n-i+1), \tag{7-16}$$

and the following (2×2) linear systems must be solved

$$\left.\begin{aligned} z_{ii}w_{ji} + z_{n-i+1,i}w_{i,n-i+1} &= a_{ji} \\ z_{i,n-i+1}w_{ji} + z_{n-i+1,n-i+1}w_{j,n-i+1} &= a_{j,n-i+1} \end{aligned}\right\} \tag{7-17}$$

for $j = (i+1)(1)(n-1)$, to give the unknown quantities $w_{ji}, w_{j,n-i+1}$, *forj* $= (i+1)(1)(n-1)$; finally, the modified a_{ij}'s are evaluated by the formula

$$a_{kl} = a_{kl} - w_{ki}z_{il} - w_{k,n-i+1}z_{n-i+1,l}, \quad for\ k,l = (i+1)(1)(n-i). \tag{7-18}$$

(ii) Solution of System (7-11)

In this part of the butterfly algorithm we have the analogous modification of $(n-2i)b_i$'s (moving, at each computational step i, one row inwards from the *top* and *bottom* of the r.h.s. vector b). The solution of the linear system given by Eq. (7-11) can be obtained in $\lfloor \frac{1}{2}(n-1) \rfloor$ steps, with the evaluation procedure carried out in pairs from the top and bottom of the vector y, i.e., y_1 and y_n are evaluated first, then y_2, y_{n-1}, and so on. To typify the computational process, in general, at the ith *stage*$($*for* $i = 1, 2 \cdots \lfloor \frac{1}{2}(n-1) \rfloor)$ we have

$$y_i = b_i, \qquad y_{n-i+1} = b_{n-i+1}, \tag{7-19}$$

and

$$b_j = b_j - w_{ij}y_i - w_{j,n-i+1}y_{n-i+1}, \quad for\ j = (i+1)(1)(n-1), \tag{7-20}$$

and then we proceed to the next stage.

(iii) Solution of System (7-12)

For the final solution of x of the system of Eq. (7-1), we have to solve at every stage a (2×2) linear system[a] and to modify all the outer-positioned y_i's (moving, at each computational step i, one row outwards to the *top* and *bottom* of the r.h.s. vector y). For the solution of the linear system of Eq. (7-12), we distinguish the cases that n is an *odd* or *even* number. If n is *odd*, then we can find that

$$x_l = \frac{y_l}{z_{ll}}, \quad for\ l = \frac{1}{2}(n+1) \tag{7-21}$$

[a]With an exception for the case that n is odd

and, in preparation for the next stage, we compute

$$y_j = y_j - x_l y_{jl}, \quad for\ j = 1(1)n\ and\ j \neq \frac{1}{2}(n+1). \tag{7-22}$$

The remaining elements of the vector x can again be evaluated in pairs by solving $\frac{1}{2}(n-1)$ sets of (2×2) linear systems in $\frac{1}{2}(n-1)$ distinct stages. In general, at the ith *stage* we solve the following (2×2) linear systems:

$$\left.\begin{array}{l} z_{ii}x_i + z_{i,n-i+1}x_{n-i+1} = y_i \\ z_{n-i+1,i}x_i + \\ z_{n-i+1,n-i+1}x_{n-i+1} = a_{n-i+1} \end{array}\right\}, for\ j = (l-1)(-1)1, \tag{7-23}$$

to compute x_i and x_{n-i+1}. We then set

$$y_j = y_j - x_i z_{ji} - x_{n-i+1}z_{j,n-i+1}, \quad \begin{array}{l} for\ i = (l-1)(-1)1, \\ j = 1(1)(i-1)\ and \\ (n-i+2)(1)n, \end{array} \tag{7-24}$$

and proceed to the next stage. On the other hand, if n is *even*, then all the components of the vector x are found in pairs. To find all the pairs, the linear system of Eq. (7-23) and the Eq. (7-24) are executed, respectively, $for\ i = \frac{1}{2}n(-1)1$.

7.3. Multilayered Dewavefront Array Processor Meshes

The software implementation of the above linear system solver on a simulated dewavefront mesh, is discussed in Bekakos [2], where each of the three distinct phases was implemented separately. At the beginning of each phase the appropriate mesh (consisting of $(n-2)^2$ processors) is initialized with the appropriate data (Evans *et al.* [12]). According to the dewavefront concept instead of having a single wave originating from the top northwest cell and traveling downwards, we have an additional concurrent wave originating from exactly the opposite bottom southeast cell. Each of the opposite computational wavefronts will have the capability to activate the successor neighbouring processors in its own propagating direction. The experimental results obtained from the software implementation (Bekakos [2]) exhibited a significant increase in speedup as the size of the matrix and the size of the dewavefront mesh were increased, respectively.

The new Pipelined + Shared memory M-DEWAP architecture [(P+S)M-DEWAP] introduced herein, is designed to exploit the overlapping capabilities inherent in the algorithmic procedure of the three distinct computational phases described previously. Each computational phase, normally, is carried out independently; however, since the elements which participate, for example, in the first computational phase, are the same as those which are used in the second computational phase, these two phases can be executed concurrently.

Every processor in this architecture, in accordance with Huygen's wavefront principle, has an asynchronous *waiting* capability. The computational wavefronts are similar to electromagnetic wavefronts, since each processor acts as a secondary source and is responsible for the propagation of the wavefront through the activation of its successor neighbouring processors in its own propagation direction. In addition, wave propagation implies localized data flow and control.

The pipelining is feasible because the wavefronts of two successive recursion stages, in each propagating direction, will never intersect, thus avoiding any contention problems. The *confrontation* stage of the opposite traveling wavefronts on the same cell can be resolved by using appropriate arithmetic logic unit latches or dedicated I/O control latches.

Finally, the estimation of the overall time-complexity, in terms of the number of time-steps required, for the VLSI implementation of the butterfly method, is a very difficult task due to the distinction of this direct method into several different phases and the pipelined approach followed.

The proposed machine consists of two (2×2) linear system solvers implementing Cramer's rule, which are used to produce the w_{ij} elements and the final x_i, and three dewavefront square meshes interconnected via shared memory circuits, while pipelined techniques are adopted for the execution overlapping between the two meshes used for the second and third computational phases.

More analytically, a shared memory separated into two partitions, a high and a lower memory partition, interconnects the first and third dewavefront meshes, while several other shared memory modules (depending on the number of processors which consist of each mesh) interconnect the first and second dewavefront meshes. The computational activities involved in the second and third meshes are carried out in a manner that exploits these overlapping capabilities inherent in the algorithmic procedure. These two meshes are additionally provided with their own local memory modules depending again on the number of utilized processors.

The total number of processors involved in the proposed architectural design has been increased to $3 \times (n-2)^2$, instead of $(n-2)^2$ used by the simple dewavefront machine. The functional description of the new (P+S)M-DEWAP architecture, as it was described above, is introduced in Fig. 7-1.

For exemplification purposes concerning the new machine's functional capabilities, we consider again the problem given by Eq. (7-1), where A is a (5×5) compact dense matrix. In this case, the number of processors involved in each of the three dewavefront meshes will be nine. The operational description of the proposed machine is analyzed in Fig. 7-2.

At first all the Processing Elements (PEs) in the meshes are initialized concurrently, except for the PEs $(1,3)$ and $(3,1)$, in a matrix elements like notation, of the third mesh, which will be initialized during the computational process. Then, at every operational stage, the elements w_{ij} are computed on the circuit of the linear system solver which implements Cramer's rule, using the elements z_{ij}, and are pipelined to the corresponding shared memory modules of the first and second meshes. The elements of matrix Z are stored into the high partition of the shared memory and the local memory modules of the second mesh are concurrently initialized by the values of the intermediate vector y. An appropriate electronic circuit controlling the shared memory will make a copy of the elements of the high memory partition into the lower memory partition, but in different memory addresses, so that they will be available for use by the third dewavefront mesh.

During the computational activity which takes place in the second mesh, the values b_2 and b_4 are computed and pipelined into the third mesh for the initialization of the dormant so far PEs $(1,3)$ and $(3,1)$. When the computational activity of the first two meshes has been completed the local memory modules of the third mesh will already be initialized by the values of vector x and by the appropriate values of matrix Z stored in the lower partition of the shared memory. The final solution of the problem will be produced with the conclusion of the

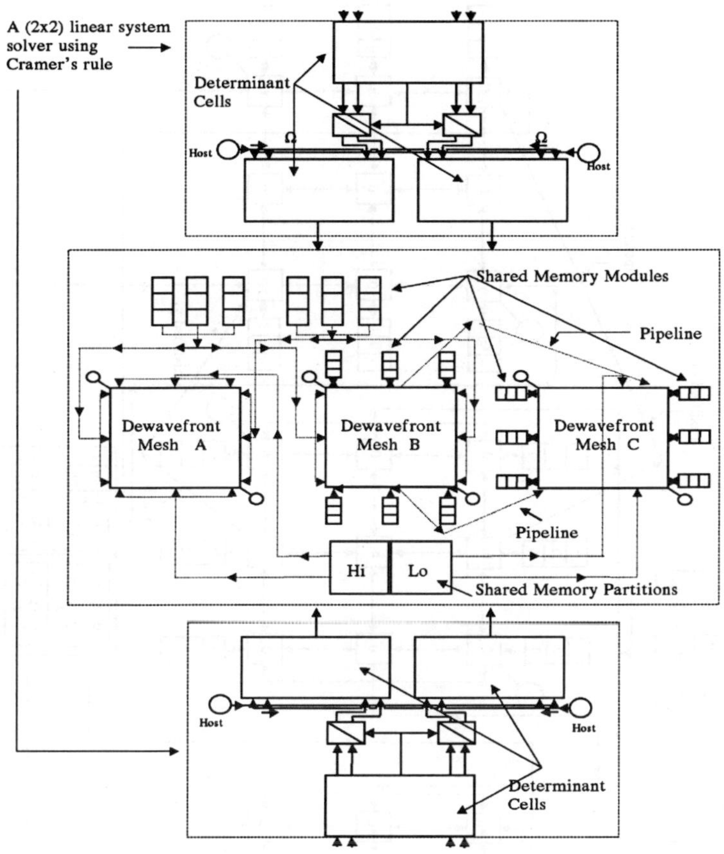

Ω : Delay, ◯ : Host

⊟ : Shared Memory Modules

a b   time

t   read a,b,e

e   t+1   c=b/e, d=a/e

c d

: Bidirectional

Fig. 7-1. Functional description of the (P+S)M-DEWAP.

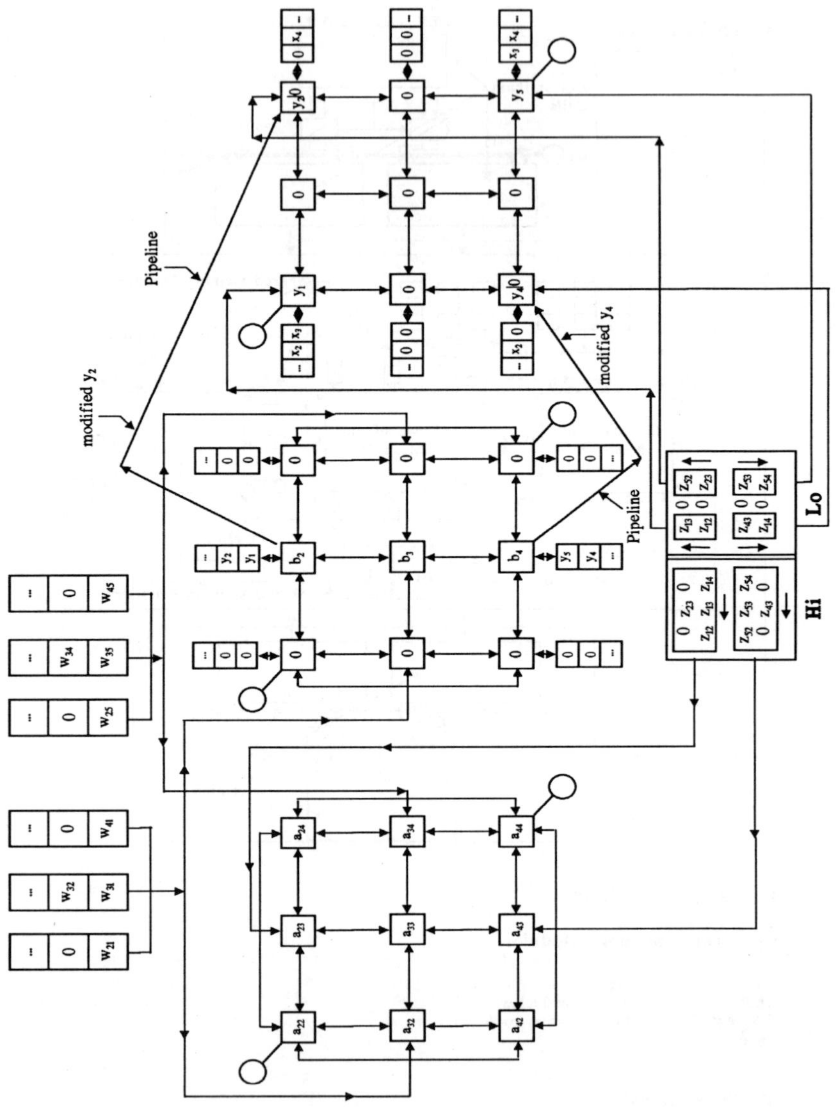

Fig. 7-2. Operational description of the (P+S)M-DEWAP.

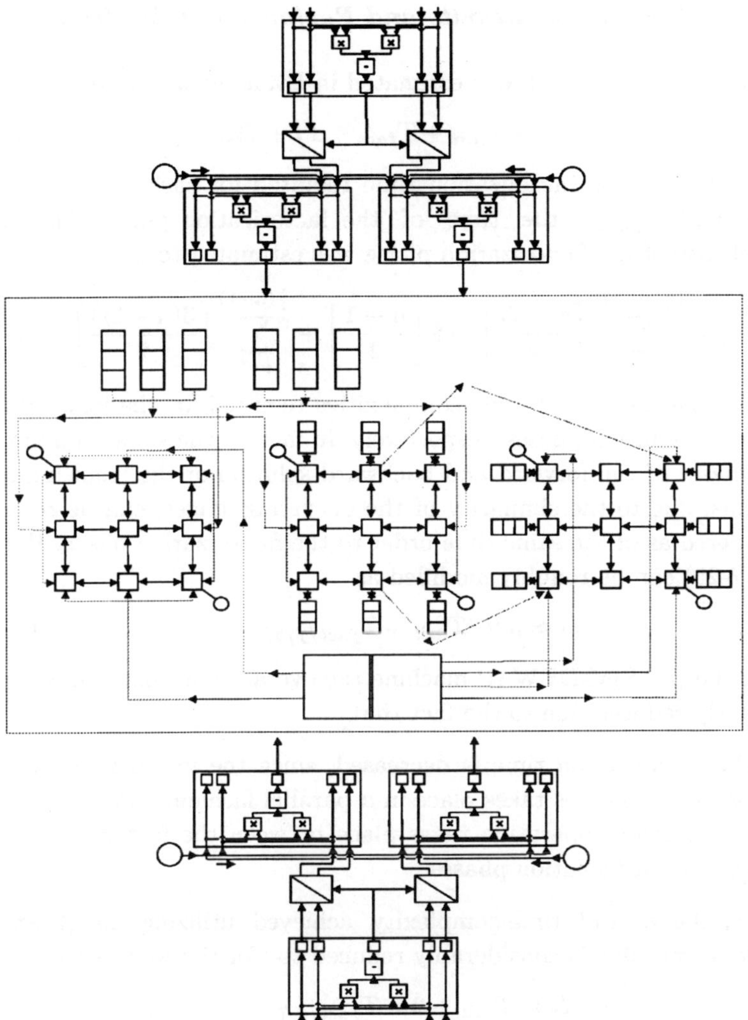

Fig. 7-3. Hardware Implementation of the (P+S)M-DEWAP.

computational activity in the third mesh. The hardware layout of the proposed (P+S)M-DEWAP architecture is depicted in Fig. 7-3.

7.3.1. *Experimental Results and Performance Analysis*

The overall time-complexity estimated in Evans *et al.* [12] was

$$T_t \approx 3 \times T_{init} + T_{fact} + T_f + T_b, \tag{7-25}$$

where T_{init} is the initialization time required for each computational phase and T_{fact} is the time of the factorization phase. The time-complexity of the factorization phase was estimated to be

$$T_{fact} = \sum_{i=1}^{\lfloor\frac{1}{2}(n-1)\rfloor} \left\lceil \frac{n-2i}{2} \right\rceil + 5\left\lceil \frac{n-1}{2} \right\rceil + \sum_{i=1}^{\frac{1}{2}(n-1)} \left\lceil \frac{3(n-2i)}{2} \right\rceil . \tag{7-26}$$

The T_f and T_b are the time-complexities required for the forward and backward solution phases, respectively. In fact, in the worst bound case the phases of the forward and backward solution of the resulting subsystems, due to the similarity of the occurring stages, can be roughly considered as of the same time order to the factorization phase. Hence, Eq. (7-25) can be roughly modified as

$$T_t \approx 3 \times (T_{init} + T_{fact/f/b}). \tag{7-27}$$

With the (P+S)M-DEWAP machine the overall time complexity is significantly reduced due to the fact that

(i) the initialization time is decreased, since the initialization of the processor meshes takes place in a parallel fashion; and
(ii) a concurrent operation takes place between the factorization and the forward solution phases.

Hence, the overall time-complexity achieved utilizing the (P+S)M-DEWAP machine is considerably reduced to (for the worst case)

$$T_t \approx T_{init} + 2 \times T_{fact/f/b}, \tag{7-28}$$

where T_{init} and $T_{fact/f/b}$ are as previously described.

7.4. Block Generalized WZ Factorization

In the following sections the block generalized WZ factorization technique is briefly discussed (Benaini *et al.* [5]); the E^3 model is then introduced and the implementation of the WZ factorization procedure on it, along with the computational complexity of each phase of the algorithmic procedure, are investigated.

The computation of the eigenvalues of a $(n \times n)$ dense real matrix A using the block generalized WZ factorization method was proposed by A. Benaini *et al.* [5]. According to this method, the eigenvalues are computed by factorizing matrix A as WZW^{-1}, where matrices W and Z are defined by the expressions given by Eq. (7-2), (7-3).

Let us consider that the $(n \times n)$ matrices A, W, Z are decomposed into $(r \times r)$ blocks, noted by $A_{ij}, 1 \leq i,j \leq r$, $W_{ij}, 1 \leq i,j \leq r$ and $Z_{ij}, 1 \leq i,j \leq r$, respectively. For the sake of simplicity, A, W and Z are assumed to consist of uniform blocks of size m and $n = m \cdot r$.

Lemma 1: *Suppose that A is block factorized as $A = WZW^{-1}$, then the eigenvalues of A are those of matrices* $Q = \begin{pmatrix} Z_{ii} & Z_{i,r-i+1} \\ Z_{r-i+1,i} & Z_{r-i+1,r-i+1} \end{pmatrix}$ *of order $(2m \times 2m)$, for $1 \leq i \leq q$, where* $q = \left\lceil \frac{r-1}{2} \right\rceil$.

According to the sequential algorithm, $q = \left\lceil \frac{r-1}{2} \right\rceil$ stages, are required to compute all the eigenvalues of matrix A. At the beginning of each stage (*phase 1*), the following procedure is applied:

For $s = 0, 1, \ldots$, until convergence

$$C_i^{(s)} = \left(A_{i1} + \sum_{k=2}^{r-1} A_{ik} W_{k1}^{(s)} \right)^T, for 1 \leq i \leq r \tag{7-29}$$

and

$$D_i^{(s)} = \left(A_{ir} + \sum_{k=2}^{r-1} A_{ik} W_{kr}^{(s)} \right)^T, for 1 \leq i \leq r \tag{7-30}$$

Next, the system $B \cdot w = c$ needs to be solved, where

$$B = \begin{pmatrix} {}^T C_1^{(s)} & {}^T C_r^{(s)} \\ {}^T D_1^{(s)} & {}^T D_r^{(s)} \end{pmatrix}, c = \begin{pmatrix} {}^T C_2^{(s)} & \cdots & {}^T C_{r-1}^{(s)} \\ {}^T D_2^{(s)} & \cdots & {}^T D_{r-1}^{(s)} \end{pmatrix} \tag{7-31}$$

and

$$w = \begin{pmatrix} {}^T W_{2,1}^{(s+1)} & \cdots & {}^T W_{r-1,1}^{(s+1)} \\ {}^T W_{2,r}^{(s+1)} & \cdots & {}^T W_{r-1,r}^{(s+1)} \end{pmatrix} \tag{7-32}$$

End for

After the convergence (i.e., the computation of the final matrix W_i (i^{th} *stage*), it follows the update of matrix A, namely the computation of $Z_i = W_i^{-1} A W_i$ (phase 2), according to the following formulae:

$$\begin{aligned} & Z_{ij} = A_{ij} - W_{i1} A_{1j} - W_{ir} A_{rj}, && \textit{for } 2 \le i, j \le r-1 \\ & Z_{1j} = A_{1j}, Z_{rj} = A_{rj}, && \textit{for } 2 \le j \le r-1 \\ & Z_{1j} = A_{1j} + \sum_{k=2}^{r-1} A_{1k} W_{kj} && \textit{for } j = 1, r \\ & Z_{rj} = A_{rj} + \sum_{k=2}^{r-1} A_{rk} W_{kj}, && \textit{for } j = 1, r \end{aligned} \tag{7-33}$$

then proceed to the next stage. Note that, after each stage the problem size decreases by $2m$ and after q stages the matrices W and Z are produced. Also, note that, initially $Z_0 = A$ and $W_{11} = W_{rr} = I_m$ and $W_{i1} = W_{ir} = 0_m$, *for* $2 \le i \le r-1$ (I_m : $(m \times m)$ identity matrix and 0_m : $(m \times m)$ zero matrix).

7.4.1. *Wavefront implementation*

According to the algorithm, described previously, we can distinguish two different phases at each one of the q stages:

(i) Computation of the W_{ij} elements, and
(ii) update of matrix A.

Furthermore, the first phase can be subdivided into three more phases :

(i) Compute C_i and D_i values,
(ii) solve the $(2m \times 2m)$ linear systems (i.e., compute $B \cdot w = c$),
(iii) check if convergence is reached.

To construct the complete engine, for each algorithmic phase the corresponding wavefront architecture is designed with a WAPm being used as the building block. The abstract layout of E^3 is depicted in Fig. 7-4.

To exemplify the hardware implementation of the factorization procedure, a dense real matrix A of size (15×15) has been considered;

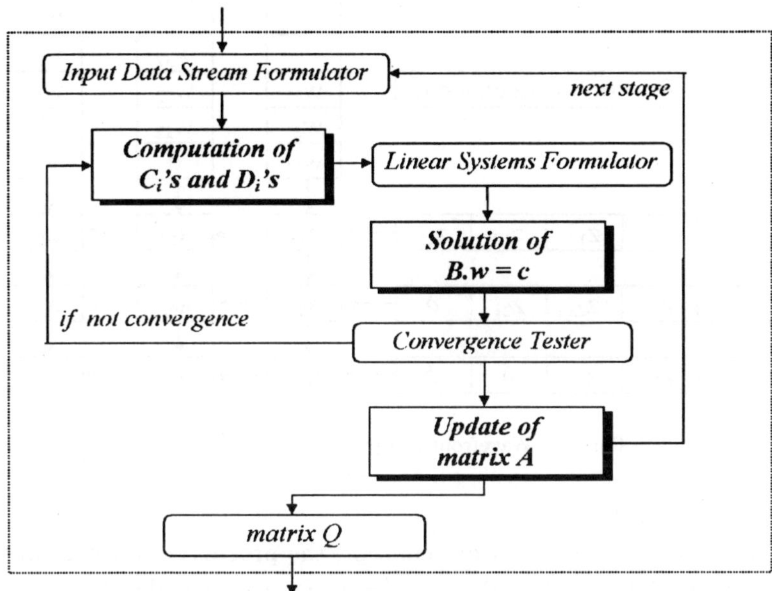

Fig. 7-4. Abstract layout of E^3.

hence, the block size is (3×3) and $r = \dfrac{n}{m} = 5$.

7.4.2. *Module - The Building Block*

The basic operation of the factorization procedure is a block multiplication and a block addition (e.g., compute C_i or update matrix A). This computation can be performed on a special-purpose orthogonal network of IPSPs (Inner Product Step Processors) adopting the wavefront concept.

Herein, this network is considered as a WAPm and it is used as the building block for larger (modular) WAPs. The number of PEs of each module equals the size of blocks, namely, $(m \times m) = m^2$, (e.g., in our case, $m = 3$ and $\#PEs = 9$).

As an example, let us consider the computation $A_{11} = A_{11} + Z_{11} * W_{11}$; the data flow of the elements of blocks W and Z into the WAP is described in Fig. 7-5. Note that, each PE is initialized by the appropriate

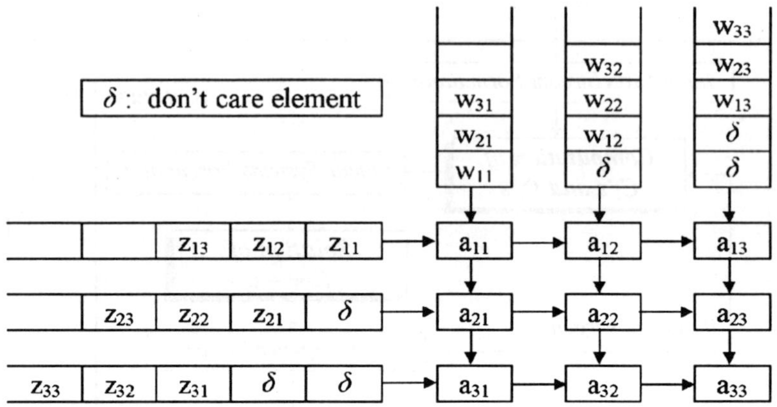

Fig. 7-5. Wavefront computations in a module.

elements of block A_{11}.

According to the wavefront concept, the process commences in the top northwest PE(1,1) as $a_{11}^{(1)} = a_{11}^{(0)} + z_{11} \cdot w_{11}$. The end of this computation activates the successor neighbours, PE(1,2), PE(2,1), which will execute in parallel: $a_{12}^{(1)} = a_{12}^{(0)} + z_{11} \cdot w_{12}$ and $a_{21}^{(1)} = a_{21}^{(0)} + z_{21} \cdot w_{11}$, while PE(1,1) will also execute in parallel: $a_{11}^{(2)} = a_{11}^{(1)} + z_{12} \cdot w_{21}$. After this computation, their corresponding successor neighbours: PE(1,3), PE(2,2) and PE(3,1) will be activated, thus creating a wave of hierarchical computations traveling down the orthogonal network of processors. The number of wavefront steps required to complete the computation is:

$$T = 3m - 1 \tag{7-34}$$

7.4.3. *Phase 1: Computation of C_i's, D_i's*

At the first stage of the algorithmic procedure ($q = 1$), the wavefront computation of C_i's and D_i's, for $1 \leq i \leq r$, is carried out. The wavefront scheme for this specific example (i.e., $n = 15$, $m = 3$ and $r = 5$), as well as the input data stream are depicted in Fig. 7-6. Note that, each shaded rectangle denotes a module and it is initialized by the corresponding blocks of matrix A (i.e., $A_{11}, A_{21}, \cdots, A_{51}, A_{15}, A_{25}, \cdots, A_{55}$).

The first row of the $(2 \times r)$ grid of modules produces the C_i's, while

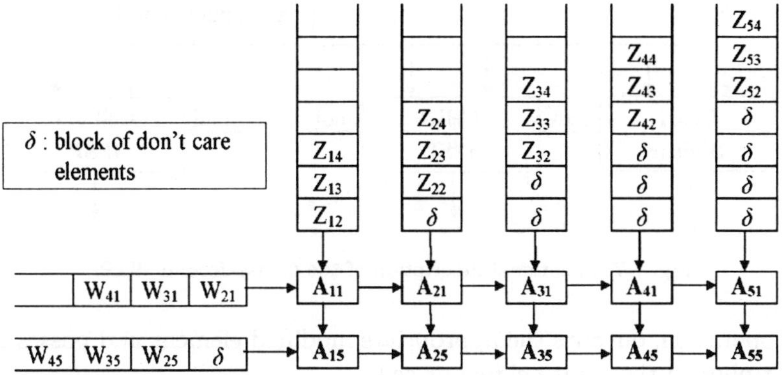

Fig. 7-6. Wavefront computation of C_i's, D_i's.

the second one produces the D_i's. The number of wavefront steps required to complete the computations is :

$$T = 2r - 1 \tag{7-35}$$

Since the number of modules required is $2r$, the number of PEs used is $2rm^2$.

7.4.4. *Phase 1: Computation of* W_{ij}*'s*

The computation of C_i's, D_i's is followed by the solution of system $B \cdot w = c$. Matrices B, w, c are of size $(2m \times 2m)$, $(2m \times (r-2)m)$ and $(2m \times (r-2)m)$, respectively. Each of the resulting $(r-2)$ linear systems is solved using the Gauss-Jordan algorithm.

For every linear system, matrix B is augmented with c_i to matrix $\overline{B}_i$ of size $(2m \times 3m)$, that is, $\overline{B}_i = (B, c_i)$, *for* $2 \leq i \leq r-1$, and a degenerated WAP (i.e., of linear form) is developed to produce W_{ij}. This array consists of $3m$ cells: One special-purpose divisor cell and $(3m-1)$ typical IPSPs (Bekakos *et al.* [1]). The only enhancement is that a register is additionally considered, attached to each cell and capable of holding one element which can be overwritten by any new occurring element.

The first cell in the linear topology is the divisor cell which produces the multipliers or the final results; the multiplier and the appropriate elements of matrix $\overline{B}_i$ input each IPSP cell which performs its typi-

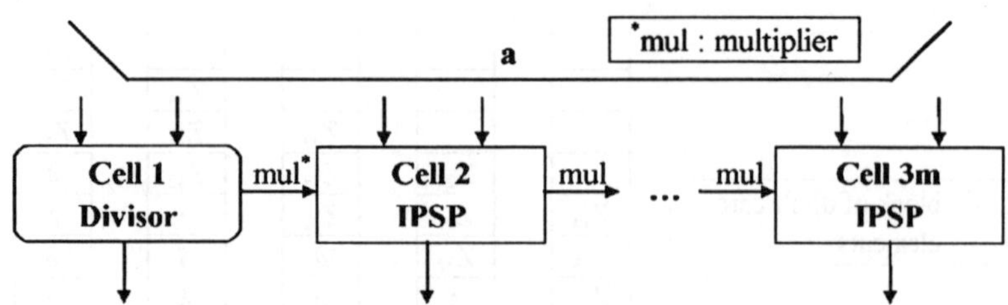

Fig. 7-7. Functional description of the Gauss-Jordan WAP.

cal operation, outputs the appropriate modified element and passes the multiplier to the right neighbour cell.

The parallel computational time complexity for the evaluation of W_{ij}'s is

$$T = 6m^2 - 2m + 1 \tag{7-36}$$

The functional description of the proposed WAP is given in Fig. 7-7, while a few consecutive snapshots of the computational procedure are described in Fig 7-8.

After the computation of W_{ij}'s the convergence tester checks the results. If convergence is not reached, then *phase 1* is repeated to refine C_i's and D_i's, and so on. Otherwise, the next phase (i.e., the update of matrix A) will commence and the computational procedure will continue until the predetermined number of stages is completed.

7.4.5. *Phase 2: Update of matrix A*

For this phase and as it can be concluded from Eq. (7-33), two different wavefront schemes are required to update matrix A (and matrix Z). The first grid needs $((r-2) \times (r-2))$ WAPms to complete the update of the inner matrix A (and Z), while the second grid needs always four (4) WAPms to produce matrix Q.

Thus, the total number of the required PEs for the carrying out of this phase is $(r2 - 2r + 8)$; the computational procedure is described in Fig. 7-9, 7-10.

The total number of steps required for each grid is :

$$1^{st} grid : T = 2r - 3 \tag{7-37}$$

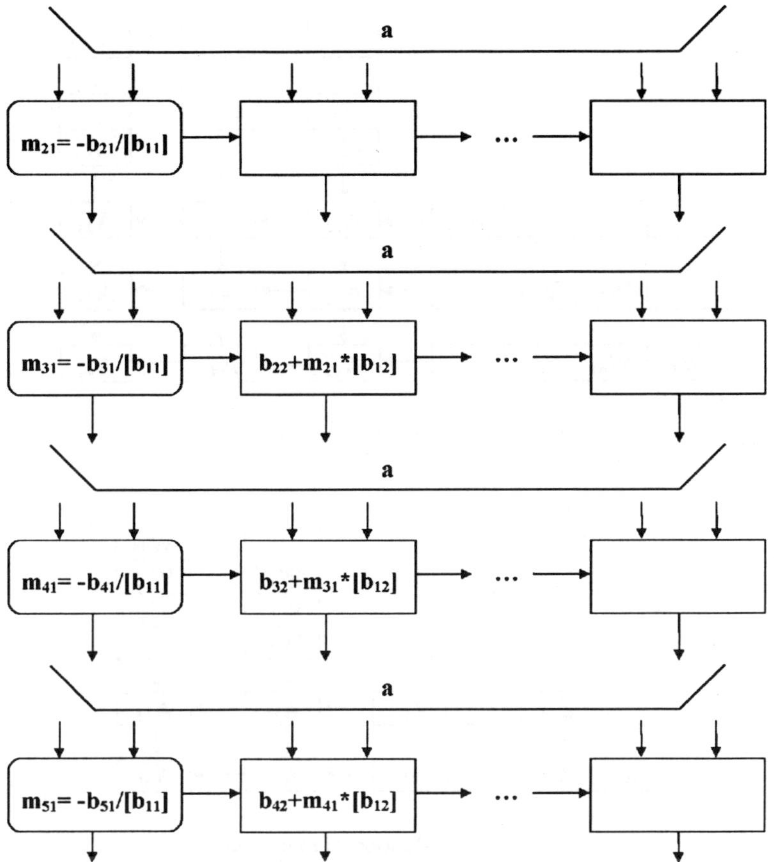

Fig. 7-8. Few consecutive snapshots of the Gauss-Jordan procedure.

$$2^{nd} grid : T = r + 1 \tag{7-38}$$

Note that, the update of the matrices in each grid is performed applying the wavefront concept. After the update, the resulted matrix Q is flushed out and the next stage of the algorithm commences.

7.5. Systolic Quadrant Interlocking Elimination - SQIE

The construction of a systolic array for the problem investigated in this

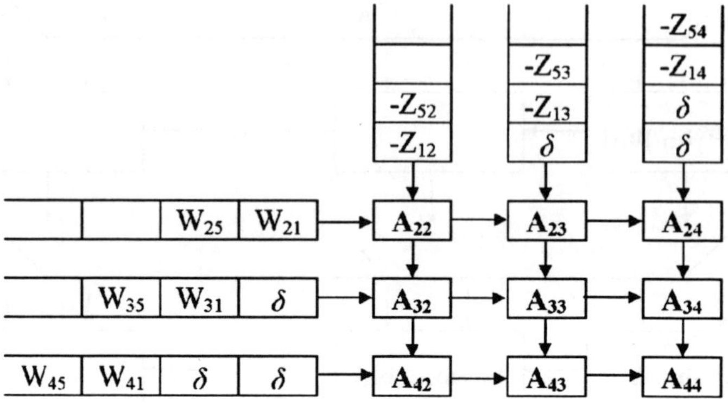

Fig. 7-9. Update of inner matrix A.

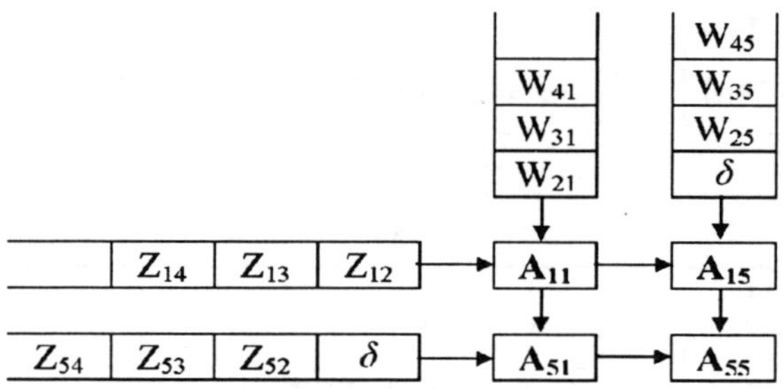

Fig. 7-10. Evaluation of matrix Q.

Chapter requires the representation of the dataflow for various (2×2) systems to be arranged in order to retain a regular communication structure. A natural type of basic cell based on (2×2) system operations suggest itself. However, the locality of data varies from stage to stage during the factorization. The first stage uses data in four corners of A, and the last stage up to four adjacent neighbours at the center of A. Consequently, a data permutation is required to smooth out these locality variations.

Now applying the QI permutation produces locally placed elements for a simplified dataflow and yield the following interesting (2×2) block partitioning.

For $n = 6$,

$$\overline{W}= \begin{array}{c} 1\\6\\2\\5\\3\\4 \end{array} \left[\begin{array}{cc|cc|cc} \multicolumn{6}{c}{\begin{array}{cccccc}1&6&2&5&3&4\end{array}} \\ 1 & 0 & & & & \\ 0 & 1 & & & & \\ \hline w_{21} & w_{26} & 1 & 0 & & \\ w_{51} & w_{56} & 0 & 1 & & \\ \hline w_{31} & w_{36} & w_{32} & w_{35} & 1 & 0 \\ w_{41} & w_{46} & w_{42} & w_{45} & 0 & 1 \end{array}\right] \quad \overline{Z}= \begin{array}{c} 1\\6\\2\\5\\3\\4 \end{array} \left[\begin{array}{cc|cc|cc} z_{11} & z_{16} & z_{12} & z_{15} & z_{13} & z_{14} \\ z_{61} & z_{66} & z_{62} & z_{65} & z_{63} & z_{64} \\ \hline & & z_{22} & z_{25} & z_{23} & z_{24} \\ & & z_{52} & z_{55} & z_{53} & z_{54} \\ \hline & & & & z_{33} & z_{34} \\ & & & & z_{43} & z_{44} \end{array}\right] \tag{7-39}$$

(Column order of both matrices: 1 6 2 5 3 4.)

where $\overline{W}$, $\overline{Z}$, and $\overline{A}$ are permuted forms of W, Z, and A.

It immediately follows by multiplying out $\overline{W}$ and $\overline{Z}$, to produce $\overline{A}$ that the QIF method is permuted form of (2×2) Block LU Factorization (BLUF). Consequently the SQIF algorithm can described by three steps:

Step (i) Permute A to $\overline{A}$
Step (ii) Pass $\overline{A}$ through (2×2) block array of Robert [19] to produce $\overline{W}$, $\overline{Z}$
Step (iii) Permute $\overline{W}$ and $\overline{Z}$ to produce W and Z.

As only simple row and column permutations are employed Steps (i) and (iii) constitute host pre- and post-processing. This allow Step (ii) to dominate computation costs when array input data is generated using pointers rather than explicit row and column interchanges in the host memory. Furthermore, the equivalence of (2×2) block schemes and QIF methods allows the use of the well known block partitioning theorems. It follows that the improved performance of QIF schemes carries over the SQIF implementation. The permutation technique also suggests that a variety of patterns other than QI structure exist. Different patterns being produced by different permutations or by selecting larger block sizes. New permutations on the other hand could yield new array but the SQIF form produces nearest neighbour data orderings which minimize communication problems and maximize efficiency. Soft-systolic arrays at present would be only way of utilizing non-local data orderings, while maintaining array efficiency.

Notice that the diagonal blocks contain extra sparsity. The above block partitioning can be termed implicit because without the partitioning lines the permuted W and Z appear to be simple L and U factors. As the point LU factorization is unique, it follows that the hexagonal (point) array to Kung *et al* [13], [14] can be applied to find the modified QIF of a matrix by applying simple pre- and post-permutations on the input and output. Likewise, the form of the permuted coupled system, described by Eq. 7-11, 7-12, corresponding to the triangular or LU factorized systems and allows triangular solver arrays to be employed to solve system of Eq. (7-1). However, the modified QIF has no advantages in speed or efficiency over the ordinary QIF (or explicit block computation) regarding the application of systolic arrays. However, this argument is based on a comparison of point and block methods using point-(IPS)[b] structured arrays. The essential feature of the modified QIF is its implicit block structured nature which allows a block structure data; thereby retaining the improved efficiency and reduced computation time of the explicit block schemes but producing point (or implicit block) structured outputs. This resolves the difficulties associated with the ordinary QIF scheme, which requires an explicitly block structured triangular solver, or the more complex LDU factorization (requiring the solution of three coupled systems) using implicit block (point) structured solvers.

7.5.1. *The QIE Systolic Array*

The implicit block structured array for the modified QIF uses the same principles as the block (2×2) block array in Robert [19] but utilizes the structure of $\overline{W}$ and $\overline{Z}$ diagonal blocks to adjust hardware and computation within each block IPS cell Megson [18]. For the development of the array we make the following simple assumptions:

(i) The $(n \times n)$ permuted input matrix $\overline{A}$ does not cause a breakdown in the factorization process (i.e., diagonally dominant or positive definite).
(ii) $n = 2m,\ for\ m < n$ (i.e., ensures an even partitioning).
(iii) $\overline{A}$ is banded with bandwidth $w = p + q - 1$.

The global structure of the array is shown in Fig 5-11 and contains

[b]inner-product step

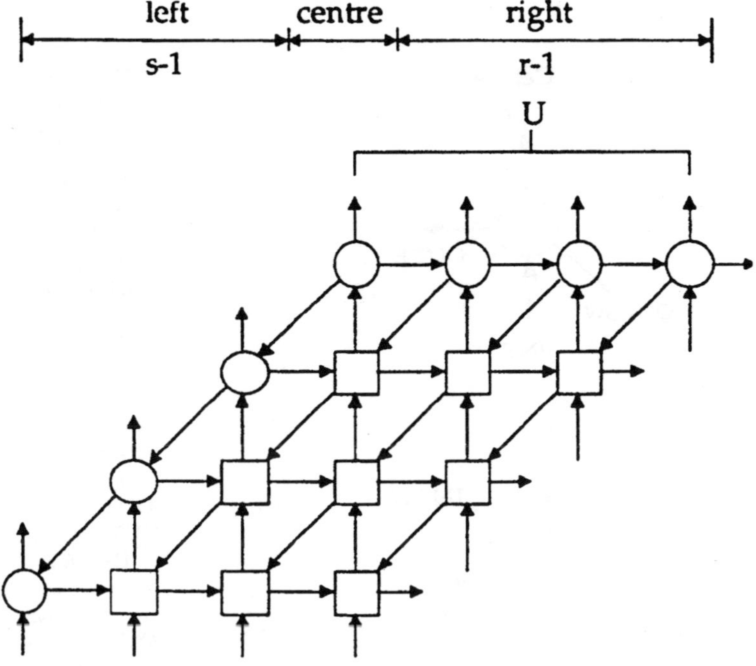

Fig. 7-11. Implicit block array for modified QIF method.

$rs(2 \times 2)$ implicit block IPS cell where,

$$p = \begin{cases} 2r - 1 \\ 2r - 2 \end{cases}, \qquad q = \begin{cases} 2s - 1 \\ 2s - 2 \end{cases} \tag{7-40}$$

In Fig. 7-12 the cell definition is given. Note that, each implicit block cell contains the equivalent of $2 \times 2 = 4$ point IPS cells and computes as it is presented in Fig. 7-13.

Each cell receives four inputs corresponding to an implicit block every four point IPS cycles, this being the longest period of any block cell. As there are $m = n/2$ implicit blocks of input data the total output time for data is $4m = 2n$ point IPS cycles. Output/computation starts after the first implicit block has reached the center processor. As each implicit block cell requires two point IPS cycles for computation data is shifted into its initial starting position in at most $2min(r, s) \leq min(p, q)$.

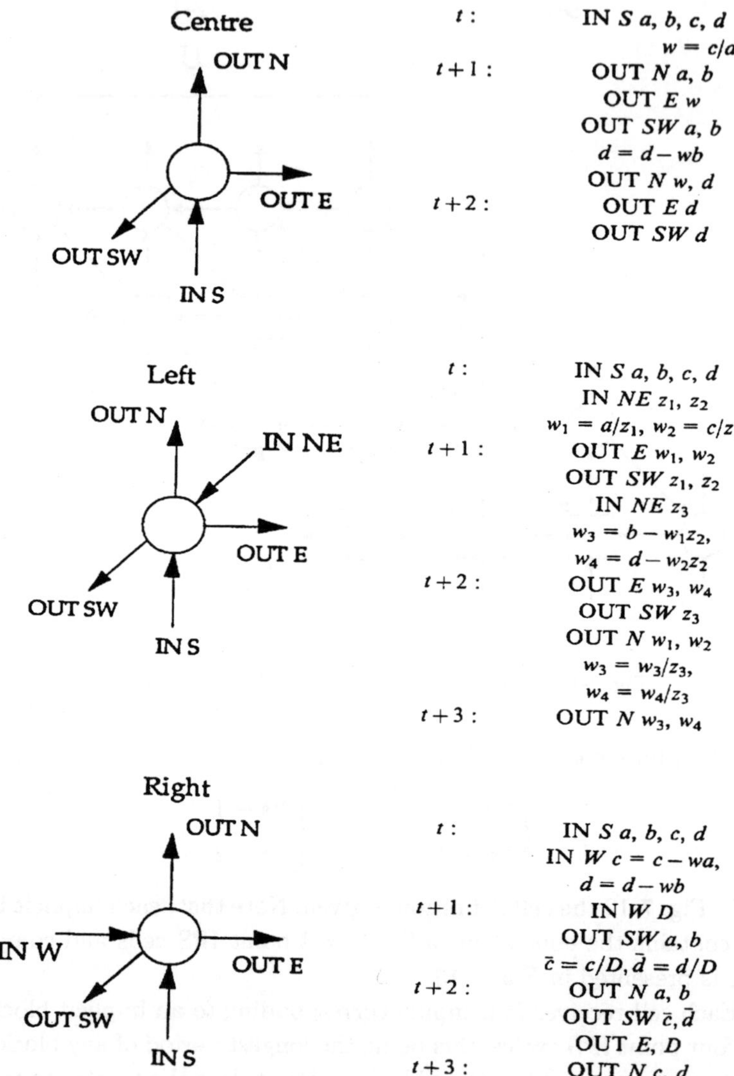

Fig. 7-12. Cell definitions.

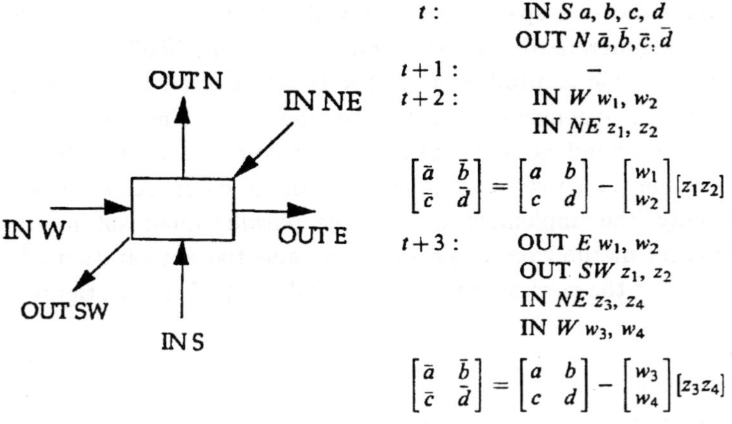

Fig. 7-13. Implicit block cell.

Hence, we have the following theorem.

Theorem 1: The modified QIF of an $(n \times n)$ matrix A whose QI permutation matrix $\overline{A}$ has bandwidth $w = p+q-1$ can be computed in $T = 2n + min(p,q)$ IPS cycles using approximately pq point IPS cells. This is identical to the result in Robert [19], and arises from the fact that the global dataflow of the two arrays are the same. Examining the computation of the implicit cells reveals that calculations occur two cycles in four and efficiency $e = 1/2$ achieved in Robert [19] is also preserved.

7.6. Conclusions

The QIF or ***butterfly*** algorithm, established by Evans ***at al.*** [7] is studied in this Chapter.

Through the use of a new (P+S)M-DEWAP architecture, especially designed to benefit from the inherent overlapping capabilities in the algorithm execution, the overall time-complexity has been significantly reduced, as compared to the simple case (Bekakos [2]), where each distinct stage of the algorithmic procedure was performed independently.

Furthermore, a special-purpose Multi-WAP engine is investigated. The block generalized WZ factorization algorithm is efficiently mapped onto this engine to produce the (2×2) matrices Q, the eigenvalues of which are the eigenvalues of the initial matrix A. The architectural design and some indicative snapshots of the computational procedure are given, while the corresponding computational complexity is discussed.

Finally, the implementation of the parallel quadrant interlocking elimination method on a systolic array and the exploitation of its efficiency over the corresponding sequential method is investigated.

References

[1] M.P. Bekakos, E.A.Lipitakis and O.B. Efremides, Parallel exploitation of multiple pipes arrangements on mesh architectures, 2nd Hellenic European Conference on Mathematics and Informatics, Greece, pp. 855-868, (1994).

[2] M.P. Bekakos, A Fast QIF Algorithm Implementation on a Shared Memory Dewavefront Machine. 5th International Conference, Applications of High-Performance Computers in Engineering, Italy, pp. 27-35, (1995).

[3] M.P. Bekakos and D.J. Evans, The Exposure and Exploitation of Parallelism on Fifth Generation Computer Systems. Int. Conf. Parallel Computing '85, (eds.) M. Feilmeir, G. Joubert and U. Schendel, Elsevier Science Pub., B.V., North-Holland, pp. 425-442, (1986).

[4] A. Benaini, and M. Drissi, Generalized WZ factorization based on eigenvalues, RR 27-93, LIB-Besancon, 1993.

[5] A. Benaini, and D. Laiymani, Generalized WZ factorization on a reconfigurable machine, *Journal of Parallel Algorithms and Applications*,**3(1)**, pp. 1-10, (1994).

[6] A. Benaini, Block Generalized WZ factorization, RR 04-94, LIB- Besancon, (1994).

[7] D.J. Evans, and M.. Hatzopoulos, A Parallel Linear System Solver. *Internat. J. Comput. Math.*, **Sect. B. 7**, pp. 227-238, (1979).

[8] D.J. Evans, and A. Hadjidimos, A Modification of the Quadrant Interlocking Factorization Parallel Method. *Internat. J. Comput. Math.*, **Sect. B 8**, pp. 149-166, (1980).

[9] D.J. Evans, A. Hadjidimos, and D., Noutsos, The Parallel Solution of Banded Linear Equations by the new Quadrant Interlocking Factorization (Q.I.F.) method. *Intnat. J. Comput. Math.*, **Sect. B 9**, pp. 151-161, (1981).

[10] D.J. Evans, Parallel Numerical Algorithms for Linear Systems, Cambridge Univ. Press, (1982).

[11] D.J. Evans, New Parallel Algorithms in Linear Algebra, Electricle de France, Bull. deirect des etude et recherche serie, pp. 61-69, (1983).

[12] D.J. Evans, and M.P. Bekakos, The Solution of Linear Systems by the QIF Algorithm on a Wavefront Array Processor. *Journal of Parallel Computing*, **7**, Elsevier Science Publischers B.V., North-Holland, pp. 111-130, (1988).

[13] H.T. Kung and C.E. Leiserson, Systolic Arrays for VLSI, *Sparse Matrix Proc*, I.S. Duff and G.W. Stewart, pp. 256-282, SIAM 1977, (1978).

[14] H.T. Kung, Notes on VLSI Computations, in Evans D.J., Parallel Processing Systems, Chap. 6, pp. 339-356. Cambridge Univ. Press, (1982).

[15] S.Y. Kung, and R.J. Gal-Ezer, Synchronous vs. Asynchronous Computation in VLSI Array Processors. Proc. SPIE Conference, Arlington, VA, (1982).

[16] S.Y. Kung, R.J. Gal-Ezer, and K.S. Arun, Wavefront Array Processor: Architecture, Language and Applications. *IEEE Trans. Comput., Special Issue on Parallel and Distributed Computers*, **31 (11)**, pp. 1054-1066, (1982).

[17] B.P. Lester, The Art of Parallel Programming, Prentice-Hall Int. Tnc., Englewood Cliffs, N. Jersey, (1993).

[18] G.M. Megson, Novel Algorithms for the Systolic Paradigm, PH.D. Thesis. Loughborough University of Technology, (1987).

[19] Y. Robert, Block LU-Decomposition of a Band Matrix on a Systolic Array, *Int. Journal Comp. Maths*, **19**, pp. 295-316, (1985).

[20] A.H. Sameh, and D.J. Kuck, Linear System Solvers for Parallel Computers, T.R. No. UIUCDCS-R-701, Department of Computer Science, University of Illinois, Urbana IL, (1975).

[21] J.M. Speiser, and H.J. Whitehouse, Architectures for Real-Time Matrix Operation, Proc. 1980 Government Microcircuits Applications Conference, Houston, pp. 19-21, (1980).

CHAPTER 8

SYSTOLIC S.O.M. NEURAL NETWORK FOR HYPERSPECTRAL IMAGE CLASSIFICATION

P. Martínez, P.L. Aguilar, R.M. Pérez and A. Plaza

Departamento de Informática, Universidad de Extremadura
Campus Universitario s/n 10071 Cáceres, Spain
E-mail: pablomar,paguilar,rosapere,aplaza @unex.es

Hyperspectral image sensor developments on the study of the Earth's surface give way to images with higher spectral and spatial resolutions. In fact, the higher the resolution, the greater the size of these images. The use of these sensors by space-borne satellite systems will provide an enormous and continuous flow of data with constraints placed on onboard storage, and data transmission bandwidth. New algorithms and computer capabilities will be necessary for the classification, compression and pre-processing of these images. In this chapter, we propose an SOM algorithm for hyperspectral classification implemented by one systolic array, to provide real-time hyperspectral compression facilities.

8.1. Introduction

The use of remote sensors to study the Earth canopy has created a great deal of expectations for applications in very different fields.

Some of these applications are not possible for the low spatial and spectral resolution of first-generation remote sensors.

Currently, hyperspectral sensors lack these restrictions, providing a large number of narrow bands that ensure sensor capabilities for narrow absorption band recognition.

Molecules and particles in land, water and atmosphere environments interact with the sun energy in the 400-2500nm spectral region through absorption, reflection and scattering processes; these occur in narrow spectral regions. Hyperspectral sensors are developed to measure spectra as images in these portions of the spectrum [1, 2] .

The amount of hyperspectral image (HI) spectral bands results in large size data sets; for example, a DAIS image with a size of 614x512 spatial pixels occupies about 140 Mbytes.

The high spatial and spectral resolution of HIs can be used for high resolution classification and/or determination of the Earth's surface constituent signature variation. This is achieved via the hyperspectral information provided by the sensor for a composite pixel spectrum. These analytical capabilities increase the HI applications.

Recent scientific research and applications are expanding HI investigations of ecology and vegetation, geology and soils, inland and coastal waters, the atmosphere, snow and ice, hydrology, biomass burning, environmental hazards [2].

Today, most hyperspectral images are taken from airborne sensors. Each image datum requires one careful schedule of the flight, regional weather, flight permission...etc. Future space-borne satellite hyperspectral sensors will provide access to spectral images from all regions of the Earth with multi-temporal coverage.

The new sensors will present significant challenges to the measurement of high quality spectra from space.

One of the bottlenecks of these new sensors is the enormous and continuous flow of data that is produced [3]. Data compression becomes increasingly important in these applications for two reasons:

- Onboard storage.
- Data transmission bandwidths.

New special computer architectures must be designed to aid massive compression tasks. This paper discusses the systolic implementation of one unsupervised classification algorithm performed by an SOM neural network.

The design is based on the contiguous flow of multidimensional data, using the parallel capabilities of the systolic arrays to provide real time computing facilities.

8.2. SOM Neural Network for Hyperspectral Analysis

Most algorithms in conventional multi-spectral images cannot be used with hyperspectral formats due to the following reasons.

1. Great size of the hyperspectral images (30 times larger than a Landsat TM image of the same spatial size).

2. High dimensionality of the images.
3. New Sub-pixel analysis possibilities.
4. High discrimination capabilities to resolve classes
5. Difficulties to use training data.

New methods should be applied for hyperspectral image classification; they should consider the mentioned features, and should be capable of exploiting both high dimensional feedback and spatial information.

The use of neural networks techniques for hyperspectral image processing have increased in the past few years.

The neural network approach has the following advantages [4]:

- Simplicity.
- A lot of parameters for adjusting performance.
- Intuitive method
- Intrinsic parallelism (different degrees of freedom for implementations)
- Easy VLSI implementation.
- Viable for high dimensional data.

Unsupervised classification algorithms do not require ground-truth data. They are namely interesting in remote sensing due to the great cost of these data in Earth canopy studies.

Unsupervised clustering leads to problems with applications in many areas. Given a set of N data points in a feature space of D dimensions (x1,x2,x3.....xD) $\in$ RD, we wish to characterize the data as belonging to a K cluster, where K must be obtained from the statistics of the image data without ground truth information. This clustering is based on distance metrics. Various unsupervised algorithms can be used to classify hyperspectral data: ISODATA, K-means. The accuracy of these unsupervised algorithms is usually very low [5].

One of the most useful neural network (NN) algorithms is the Self-Organizing NN, or Self Organized Map (S.O.M), proposed by T. Kohonen, with applications in various signal processing tasks. The SOM neural network is an unsupervised classification method, widely used for image treatment. One of the most interesting SOM characteristics is the creation of topologic maps [6].

Some reasons for using S.O.M. in hyperspectral analysis have been described by [7]:

- Avoiding the need to degrade data

- Providing speed (when implemented in hardware as massively parallel algorithms).
- Surpassing conventional classification algorithm performance.
- Good performance for large real-life tasks.

In the following section, we design one S.O.M. to classify HIs.

8.3. Topology of the Proposed Neural Network

Our proposed network architecture is depicted in Fig. 8-1 [8]. In our case, N corresponds to the number of channels of the hyperspectral image, and M is the number of classes or prototypes to be extracted by the network. M depends on image complexity and must be carefully selected according to certain metrics. The weight matrix W has one weight for the connection of each input neuron (channel) with each prototype neuron.

In the classification phase, the input signals x are projected on the feature space by the feed-forward connections W; each neuron produces a selective response to the input signals. In the learning phase, lateral and feedback output layer connections produce excitatory or inhibitory effects depending on the distance from the neuron to the winning neuron [6]. These weights are used to determine the Wi classification prototype for each output neuron.

8.4. SOM Training Algorithm

There are five basic steps involved in the training algorithm. These steps are repeated until the topological map is completely formed:

a) Initialization. Choose random values for the initial weight vectors $\boldsymbol{w}_i(0)$, i = 1,2,...,*M.* It is desirable to keep the magnitude of the weights small.

b) Sampling. Choose an input pattern ***x(n)*** belonging to the pixel of the hyperspectral image. The selection is done randomly.

c) Similarity Matching. Find the best-matching (winning) neuron i^* at time t, using the minimum-distance criterion, as shown in the following equation:

$$i^*[x(n)] = \min_j dist\{x(n), w_j(t)\} \qquad j = 1,2,\cdots,M \qquad \text{(8-1)}$$

where *dist (i*,i)* is the Euclidean distance.

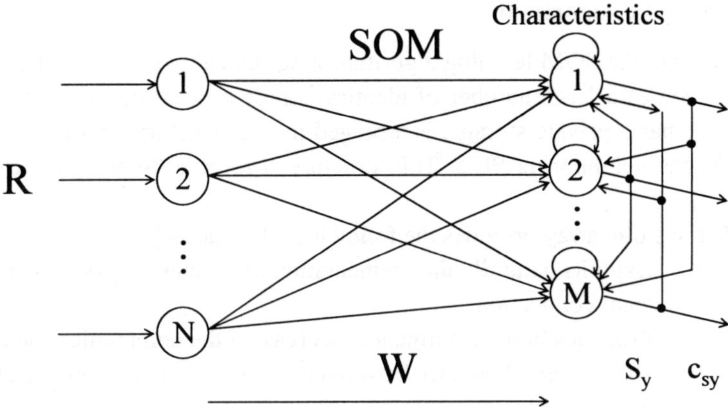

Fig. 8-1. SOM neural network topology including weight matrix W and lateral and feedback connections.

d) Learning. Adjust the synaptic weight vectors of all neurons, using the update formula (8-2), where η(t) is a learning-rate parameter, and γ is a Gaussian neighborhood function centered around the winning neuron. The size of the neighborhood is determined by a parameter σ(t) (see Eq. (8-3)).

$$w_i(t+1) = w_i(t) + \eta(t)\gamma\left(t, i, i^*[x(n)]\right)\left(x(n) - w_i(t)\right) \tag{8-2}$$

Among the different choices made for the involved parameters, we follow the approach in [9], thus coming up with:

$$\eta(t) = \frac{1}{t} \qquad \gamma\left(t, i, i^*[x(n)]\right) = e^{\frac{dist(i^*,i)^2}{\sigma(t)}} \qquad \sigma(t) = \left(\frac{\sigma_0}{t}\right)^2 \tag{8-3}$$

where σ_o is the initial width.

e) Continue from step b) until no noticeable changes in the weight space are observed, or until the maximum convergence time is achieved.

8.5. Systolic Algorithm

To alleviate the SOM learning algorithm described above, one solution is the connection of a large number of identical single processing elements (PEs). Each PE has a private storage, is arranged in one structure, and is connected only to the neighboring PE's. This structure is referred to as Systolic Array [10, 11].

The systolic array provides the following advantages:

- Exploits naturally the segmentation of regular networks with local connections.
- Produces high performance, decreasing the communication cost.
- Offers a good balance between computation and communication for array operations.

In order to design the systolic linear array for the implementation of the SOM algorithm described above, the first step is the use of one regular algorithm with local dependences, equivalent to the algorithm described in Section 4:

```
For j=1 to M
   s(j,0)= 0
   For  i=1 to N
        x(0,i)=x_i
        x(j,i)=x(j-1,i)
        s(j,i)= s(j,i-1)+(x(j,i)-w(j,i)) *(x(j,i)-w(j,i))
   End For
End For
```

From these sequential algorithms, we can obtain the dependence graph with connected MxN nodes, as shown in Fig. 8-2 [12].

In order to establish the correspondence between the dependence graph and the corresponding systolic algorithm, we develop a bijection T, going from the original algorithm index set to the systolic index set:

$$T = \begin{bmatrix} 11 \\ 10 \end{bmatrix} \tag{8-4}$$

According to this bijection, we should have one graph projection in the east-west direction. For these reasons, the difference between the input pattern and the weight vector s(j) of the neuron j will remain static for the corresponding EPj. The input pattern will be displaced from one EP to its right neighbor. The total number of cycles in the retrieving phase will be M+N.

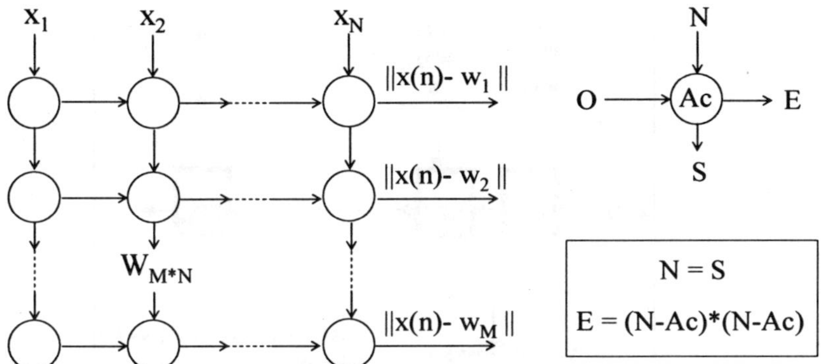

Fig. 8-2. Dependence graph to compute similarity between the input pattern and the weight vectors.

The data movements within the systolic algorithm for distance computation, are the following:

- The input data ***x(n)*** are introduced in the first EP and are propagated to the other EPs sequentially.
- The weight data are stored in the EPs row by row.
- When the ***x(n)*** value arrives at the j EP, the square of the difference between the stored weight on the EP and the ***x(n)*** is computed, and the partial sum s_j is stored on the same EP.
- The control unit changes the EP mode to read its weight in the following step, and is compared with the other s_j to obtain the winning neuron.

Now, we describe the components of each EP:

- Memory: Each EP stores one row of the weight matrix.
- Communication: The data moves along one direction between neighbor EPs (E=W).
- Processing: Each EP supports all arithmetic operations including subtraction and accumulation.

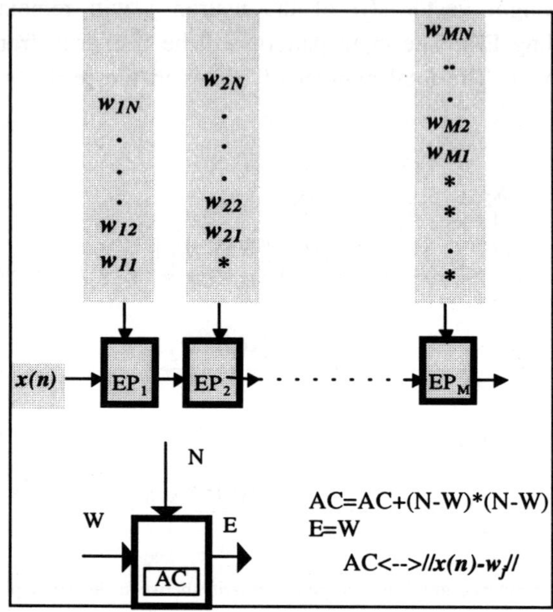

Fig. 8-3. Systolic algorithm for computing the distance.

When all the distances are computed, we can obtain the winning neuron by searching for the EP with the lower AC value.

The same scheme can be used with this purpose when performing a comparison operation such as the one described in Fig. 8-4.

We must include one communication between the last EP and the first one.

Each EP must include one comparator circuit.

When the M cycles are completed, the output of the last EP offers the minimum distance computed in the previous phase. By comparing this output with the accumulators of each EP, we obtain the winning neuron and begin the weight upgrading phase.

Before changing the weights, the η factor (with the same value for each EP) will be computed. The neighbor function γ will be initialized to 1 for the winning neuron and with lower values for the others neurons, according to the distance from this winning neuron. The γ values are computed in advance, and are stored in one memory array.

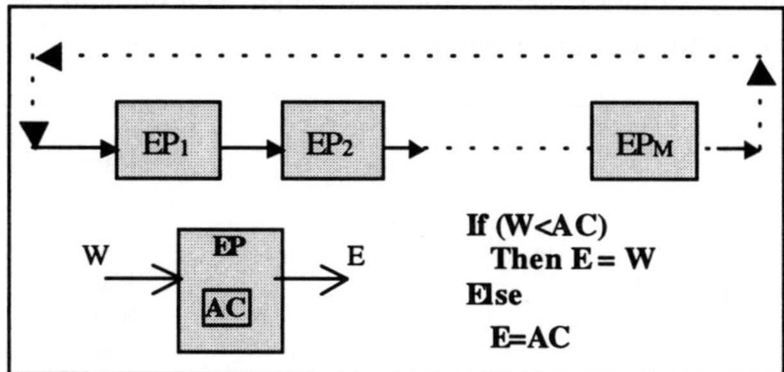

Fig. 8-4. Linear systolic scheme to determine the lower distance. (AC stores the distances previously computed).

In order to compute the weights according to Eq. (8-2), the following steps will be performed:

1. The x values stored in the first EP in the distance computation phase will be propagated to the other EPs.

2. When x_i arrives at the j element, the difference between the x_i value and the weight value w_{ji} is obtained and multiplied by the η and γ values of this EP. In this way, we obtain Δw_{ji} by adapting the weight of the j element.

Fig. 8-5 shows the scheme of this process. By following the previous systolic description, we can conclude:

- The systolic SOM implementation has M EPs.
- Each EP will be programmable to change functionality in the different phases.
- The memory for each EP must store one row of the weight matrix, and one FIFO is required to recycle the x data.
- The data are transmitted in a single direction between two neighbor EPs. To compute the winning neuron, we include one connection from the last neuron to the first.
- Each EP supports all the arithmetic processing capabilities (subtractions, comparisons, and accumulations).

The classification process includes resolution of the winning neuron, achieved by computation of the minimum distance.

In order to classify hyperspectral vectors provided by the DAIS sensors with 220 bands, the proposed neural network has one input neuron for each channel (220 neurons), and the number of output neurons is fixed at 17. The systolic array includes one EP for each output neuron, and each EP has 220 memory cells to store the weights.

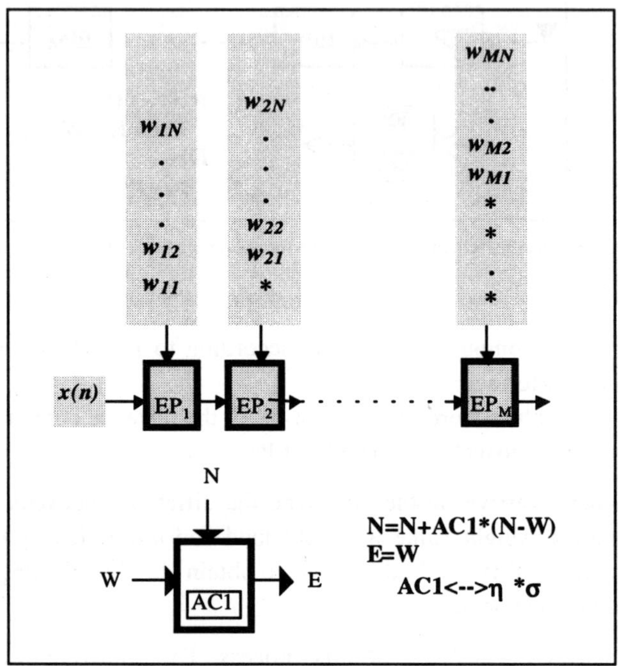

Fig. 8-5. Systolic scheme for the weight upgrading phase.

8.6. Conclusions

This chapter describes the design of one linear systolic array to implement one SOM neural network. The aim is to carry out unsupervised classification of hyperspectral images. The proposed architecture reduces classification time by using M EP's to perform parallel computations with N-element vectors. For the retrieving phase, the number of sequential operations is 4 MxN. The systolic array reduces this number to M+N. For the weight modification phase, the cycle number is reduced from 5NxM cycles for each iteration to N+M cycles for each iteration. This great reduction alleviates the processing of the image pixels by allowing real-time processing.

Acknowledgements

This chapter has been conceived under the funding project Aplicación de las imágenes hiperespectrales a la vigilancia de recursos naturales (TIC 2000-0739-C04-3), Ministerio de Educación y Ciencia (Spain).

We also wish to acknowledge the linguistic revision of the paper by Dr. Alejandro Curado Fuentes, from the Department of English at our Institution.

References

[1] A. F. H. Goetz, G. Vane, J. E. Solomon, and B. N. Rock. Imaging spectrometry for Earth remote sensing. *Science*, 211, pages 1147-1153, 1985.

[2] R. Green, M. Eastwood, T. Sarture, M. Aronson, J. Chippendale, J. A. Faust, B. Pavri, Ch. Chovit, M. Solis, M. Olah and O. Willians. Imaging Spectroscopy and the Airborne Visible/Infrared Imaging Spectrometer (AVIRIS). *Remote Sensing of Environments*, 65, pages 227-248, 1998.

[3] S. Subramianian, N. Gat, A. Ratcliff and M. Eismann. Real-time Hyperspectral Data Compression Using Principal Components Transformation. *Summaries of the X Annual JPL Airborne Earth Science Workshop*, Jet Propulsion Laboratory, NASA, Pasadena ,California, pages 451-460, 2000.

[4] E. Merényi, The Challenges in Spectral Images: An Introduction and Review of ANN Approaches. *Proc. European Symposium on Artificial Neural Networks, ESANN'99,* Bruges, Belgium, pages 93-98, 1999.

[5] L. O. Jiménez, A. Morales-Morel and A. Creus. Classification of Hyperdimensional Data Based on Feature and Decision Fusion Approaches Using Projection porsuit, Mayority voting and Neural Networks. *IEEE Transactions on Geoscience and Remote Sensing*, 17(3), pages 1360-1366, 1999.

[6] T. Kohonen. *Self-Organizing Maps* (2nd. ed.), Springer Series in Information Science, 1997.

[7] J. Bruske and E. Merényi. Estimating the intrinsic dimensionality of Hyperspectral images. *Proc. European Symposium on Artificial Neural Network, ESANN'99*, Bruges, Belgium, 21-23, pages 105-110, April, 1999.

[8] P. L. Aguilar, P. Martínez, R.M. Pérez. Abundance extractions from AVIRIS image based on Self-Organizing neural network. *AVIRIS'2000, Earth Science and Applications Workshop,* Jet Propulsion Laboratory, NASA, Pasadena, California.

[9] P. L. Aguilar. *Cuantificación de firmas hiperespectrales usando mapas autoorganizativos,* Tesis Doctoral, Departamento de Informática, Escuela Politécnica, Cáceres, 2000.

[10] M. Zarghan. *Computer Architecture*, New Jersey, Prentice Hall, 1996

[11] R. M. Pérez, P. L. Aguilar, P. Bachiller and P. Martínez, Neural Network quantifier for solving the mixture problem and its implementation by systolic array. *Microelectronics Journal*, 30(1), pages 77-82, 1999.

[12] R. M. Pérez, P. Martínez, P. L. Aguilar y Linaje M., Arquitectura Sistólica para una Red SOM que Clasifica Imágenes Hiperespectrales, *XII Jornadas de Paralelismo*, Granada, Spain, pages 351-355, 2000.

CHAPTER 9

OPTIMIZING AND LEARNING ALGORITHM FOR FEEDFORWARD NEURAL NETWORKS AND ITS IMPLEMENTATION BY SYSTOLIC ARRAY

P. Bachiller-Burgos

Department of Computer Science. University of Extremadura
Avda/ de la Universidad s.n., 10071 Cáceres, Spain
E-mail: pilarb@unex.es

Feedforward neural networks constitute a good solution for many problems, such as classification, recognition and identification, and signal processing. However, the network designer confronts the problem of selecting an adequate structure that makes the network work well at both learning and recognition phases. When the hidden structure of the network is too large and complex for the model being developed, the network may tend to memorize input and output sets rather than learning relationships between them. Such a network may train well but test poorly when inputs outside the training set are presented. In addition, training time will significantly increase when the network is unnecessarily large and complex. To reach both goals, adequate hidden structure and learning time reduction, a new optimizing method is presented. This approach is based on the idea of selecting the optimum hidden nodes of the network during the training process, avoiding the retraining phase of the reduced-size network. The features of the proposed method favor the use of systolic array for its implementation. Exploiting the benefits of pipelined processing, a systolic architecture is presented that implements the optimizing and learning algorithm described in this chapter.

9.1. Introduction

Feedforward neural networks are capable of providing solutions to many problems in the areas of pattern recognition, signal processing, time series analysis, etc. The main disadvantage of this neural model is how to determine the appropriate network size for solving a specific task. Choosing the optimal size of a feedforward neural network is a very important issue of its design since the generalization capability of the network depends to a great extent on its hidden structure [2]. In addition, the training time increases significantly when the number of links and nodes is excessive. There are several approaches to solve the problem of determining the optimum size of a neural network. The first technique, referred to as growing algorithm, adds gradually hidden units to an initial small network until it reaches the convergence [1,4,5,14]. The second approach, known as pruning, consists of training a large network, eliminating the unnecessary links and nodes and finally retraining the reduced-size network [3,8].

The fact of using a pruning technique to find the optimal size of a feedforward neural model involves training an over-parameterized network. This implies that learning time increases significantly when its structure is too large and complex. Growing algorithms try to solve this problem making the network structure during the learning process. However, although these growing techniques could improve the learning time, they would not ensure an optimal structure.

A good neural network optimizing method should find the best structure and improve the learning time. To reach both goals an Optimizing and Learning Algorithm (OLA) is presented in this chapter. OLA modifies the Backpropagation training algorithm to find the optimal structure of the neural network during the training process. It is based on the idea of identifying iteratively the set of optimum hidden nodes and then updating only the weights connected to these nodes. In OLA the retraining process of the reduced-size network is avoided; it gives an optimal structure of the neural network and accelerates the training process.

The basic principle of network size reduction is to detect and eliminate the collinearity between the input data sets at the hidden layers. Orthogonal transformations can lead to relative decorrelation of the network information providing a good solution to the optimal set of hidden nodes selection. There have been some efforts for using orthogonal transformations in neural network modelling [8,11,13]. In particular, the singular value decomposition (SVD) [6] and the QR with column pivoting factorization (QRcp) [6] can stand out because of their numerical and computational stability. These orthogonal transformations have been used successfully in a post-training pruning method to find the optimal structure of a neural network [8]. Although both of them provide a good solution for network size reduction when the learning process has carried out, their application during network training would excessively increase learning time. Therefore, OLA needs other kind of orthogonal transformations. The properties and structure of Householder reflections allow developing many factorization methods with a great computational efficiency, so they are the best candidates for the proposed optimizing algorithm.

The regular nature of the Backpropagation algorithm makes specialist architectures, such as systolic array, ideal for its implementation. These specialist architectures provide many computation and communication benefits compared more general-purpose processors [7,9]. To achieve the architecture design, the proposed learning algorithm is divided in three phases: optimum hidden nodes computation, forward input propagation and backward error correction. The last two phases correspond to the Backpropagation algorithm, which can be implemented by a linear systolic array. The exploitation of the structure of Householder reflections allows mapping the optimum hidden nodes computing phase into a systolic array without a great effort. Thus, it is possible to design a specialist architecture for the proposed algorithm taking advantages of the features of the pipelined processing.

The rest of the chapter is organized as follows. In Section 9.2, the main properties of Householder reflections are introduced. The proposed method is presented in Sections 9.3 and 9.4. In Section 9.5 a systolic architecture design to implement the optimizing and learning algorithm is described. Experimental results on different test problems are shown in Section 9.6. Finally, Section 9.7 summarizes the main conclusions of this chapter.

9.2. Householder Reflections

Let $v \in \Re^n$ be nonzero. An nxn matrix P of the form

$$P = I - 2vv^T/v^Tv \qquad (9\text{-}1)$$

is called a Householder reflection. The vector v is called a Householder vector [6].

The properties of Householder reflections allow developing algorithms with low computational cost to solve several problems. In particular, for the neural network size optimization, the following properties can stand out:

Property 1: *Householder reflections can be used to zero selected components of a vector* [6].

Given a vector $x \neq 0$, if we want Px to be multiple of e_1 (the first column of the nxn identity matrix), then, for any $x \in \Re^n$, v must be defined as follows:

$$Px = (I - 2vv^T/v^Tv)x = x\text{-}(2v^Tx/v^Tv)v \tag{9-2}$$

Setting $v = x + \alpha e_1$ gives

$$v^Tx = x^Tx + \alpha x_1 \tag{9-3}$$

$$v^Tv = x^Tx + 2\alpha x_1 + \alpha^2 \tag{9-4}$$

If we assume $\alpha = \pm \|x\|_2$ (2-norm of the vector x), then

$$v = x \pm \|x\|_2 e_1 \Rightarrow Px = (I - 2vv^T/v^Tv)x = \pm \|x\|_2 e_1 \tag{9-5}$$

To zero the last m-1 components of a vector $x \in \Re^n$, we have to define a matrix of the form:

$$\begin{bmatrix} I & 0 \\ 0 & H \end{bmatrix} \begin{matrix} n-m \\ m \end{matrix} \tag{9-6}$$

$$\begin{matrix} n-m & m \end{matrix}$$

being I the identity matrix and H a Householder reflection that verifies:

$$H\,[x_{n-m+1}\; x_{n-m+2}\; \ldots\; x_n]^T = [c\; 0\; 0\; \ldots\; 0]^T \tag{9-7}$$

Property 2: *Given m vectors* $\in \Re^n$ $[x_1, x_2, \ldots, x_m]$*, Householder reflections are used to determine which of them are linearly independent.*

Assume that H is a product of Householder reflections which zeroes at least the last n-L components of each vector of the set $S = [x_1, x_2, \ldots, x_i]$, being L the number of linearly independent vectors of S. To determine the linear dependency among a new vector x_{i+1} and the vectors of S, a vector y has to be obtained by the product Hx_{i+1}. If x_{i+1} can be expressed by a linear combination of $x_1, x_2, \ldots, x_i$, then the last n-L components of vector y will be equal to zero:

$$x_{i+1} = \sum_{j=1}^{i} a_j x_j \tag{9-8}$$

$$Hx_{i+1} = \sum_{j=1}^{i} a_j Hx_j \tag{9-9}$$

$$y = \sum_{j=1}^{i} a_j \, [n_{j1}, ..., n_{jL}, 0, ..., 0]^T \tag{9-10}$$

$$y = [m_1, ..., m_L, 0, ..., 0]^T \tag{9-11}$$

Since H is an orthogonal matrix, the equality $||y||_2 = ||x_{i+1}||_2$ holds. So if the 2-norm of vector x_{i+1} can be computed using only the first L components of y, the linear dependency among x_{i+1} and the vectors of S is verified.

If Eq. (9-11) does not hold, matrix H has to be updated in order to determine linear dependencies among new vectors and the vectors of $S' = [x_1, x_2, ..., x_{i+1}]$. Now, matrix H has to zero at least the last n-L-1 components of the vectors of S', so a new Householder reflection H' to zero the last n-L-1 components of y is computed and matrix H is updated by the product $H'H$.

Property 3: *Householder updates never entails the explicit formation of the Householder matrix* [6].

Given a matrix $A \in \Re^{mxn}$ and a Householder matrix $P = I - 2vv^T/v^Tv \in \Re^{mxm}$ then

$$PA = (I - 2vv^T/v^Tv)A = A - vw^T \tag{9-12}$$

where $w = (2/v^Tv)A^Tv$

Thus, a Householder update of a matrix involves a matrix-vector multiplication and an outer product update, which entails 4*mn* floating point operations (flops).

The same operations can be applied for the Householder update of a vector $x \in \Re^m$:

$$Px = (I - 2vv^T/v^Tv)x = x - \beta v \tag{9-13}$$

where $\beta = 2\, v^Tx/v^Tv$

Thus, instead of computing a matrix-vector multiplication, the Householder update of a vector can be computed by dot products and a scalar-vector product update, which involves about 4*m* flops.

Many applications of Householder reflections implicate computing products of several Householder matrices:

$$Q = Q_1Q_2...Q_r \qquad Q_j = (I - 2v_jv_j^T/v_j^Tv_j) \tag{9-14}$$

It is usually not necessary to compute Q explicitly even if it has to be used in subsequent calculations. For instance, assuming m vectors $w_i \in \Re^m$ and the matrix $Q \in \Re^{mxm}$ of Eq. (9-14), the products $Q^T w_i$ can be implemented as follows:

```
for i = 1:m
    for j = 1:r
        β = (2v_j^T w_i/v_j^T v_j)
        w_i = w_i - βv_j
    end
end
```

This involves about $4rm^2$ floating point operations. If Q is explicitly represented, $Q^T w_i$ has to be computed as follows:

```
Q = I
for j = 1:r
    w = (2Q^T v_j/v_j^T v_j)
    Q = Q - v_j w^T
end
for i = 1:m
    w_i = Q^T w_i
end
```

This implementation involves $4rm^2 + m^3$ flops, which shows the computational benefits of exploiting the structure of Householder reflections.

All these properties allow developing a robust method for subset selection, which is the basis of neural network size-reduction.

9.3. Selection of the Optimal Hidden Structure

Detecting and eliminating collinearity between the input data sets at different layers of the network is the basic principle of its size reduction [8]. A node is redundant if net inputs at the subsequent layer keep invariant when that node has been eliminated. Hence, if a node i is redundant, a weight combination (w'_{lj}) exists, which makes the network works in the same way as it does before eliminating that node:

$$\sum_{j=1}^{M} h_j w_{lj} = \sum_{\substack{j=1 \\ j \neq i}}^{M} h_j w'_{lj} \qquad \text{for each node } l \text{ of the subsequent layer} \qquad (9\text{-}15)$$

where h_j is the vector formed by the outputs of the node j for all the training patterns and M is the number of nodes in the layer where i belongs to. Simple algebraic manipulations yield

$$\sum_{\substack{j=1\\j\neq i}}^{M} h_j w_{lj} + h_i w_{li} = \sum_{\substack{j=1\\j\neq i}}^{M} h_j w'_{lj} \tag{9-16}$$

$$h_i = \sum_{\substack{j=1\\j\neq i}}^{M} h_j \frac{(w'_{lj} - w_{lj})}{w_{li}} \tag{9-17}$$

If a new weight combination (w'_{lj}) can be found to preserve the original network behaviour, then Eq. (9-17) can be expressed as follows:

$$h_i = \sum_{\substack{j=1\\j\neq i}}^{M} c_j h_j \tag{9-18}$$

being $cj = (w'_{lj} - w_{lj})/w_{li}$.

Hence, a node i is redundant if its output vector h_i is a linear combination of the output vectors of the remaining nodes belonging to the layer of i.

The proposed selection process uses the properties of Householder reflections to compute the optimal set of hidden nodes. The output information for the training patterns of each node of the hidden layer is grouped forming vectors. Then, linear dependencies among them are determined using Householder reflections. The final structure of the network will be composed by the nodes whose output vectors are linearly independent. The selection algorithm starts with a NxMxK neural network and P training patterns and proceeds as follows:

(i) Initialize L to 0 and H as the identity matrix, being L the optimum number of hidden nodes and $H \in \Re^{PxP}$ an orthogonal matrix that zeroes at least the last P-L components of the output vectors of the optimum hidden nodes.

(ii) For every hidden node ($1 \leq i \leq M$):

(a) Obtain the vector $h_i \in \Re^P$ formed by the outputs of the current hidden node for all the training patterns.

(b) Compute the vector $y = Hh_i$.

(c) If vector h_i is a linear combination of the linearly independent vectors (h_j's) determined in previous iterations, then y will have its last P-L components equal to zero and, therefore, the equality of Eq. (9-19) is verified. In such case, the node i is redundant.

$$\|h_i\|_2 = \sqrt{\sum_{m=1}^{L} y_m^2} \tag{9-19}$$

(d) If Eq. (9-19) does not hold, the i-th hidden node can not be considered redundant and it is necessary to update matrix H in order to determine more redundant hidden nodes. A Householder matrix H' that zeroes the last P-L-1 components of y is computed, H is updated by the product $H'H$ and the number of optimum nodes is increased ($L = L$+1).

The property 3 of Householder reflections can be applied to reduce the number of operations of the proposed selection algorithm. Since H is a product of L Householder matrices, it can be considered as a sequence of L Householder vectors. Thus, in step (d) it is not necessary to update matrix H, but only to compute a new Householder vector to zero the last P-L-1 components of vector y. In addition, the product Hh_i of step (b) is computed as L Householder update operations, which reduces the number of operations required to compute vector y:

$$Hh_i = \left(I - 2\frac{v_L v_L^T}{v_L^T v_L}\right)\cdots\left(I - 2\frac{v_1 v_1^T}{v_1^T v_1}\right)h_i \tag{9-20}$$

where each v_i has the form:

$$v_i = \begin{bmatrix} \underbrace{0 \quad \cdots \quad 0}_{i-1} & v_{ii} & \cdots & v_{iP} \end{bmatrix}^T \tag{9-21}$$

9.4. The Optimizing and Learning Algorithm

When a pruning technique is used to compute the optimal set of hidden nodes of a given network, it is always necessary to train an over-parameterized network. The training time increases significantly when the hidden structure of the network is too large and complex for the model that is being developed. In order to accelerate the training process while an optimum structure for the network is obtained, it is proposed the optimizing and learning algorithm (OLA). This method applies the selection algorithm explained in the previous section. The idea consists of pruning each output of the hidden nodes that have been considered redundant at the current iteration during the training process, and then updating only the weights of the links connected to those hidden units whose connections have not been pruned.

Two different sets of hidden nodes are considered in the algorithm. The first one is the set O, composed by the nodes whose output vectors are linearly independent. The other one is the set R, which contains the redundant nodes. Each node belonging to R has an output vector that is a linear combination of the output vectors of the hidden units of O.

The first step of every iteration of the training process is to compute the sets O and R using the proposed selection method. Then the connection weights have to be updated by the Backpropagation algorithm [15]:

$$w_{ji} = w_{ji} + \alpha\delta^{h}_{pj} f'(Net^{h}_{j})x_{pi} \tag{9-22}$$

$$v_{kj} = v_{kj} + \alpha\delta^{o}_{pk} f'(Net^{o}_{k})h_{pj} \tag{9-23}$$

where w_{ji} is the weight from the i-th input node to the j-th hidden node, v_{kj} is the weight from the j-th hidden node to the k-th output node, δ^{h}_{pj} is the error of the j-th hidden unit for the pattern p, δ^{o}_{pk} is the error of the k-th output unit for the pattern p, x_{pi} is the value of the i-th element of input pattern p, h_{pj} is the j-th hidden output and f' is the derivative of the activation function for the concerned node.

From the definition of redundant node, explained in Section 3, it can be shown that there is no need for updating all the weights. The output vectors of the nodes of R are obtained by linear combinations of the output vectors of the nodes of O. This implies that output weights (v_{kj}) of the nodes of R can be considered as 0 and there is no need to update them because they have no influence in the final behavior of the network. Taking this into account, for any node j of R their weights from the previous layer (w_{ji}) keep invariant since the error for such node (δ^{h}_{pj}), defined in Eq. (9-24), is zero. In sort, the only weights that have to be updated are those connected to the nodes of the set O.

$$\delta^{h}_{pj} = \sum_{k} \left(\delta^{o}_{pk} \, v_{kj} \right) \quad \forall k \in \text{output nodes} \tag{9-24}$$

More explicitly, assuming a NxMxK network and P training patterns the optimizing and learning algorithm proceeds as follows:

(i) Compute the sets O and R using the selection process of section 3:

- Set O: hidden nodes whose output vectors are linearly independent.
- Set R: hidden nodes whose output vectors are linear combinations of the output vectors of the nodes of O.

(ii) Prune the outputs of the nodes of R, taking their weights to the subsequent layer as 0. Update the weights connected to the nodes of the set O for the P training patterns using the Backpropagation algorithm.

(a) Compute network outputs considering only the optimum hidden nodes.

(b) Update the weights connected to the nodes of O.

(iii) Go to step (i) until the network reaches the convergence.

Remarks of the algorithm:

- The initial number of hidden units depends on the specific problem to solve. However it will be always less or equal than the number of training patterns.
- Once the network is trained, the hidden nodes belonging to the set R can be eliminated, since their weights to the subsequent layer are zero.
- In case of a network with more than one hidden layer, it is necessary to compute the set of optimum hidden nodes for each hidden layer. Then the weights connected to those nodes are updated using the Backpropagation algorithm as in step (ii).

9.5. Implementation of OLA by Systolic Arrays

The iterative features of the optimizing and learning algorithm make systolic array a good architecture for its implementation. Four blocks compose the proposed architecture (Fig. 9-1).

The *Control Block* takes charge of synchronizing the remaining blocks of the architecture. It stores the input patterns and output targets to be sent to the *Learning Block* and computes the output error produced in each iteration of the learning process. The *Learning Block* implements the Backpropagation algorithm and sends the hidden layer outputs to the *Hidden nodes information Block.* This network information is used by the *Optimum hidden nodes computing Block* to decide which processing elements of the *Learning Block* must work and which of them are inactive. Next, the main blocks of the proposed architecture are described in detail.

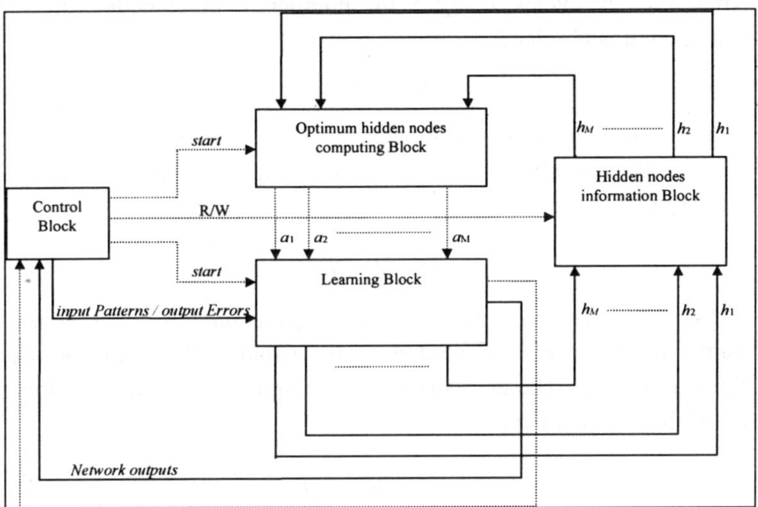

Fig. 9-1. General architecture for OLA.

9.5.1. *The Learning Block*

The internal implementation of the *Learning Block* is shown in Fig. 9-2. Assuming a feedforward neural network with M nodes in its single hidden layer, the *Learning Block* can be implemented by a linear systolic array with M processing elements locally connected and a non-linear unit that generates the output value of each neuron of the last layer. Each processing element PE_i stores the weights of the links connected to the i-th hidden node and works in two different ways depending on the value of the signal a_i. If a_i is inactive, the processing element works asynchronously propagating the input data to the next PE without performing any computation. Otherwise, it carries out the following phases, assuming the PE description of Fig. 9-3:

- *Computation of the hidden outputs*: The input pattern is propagated through the PEs by the input I_1. At the j-th clock cycle

$$Net_i^h = Net_i^h + I_1 w_{ij}$$

$$O_1 = I_1$$

After N clock cycles, being N the number of input nodes, the hidden node output is obtained:

$$h_i = f(Net_i^h)$$

- *Computation of the network outputs*: The net value of each output node is partially computed by each PE and propagated to the non-linear unit. At the j-th clock cycle:

$$O_1 = I_1 + h_i v_{ji}$$

- *Update output layer weights and compute hidden nodes error*: The output nodes error is computed by the *Control Block* and propagated through the PEs. The net value of each output is obtained from the input I_2. At the j-th clock cycle:

$$\delta_i^h = \delta_i^h + v_{ji} I_1$$

$$v_{ji} = v_{ji} + \alpha I_1 f'(I_2) h_i$$

$$O_1 = I_1$$

$$O_2 = I_2$$

- *Update the hidden layer weights*: The input pattern is propagated through the input I_1. At the j-th clock cycle:

$$w_{ij} = w_{ij} + \alpha \delta_i^h f'(Net_i^h) I_1$$

$$O_1 = I_1$$

The start signal is propagated through all PEs and sent back to the *Control Block* through the last PE. This communication process allows the *Control*

Block to know the number of optimum hidden nodes (i.e. the number of active processing elements) in order to synchronize the remaining blocks of the architecture.

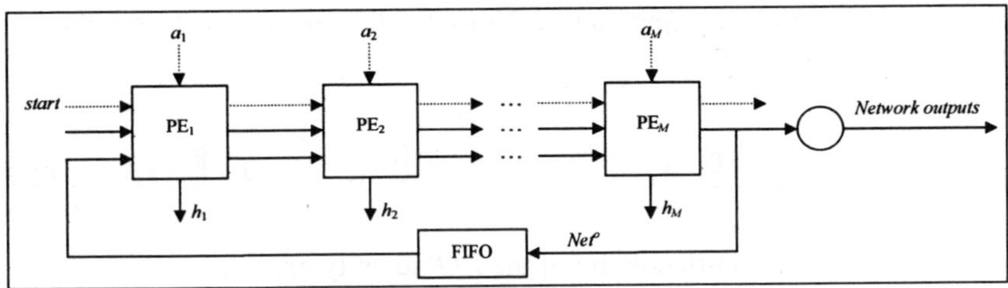

Fig. 9-2. Learning Block.

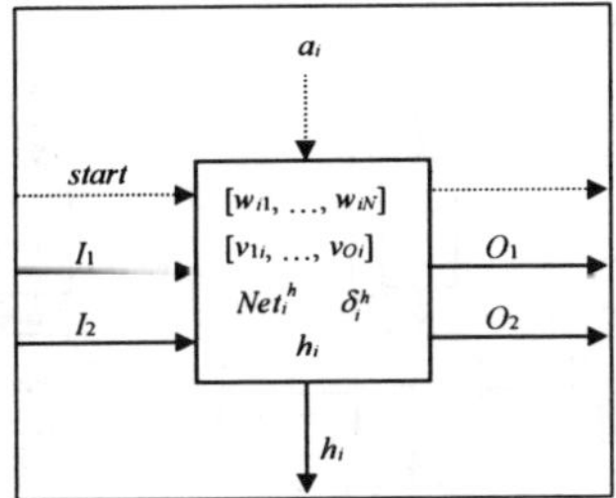

Fig. 9-3. Processing element description.

9.5.2. *The Optimum Hidden Nodes Computing Block*

From the hidden layer information, the *Optimum hidden nodes computing Block* has to establish which processing elements of the *Learning Block* have to be active. This involves computing, for each hidden vector h_i, a vector y as:

$$y = \left(I - 2\frac{v_L v_L^T}{v_L^T v_L} \right) \cdots \left(I - 2\frac{v_1 v_1^T}{v_1^T v_1} \right) h_i \tag{9-25}$$

where $L < i$ is the current number of optimum hidden nodes and each v_i has the form:

$$v_i = \left[\underbrace{0 \ \cdots \ 0}_{i-1} \ v_{ii} \ \cdots \ v_{iP}\right]^T \tag{9-26}$$

We want to verify if the last *P-L* components of *y* are equal to zero. If not, it is necessary to compute a new Householder vector v_{L+1} to zero the last *P-L*-1 components of *y*:

$$v_{L+1} = \left[\underbrace{0 \ \cdots \ 0}_{L} \ y_{L+1} \ \cdots \ y_P\right]^T - \left\|\left[y_{L+1} \ \cdots \ y_P\right]\right\|_2 e_{L+1} \tag{9-27}$$

where e_{L+1} is the $(L+1)$-th column of the *P*x*P* identity matrix.

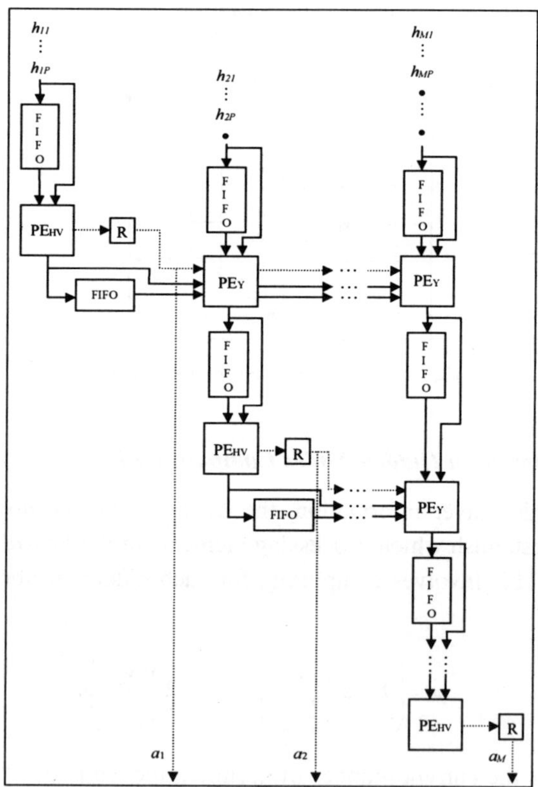

Fig. 9-4. Optimum hidden nodes computing block.

To make easy the architecture design, instead of considering the Householder vectors of the form of Eq. (9-26), each vector v_i is represented as:

$$v_i = \begin{bmatrix} v_{ii} & \cdots & v_{iP} & \underbrace{0 \ \cdots \ 0}_{i-1} \end{bmatrix}^T \tag{9-28}$$

and y is computed as:

$$y = \begin{bmatrix} y_{L+1} & \cdots & y_P & \underbrace{0 \ \cdots \ 0}_{L} \end{bmatrix}^T \tag{9-29}$$

Thus, it is not necessary to know the current number of optimum hidden nodes, but only to verify if the 2-norm of y is zero. If not, the new Householder vector can be computed by:

$$v_{L+1} = y - \|y\|_2 e_1 \tag{9-30}$$

In this way, a new Householder vector can be obtained without taking into account the value of L.

To achieve this process a triangular systolic array is proposed (Fig. 9-4). Each hidden vector h_i is propagated through a column of the array to establish the activation mode of the corresponding processing element of the *Learning Block*. Two different kind of PEs are considered in this array:

- PE_{HV}: This is the last processing element of each column of the array. It computes a new Householder vector by the Eq. (9-30) and verifies $\| y \|_2 \approx 0$. The result of this comparison (a_i) is stored in a register (labeled with R), which indicates if the corresponding hidden node is optimum or not. The resulting Householder vector and the activation signal a_i are propagated through the PEs situated in the same row.

- PE_Y: This processing element computes partially the vector y associated with a hidden vector. Assuming the PE_Y situated at the position (i, j) of the array, it works as follows:

  $\beta = 2v_i^T y^j_i / v_i^T v_i$
  if $a_i = 1$ **then**
  $\quad s = y^j_i - \beta v_i$
  $\quad y^j_{i+1} = [s_2 \dots s_P \ 0]^T$
  else

$$y^j_{i+1} = y^j_i$$

end

The activation signal a_i is received after the Householder vector v_i. This is the reason why the value of β is computed without depending on the value of a_i. Once this signal has been received, the output y_{ji+1} is computed. The activation of a_i indicates that the hidden node i is optimum and, hence, it is necessary to compute y_{ji+1} as a Householder update of the vector y_{ji}, in order to check the dependency between h_i and h_j. However, if a_i is inactive, the node i is redundant and y_j can not be modified. In such case, the input vector y_{ji} is propagated to the next PE without making any modification.

In this way, the *Optimum hidden nodes computing Block* establishes the activation mode of the processing elements of the *Learning Block*. Once this process has finished, the *Control Block* sends the start signal to the *Learning Block* to update the weights stored in the active processing elements.

9.6. Experimental Results

To test the effectiveness of the proposed optimizing algorithm, several software simulations have been carried out over different test problems. In all the presented experiments, three-layer feedforward neural networks have been considered. For each test problem, several networks with different number of hidden nodes were trained by the optimizing and learning algorithm and the Backpropagation method.

To evaluate the proposed method four parameters have been considered:

- Learning time required for the proposed optimizing method in comparison with the Backpropagation algorithm. Though this measure can only be considered for a software implementation, it gives us an idea of the number of operations required for both methods.
- Number of iterations spent for both learning methods to reach the convergence of the network. This parameter measures the speed of both algorithms without taking into account the implementation.
- Optimum number of hidden nodes obtained by OLA in comparison with the reduction provided by the pruning after training method based on SVD and QRcp [8].

- Generalization provided by networks trained with the proposed method with regard to the generalization obtained by networks trained using the Backpropagation algorithm.

Three examples are studied to evaluate the optimizing and learning algorithm. One is the time-series generated by the Mackey-Glass (MG) equation [12] and the other two are the parity and symmetry problems [15].

9.6.1. *Mackey-Glass Time Series*

A system is said to be chaotic if its evolutionary trajectory is generated by a deterministic mechanism, but it is very sensitive to the initial conditions of the system [16]. Since under certain conditions a chaotic system behaves randomly, the identification of such system is difficult. Under those conditions, a model capable of identifying the underlying deterministic mechanism can greatly improve system performance, predictability and control [8].

The discrete time representation of the Mackey-Glass equation is given by

$$x(k+1) - x(k) = \alpha x(k-\tau)/(1+ x^{\gamma}(k-\tau)) - \beta x(k) \quad (9\text{-}31)$$

Consider the series generated with α=0.2, β=0.1, γ=10 and τ=17. This combination generates a quasiperiodic time series, where a quasiperiodic process is a linear combination of several periodic processes.

The objective is to model the Mackey-Glass series to produce ahead predictions. The Mackey-Glass series $\{x(k)\}$ can be expressed as

$$x(k+p) = f(x(k), x(k-\tau), x(k-2\tau), ..., x(k-(N-1)\tau)) \quad (9\text{-}32)$$

where p is the prediction time, which is chosen according to the need for long-term or short-term prediction, and N is generally between four and eight [10,16]. Proofs were made choosing N=6, so a six-input neural network is considered where $x(k)$, $x(k-\tau)$, ...,$x(k-5\tau)$ are used as the inputs and $x(k+p)$ is used as the output, taking a prediction time of 5.

Simulation results have been obtained from several neural networks with different number of hidden units using 300 data sets for training. For each of those neural networks both, the Backpropagation algorithm and the optimizing and learning algorithm (OLA), have been applied. When the proposed method is applied, a reduced 6x2x1 network is obtained.

Fig. 9-5 and Fig. 9-6 show the training and iteration times using different number of hidden nodes in the proposed and the original Backpropagation algorithms. As it can be seen, although training time increases for large networks in both algorithms, the optimizing and learning method provides better results than the Backpropagation algorithm, even when the optimum number of hidden nodes is near to the initial number of hidden nodes.

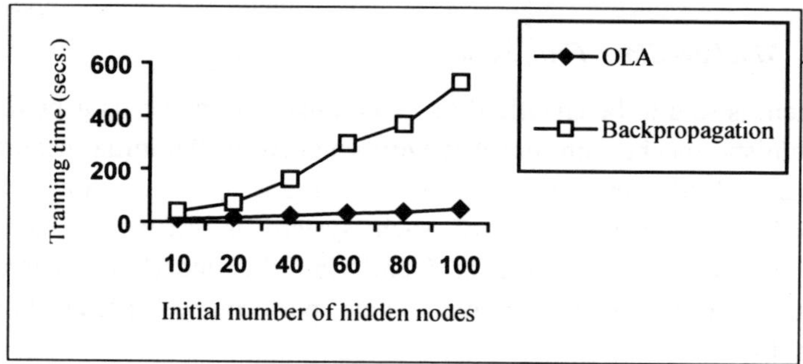

Fig. 9-5. Training time for several networks with different number of hidden nodes.

Fig. 9-7 shows the number of iterations required for network training. From the results obtained we can observe that small networks generally need less number of iterations than large networks to reach the desired convergence. However, the number of iterations used by OLA keeps almost invariant, providing better results than the Backpropagation algorithm in most of the experiments carried out.

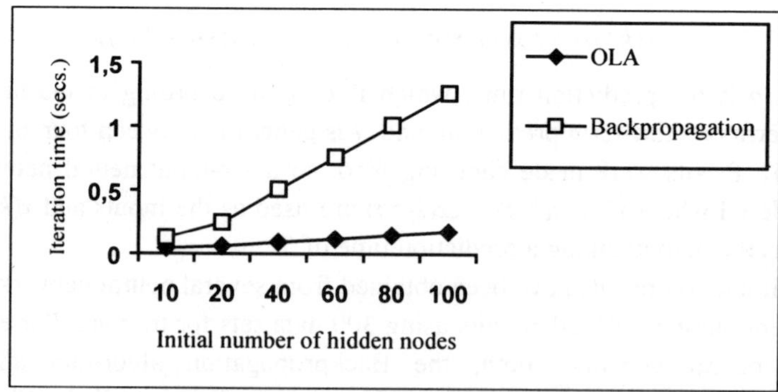

Fig. 9-6. Iteration time for MG series using networks with different number of hidden units.

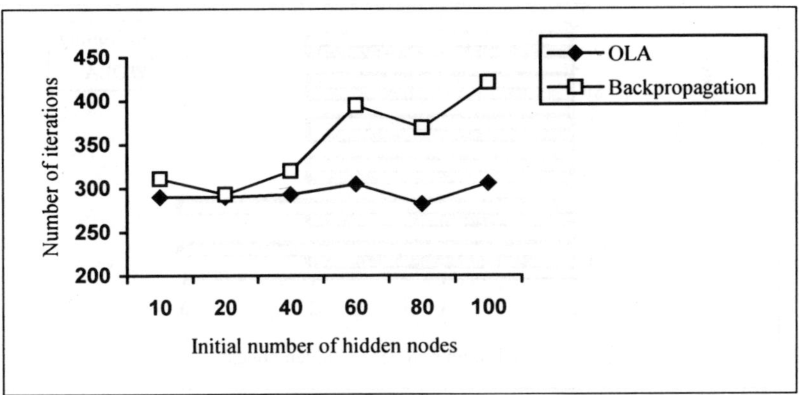

Fig. 9-7. Number of iterations required for the Backpropagation algorithm and the proposed method.

To assess the efficiency of the proposed method, the Mackey-Glass time series have been generated with a 6x100x1 neural network. Tests were made comparing a network trained using the Backpropagation algorithm with another network obtained by OLA. The results of Fig. 9-8 show that the performance of both methods is equally good.

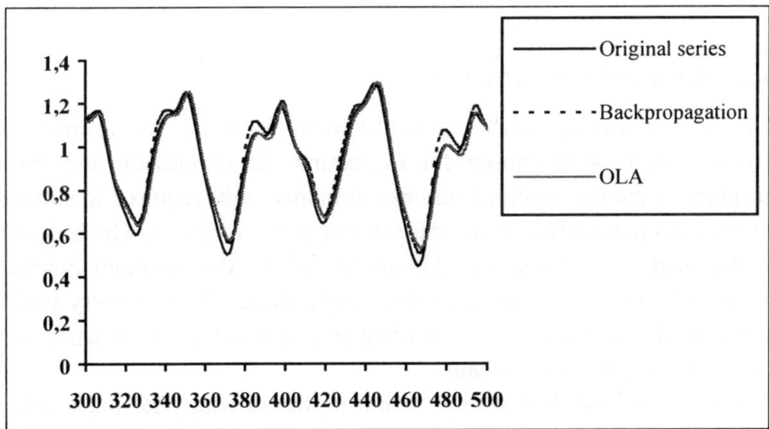

Fig. 9-8. Mackey-Glass series modelled using the Backpropagation algorithm and the optimizing and learning method.

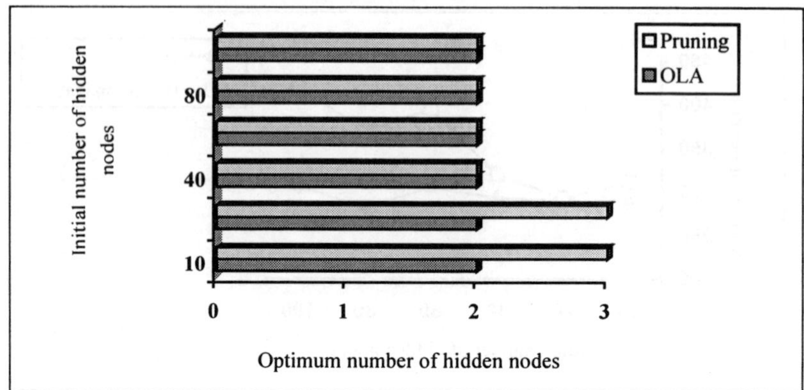

Fig. 9-9. Optimum number of hidden nodes obtained by OLA and the pruning after training method.

As results of the experiments, Fig. 9-9 exhibits the stability of OLA. Changing the initial number of hidden units between 10 and 100 the reduction results have not varied in OLA, though the pruning after training method shows some variations. Hence, this figure underlies that the initial number of nodes belonging to the hidden layer has no influence in the final network obtained by OLA. This means that the presented approach is suitable for networks with an initial large size.

9.6.2. *Parity and Symmetry Problems*

Parity and symmetry tasks are well-known problems for testing neural networks. They were chosen for evaluating the optimizing and learning algorithm, as the near-optimal number of hidden units required to solve both problems is known [15]. In the case of the parity problem, which detects if an input pattern contains an odd number of 1s, the optimum number of hidden nodes is equal to the number of input units. The symmetry problem, which classifies input strings as to whether or not they are symmetric, can be solved with only two hidden units.

Tests were made for several neural networks with different number of hidden units (from 10 to 60). Each network was trained using the optimizing and learning algorithm and the Backpropagation method (BP). Networks trained with the Backpropagation algorithm were pruned using SVD and

QRcp. There have been considered data of size 6 for both problems, using 40 patterns for training and the remaining 24 for validation.

Table 9-1. Numerical results for the parity problem.

	Training time (secs.)		Number of iterations		Optimum number of hidden nodes		Recognition rate (%)	
Initial size	***OLA***	***BP***	***OLA***	***BP***	***OLA***	***Prun-ing***	***OLA***	***BP***
10	15	29	1148	1551	6	9	87,5	82,81
20	19	64	1184	1802	6	11	85,94	76,56
30	29	87	1437	1593	6	13	85,94	79,68
40	35	147	1473	1961	7	13	84,37	81,25
50	40	139	1491	1522	7	13	84,37	81,25
60	46	195	1475	1839	7	13	84,37	73,43

Table 9-2. Numerical results for the symmetry problem.

	Training time (secs.)		Number of iterations		Optimum number of hidden nodes		Recognition rate (%)	
Initial size	***OLA***	***BP***	***OLA***	***BP***	***OLA***	***Prun-ing***	***OLA***	***BP***
10	8	18	1031	1117	4	7	92,19	85,94
20	13	44	1191	1404	4	8	92,19	84,37
30	14	56	1122	1209	4	9	95,31	84,37
40	16	76	1053	1201	4	10	93,75	84,37
50	21	97	1165	1233	4	9	95,31	84,37
60	21	136	1032	1505	4	9	95,31	87,5

The results of the experiments are summarized in Table 9-1 for parity and Table 9-2 for symmetry. As it can be observed, the time required to train the different networks by OLA is always smaller than the training time spent by Backpropagation. Moreover, trials show certain degree of independence on the number of iterations needed by OLA with regard to the initial number of hidden units. The results also exhibit the stability of OLA in terms of reduction. The proposed method always provides a near-optimal network, while the pruning after training technique shows a less suitable behavior, which depends on the initial size of the network. The last two columns of tables 1 and 2 show the recognition rates provided by the different size networks. Those values were measured as the proportion of examples for which network output differed from the corresponding target by less than 0.3. As it can be observed, recognition results show the improvement of the generalization capability when the networks are trained by OLA.

9.7. Conclusions

In this chapter, a method for optimizing the size of feedforward neural networks during the training process has been presented. The key idea of this approach consists of computing iteratively the optimum hidden nodes of the network and then updating only the weights connected to these nodes by the Backpropagation algorithm.

Householder reflections are applied to compute the optimal network size. This orthogonal transformation leads to relative decorrelation of network information using few operations, which allows computing the optimal set of hidden nodes during the training process without increasing significantly the training time.

The design of a systolic architecture that implements the proposed learning algorithm has been described. This architecture exploits the pipelined processing to maximize performance.

From experimental results, the benefits of using the optimizing and learning algorithm are twofold. Firstly, a significant reduction on training time can be observed and secondly an optimal hidden structure is provided at the end of the training process, avoiding any retraining phase. In addition, the number of iterations according to the initial hidden layer size is almost invariant. This implies that the presented approach could provide good results with some independence on the employed implementation techniques.

References

[1] T. Ash. Dynamic node creation in backpropagation networks. *Connection Sci.*, 1(4), pages 365-375, 1989.

[2] E. B. Baum and D. Haussler. What size net gives valid generalization?. *Neural computation*, 1, pages 151-160, 1989.

[3] G. Castellano, A. M. Fanelli and M. Pelillo. An iterative pruning algorithm for feedforward neural networks. *IEEE Transactions on Neural Networks*, 8(3), pages 519-531, 1997.

[4] S. E. Fahlman and C.Lebiere. The cascade-correlation learning architecture. *Advances in Neural Information Processing*, Ed. San Mateo, pages 524-432, 1990.

[5] S. I. Gallant. Optimal linear discriminants. *Proc. 8th Int. Conf. Pattern Recognition*, Paris, France, pages 849-852, 1986.

[6] G. H. Golub and C. F. Van Loan. *Matrix computations*. Baltimore, MD: John Hopkins Univ. Press, 1996.

[7] D. Hammerstrom. A highly parallel digital architecture for neural network emulation. *VLSI for Artificial Intelligence and Neural Network*, J. G. Delgado-Frias, W. Moore (eds.), pages 357-366, Plenum Press, 1991.

[8] P. Kanjilal and N. Banerjee. On the application of orthogonal transformation for the design and analysis of feedforward networks. *IEEE Transactions on Neural Networks*, 5(5), pages 1061-1070, 1995.

[9] S.Y. Kung and J.N Hwang. A unifying algorithm/architecture for artificial neural networks. *Proceedings of the International Conference on Application Specific Signal Processors*, pages 2505-2508, Edinburg, Scotland, 1989.

[10] A. Lapedes and R. Farder. *Nonlinear signal processing using neural networks*. Los Alamos National Lab. Tech. Rep. LA-UR-2662, 1987.

[11] A. Levin, T. K. Leen and J. E. Moody. Fast pruning using principal components. *Advances in Neural Information Processing Systems 6*, J. D. Cowan, G. Tesauro, and J. Alspector, Eds. San Francisco, CA: Morgan Kaufman Publishers, 1994.

[12] M. C. Mackey and L. Glass. Oscillations and chaos in physiological control systems. *Sci.*, 197, pages 287-289, 1977.

[13] H. A. Malki and A. Mogghaddamjoo. Using the Karhuren-Lorve transformation in the backpropagation training algorithm. *IEEE Transactions on Neural Networks*, 2(1), pages 162-165, 1991.

[14] M. Mézard and J. P. Nadal. *Learning in feedforward layered networks: The Tiling algorithm*. J. Phys. A, 22, pages 2191-2204, 1989.

[15] D. E. Rumelhart, G. E. Hinton and R. J. Williams. Learning internal representation by error propagation. *Parallel distributed processing*, 1, pages 318-362, D. E. Rumelhart and J. L. McClelland (Eds.), Cambridge, MA: MIT Press, 1986.

[16] M. F. Tenorio. Self-organizing network for optimum supervised learning. *IEEE Transactions on Neural Networks*, 1(1), pages 100-110, 1990

CHAPTER 10

PARALLEL ANN ARCHITECTURE FOR FUZZY PATTERNS

D. Zhang

The Centre for Multimedia Signal Processing and Department of Computing
Hong Kong Polytechnic University, Kowloon, Hong Kong
Tel: (852) 2766 7271 Fax: (852) 2774 0842
E-mail: csdzhang@comp.polyu.edu.hk

S.K. Pal

Machine Intelligence Unit Indian Statistical Institute
203, B.T. Road, Calcutta 700035, India
E-mail: sankar@isical.as.in

In this chapter, a system design methodology for fuzzy clustering neural networks (FCN) is presented. This methodology emphasizes coordination between FCN model definition, architectural description, and systolic implementation. Two mapping strategies both from FCN model to system architecture and from the given architecture to systolic arrays are described. In Section 10.2, an FCN model with the direct fuzzy competitive learning using a membership function for fuzzy clustering is presented. For mapping the FCN model onto parallel architecture, we have designed several processing cells and discussed their hardware complexity in Section 10.3. There also includes a comparison with FCM architecture with respect to hardware complexity. Section 10.4 describes some mapping strategies and typical systolic structures based on which the FCN architecture can be systematically implemented by the corresponding SAs. In Section 10.5, the contribution of the chapter is briefly summarized.

10.1. Introduction

For a set of data points in Euclidean space, a "hard clustering problem" is defined when each point is restricted to one of a given number of clusters, whereas a "fuzzy clustering problem" is defined when each point belongs to every cluster with some degree of membership. Historically, the latter problem evolved from the former. Ruspini [1] first introduced the problem of fuzzy clustering for which the objective is to determine the fuzzy classification of each pattern by minimizing some suitably defined function. Dunn [2] defined the first generalization of the conventional minimum-variance hard clustering. Bezdek [3] generalized Dunn's work into a family of fuzzy clustering problems. He developed an algorithm known as fuzzy c-means (FCM) algorithm and gave a comprehensive treatment of the problem. Some modified versions of the fuzzy clustering problem have also been proposed [4-6].

The problem has then been addressed in neural network research. A number of approaches for designing fuzzy clustering neural networks (FCN) have been considered. Simpson [7] has discussed on fuzzy min-max neural networks in fuzzy clustering. Pal, Bezdek and Tsao [8] have developed a fuzzy clustering network based on Kohonen network. Mitra and Pal [9] have described a self-organizing neural network which is capable of handling fuzzy input and of providing fuzzy classification. Jou [10] has implemented fuzzy clustering using fuzzy competitive learning networks. These FCN models can be specified by their network topology, node characteristics, and fuzzy competitive rules. The network topology defines how each node connects to other nodes. The node characteristics define the function which combines the various inputs and weights into a single quantity as well as the function which then maps this value to an output. The fuzzy competitive rule specifies an initial set of weights and indicates how weights should be adapted during use to accomplish the task at hand.

All these approaches are concerned with algorithms; their behaviours and characteristics are primarily investigated by simulation on general purpose computers. The fundamental drawback of such simulators is that the spatio-temporal parallelism inherent in the processing of information using neural networks is lost entirely or partly. Another drawback is that the computing time of the simulated network, especially for large associations of nodes tailored to application-relevant tasks, grows to such orders of magnitude that a speedy acquisition of neural "know-how" is hindered or made impossible. This makes the actual fuzzy clustering applications difficult to implement in real time. Therefore, it is essential to implement FCN in a VLSI medium.

However, it cannot be assumed that FCN models developed in computational neuroscience, at a high level, are directly implementable in silicon. This is because the technology, the physical devices and the circuits severely limit the performance of integrated FCN. Systolic array (SA) can offer flexibility, programmability, and precision in computation, coupled with the advantages of large pipelined throughput and local interconnections in VLSI implementation [13, 17-20]. So an efficient FCN system can be realized by considering the special features and aspects of the fuzzy clustering technology, designing their special purpose architectures, and mapping the neural networks onto the corresponding systolic arrays. As a result, an FCN system design methodology (See Fig. 10-1) is proposed here. The effectiveness of the methodology is illustrated by designing a parallel FCN model, developing the corresponding special-purpose architecture and building the systolic arrays suitable for VLSI implementation.

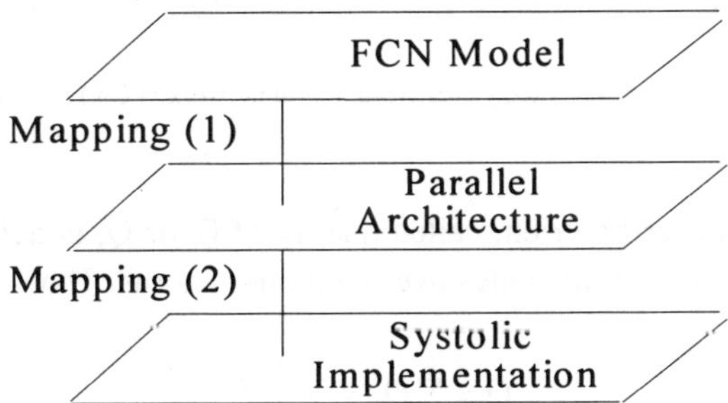

Fig. 10-1. A system design methodology for FCN implementation.

10.2. Fuzzy Clustering Neural Networks (FCN) Model

A basic FCN model using clustering competitive network is illustrated in Fig. 10-2. Each node represents a fuzzy cluster and the connecting weights from the inputs to a node represent the exemplar of that fuzzy cluster. The square of the Euclidean distance between the input pattern and the exemplar is passed through a Gaussian nonlinearity. The output of the node, therefore, represents the closeness of the input pattern to the exemplar. The degree of possibility that each input pattern belongs to different fuzzy clusters is calculated in the final membership level.

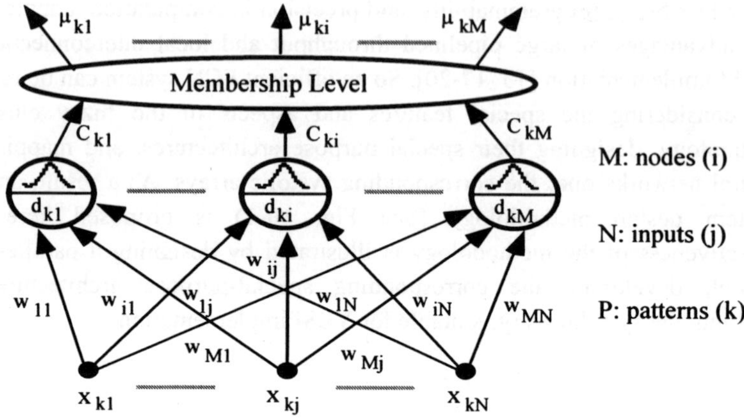

Fig. 10-2. Fuzzy clustering neural network (FCN) model.

A fuzzy cluster criterion, called quality of fit or Q, is defined as the sum of all output values of the nodes over all input patterns. That

$$Q = \sum_{k=1}^{P} Q_k = \sum_{k=1}^{P} \sum_{i=1}^{M} C_{ki} \tag{10-1}$$

where P and M are the respective numbers of input patterns and nodes, and C_{ki} is the output of node i when the input pattern is $x_k = (x_{k1}, x_{k2}, ..., x_{kN})$,

$$C_{ki} = \exp\left(-\frac{(d_{ki})^2}{2\sigma^2}\right). \tag{10-2}$$

The weight vector connecting the inputs to node i is $w_i = (w_{i1}, w_{i2}, ..., w_{iN})$. The Euclidean distance between w_i and x_k is defined as

$$(d_{ki})^2 = \sum_{j=1}^{N} (x_{kj} - w_{ij})^2 . \tag{10-3}$$

The weight vectors, w_i ($i = 1, 2, ..., M$), can also be viewed as the parameters of the Gaussian functions that determine their locations in the input space. Since the fuzzy clusters are high concentrations of the input patterns in the input space, locating the weight vectors at or close to the centers of these concentrations will insure a maximum for the quality of fit criterion.

This is clearly an optimization problem where the objective function is the Quality of fit and the variables are the coordinates of the centers of the Gaussian functions, i.e. the weight vectors. The change in the weight on the objective function (Eq. (10-1)) is

$$\Delta w_{ij} = \eta \frac{\partial Q}{\partial w_{ij}} = \eta \frac{\partial}{\partial w_{ij}} \left\{ \sum_{k=1}^{P} \sum_{i=1}^{M} C_{ki} \right\} \tag{10-4}$$

This is equal to:

$$\Delta w_{ij} = \eta \sum_{k=1}^{P} \frac{\partial C_{ki}}{\partial w_{ij}} \tag{10-5}$$

where the second summation, $\sum_{i=1}^{M}$, was dropped since w_{ij} appears only. Note that

$$\frac{\partial C_{ki}}{\partial w_{ij}} = \frac{\partial C_{ki}}{\partial d_{ki}} \frac{\partial d_{ki}}{\partial w_{ij}} \tag{10-6}$$

and

$$\frac{\partial C_{ki}}{\partial d_{ki}} = \frac{-1}{2\sigma_i^2} C_{ki} \tag{10-7}$$

$$\frac{\partial d_{ki}}{\partial w_{ij}} = -2(x_{kj} - w_{ij}) \tag{10-8}$$

This means

$$\Delta w_{ij} = \eta \sum_{k=1}^{P} \frac{1}{\sigma^2} C_{ki} (x_{kj} - w_{ij}) \tag{10-9}$$

where η is a constant of proportionality and σ^2 is the variance of the function of the node. It is clear that this formulation utilizes local information available at the weight itself and the node it is connected to. There is no need for an external orientation subsystem to decide which weights are to be increased since all the weights are adapted after accumulating the errors over all input patterns. The amount of change in the weight is a function of the distance between the input pattern and the weight vector. Thus, patterns belonging to different fuzzy clusters do not disturb the weights significantly, and patterns very close to the exemplar will not disturb its convergence.

To introduce a fuzzy competition mechanism between the nodes, a membership function for fuzzy clustering is required. In other words, a partition of the input pattern space, X = {x1, x2, ..., xP} ⊂ ℜN, into fuzzy clusters, Zi (i = 1, 2, ..., M), is associated with the membership functions μZi : X → [0,1]. Assignment of input patterns to different clusters can be given in terms of a fuzzy cluster membership matrix U = [ki], where the element ki denotes the degree of possibility that the input pattern k belongs to the fuzzy cluster i,

$$\mu_{ki} = \mu_{Z_i}(x_k) = \frac{C_{ki}}{Q_k} \tag{10-10}$$

It is evident that the elements of *U* are subject to $P > \sum_{k=1}^{P} \mu_{ki} > 0$ and $\sum_{i=1}^{M} \mu_{ki} = 1$, where $1 \leq i \leq M$ and $1 \leq k \leq P$.

Using the fuzzy cluster membership element ki to participate in the corresponding weight change, a fuzzy competitive learning update rule which moves the weight vectors towards their respective fuzzy cluster centers can be represented as

$$\Delta w_{ij} = \eta \sum_{k=1}^{P} \frac{1}{\sigma^2} C_{ki} \left(x_{kj} - w_{ij}\right) \mu_{ki} \tag{10-11}$$

We can obtain a variation of this learning algorithm from Eq. (10-1) by letting

$$\Delta_k w_{ij} = \eta \frac{\partial Q_k}{\partial w_{ij}} \tag{10-12}$$

Note that the direction of the gradient only guarantees a locally increasing direction. To avoid instability of the algorithm, the step taken in the direction is usually chosen to be very small by the control parameter, η. Thus, if η is sufficiently small, $\Delta w_{ij} \approx \sum_{k=1}^{P} \Delta_k w_{ij}$ This means the change in weights will be approximately equal to Δw_{ij} if the weights are updated immediately after the presentation of an input pattern. Considering the approximation, Eq. (10-11) becomes

$$\Delta_k w_{ij} = \eta \frac{1}{\sigma^2} C_{ki} (x_{kj} - w_{ij}) \mu_{ki} \tag{10-13}$$

A number of experiments have been conducted to investigate various aspects of both the fuzzy competitive learning algorithm and the FCN model [4-10].

Simulations show that the clustering results of the FCN can be similar to ones of the FCM [22]. However, unlike the FCM algorithm where the selection of the number of cluster or nodes, M, is critical to the results, the FCN requires an overestimate of the actual number of clusters. The FCN always converges to the natural number of clusters while the FCM will force exactly M clusters.

10.3. Parallel Architecture

10.3.1. *Mapping Strategies*

An FCN model is usually specified by some model functions with different processing phases, such as fuzzy competitive rule using membership level. The fuzzy competitive rule indicates an initial set of weights and how weights should be adapted during use to accomplish the task at hand.

An FCN architecture is, however, specified by its network topology and node characteristics. The network topology defines how each node is connected to other nodes. The node characteristics define the function which combines the various inputs and weights into a single quantity as well as the function which then maps this value to an output.

Considering parallel network topology, some mapping strategies from an FCN model to the architecture are defined in the following (See Fig. 10-3):

Model Structure Mapping:

In the FCN model, each function, like competitive and membership function, is mapped as an independent processing layer and their connection patterns within and between layers are defined. This means that all model functions will be arranged into corresponding layers.

Processing Phase Mapping:

Often two processing phases, searching and learning, are given in the FCN model. They are able to be performed in architecture design by special control paths, i.e., feedforward and feedback path, respectively.

Computing Unit Mapping:

Depending on different functional arithmetic types in each processing layer, such as weight computing and pattern summing, their processing cells can be built to achieve the given input / output functions.

10.3.2. *FCN Architecture: Processing Cells*

Based on these mapping strategies, the model described in Section 10.2 can be mapped onto its FCN architecture. Two kinds of operations, the competitive function in Eq. (10-2) and the membership function in Eq. (10-10), are defined by each independent layer in feedforward processing. Notice that an adder between these two layers is needed to generate Q_k in Eq. (10-1). However, the fuzzy competitive learning update rule in Eq. (10-11) can be arranged in feedback path, since some parameters required for changing cluster center come from the feedforward processing.

The FCN architecture comprising three kinds of processing cells, including weight, node, and output, plus adder (Σ), is obtained in Fig. 10-4. We have embedded all functions from the FCN model in this architecture so that on-line learning and parallel implementation are feasible.

Both node cell and adder (Σ) can be achieved by many current approaches [11-13]. Thus, the following discussion will focus on the other two cells. Their structures are defined in Fig. 10-5, 10-6, where a circle indicates an operator and a square a memory.

The output processing cell is used to obtain both the membership element, μ_{ki}, in Eq. (10-10) and the partial product for the fuzzy competitive learning update rule in Eq. (10-13). The two inputs, C_{ki} and Q_k, come from the outputs of node i and the adder (Σ) when the input pattern is $\boldsymbol{x}_k$. The output of the partial product is

$$S_{ki} = AC_{ki}\mu_{ki} \tag{10-14}$$

where A is a control parameter for the operator. The output cell can be built using multiplier and divider operators (See Fig. 10-5). A weight cell is used to store and change weight value as well as implement the related arithmetic (See Fig. 10-6). It is mainly composed of four different memory elements, i.e., accumulation memory ($\Sigma\Delta_k w_{ij}$), weight memory (w_{ij}), difference memory (g_{ijk}) and its square memory ($g^2{}_{ijk}$), and five operators, including adders, subtracter, multiplier and the unit that generates square of number. In the feedforeward processing, an input, x_{kj}, is given and the outputs of the cell are represented as:

$$g_{ijk}^2 = (x_{kj} - w_{ij})^2 \tag{10-15}$$

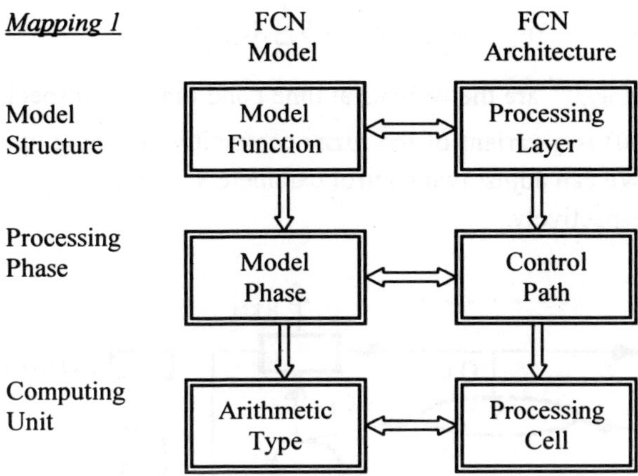

Fig. 10-3. Mapping strategies from FCN model to architecture.

In the feedback processing, the input S_{ki} is to come from the output cell and its corresponding arithmetic to implement the update rule in Eq. (10-13) is

$$\Delta_k w_{ij} = BS_{ki} g_{ijk}, \tag{10-16}$$

where B is the control parameter of the multiplier in the cell, and g_{ijk} is obtained in the previous processing and stored in the difference memory (g_{ijk}). We use the update rule in Eq. (10.11) in batch mode: the changes Δ_k over all input patterns are accumulated before updating the weight. That is

$$\Delta w_{ij} = \sum_{k=1}^{P} \Delta_k w_{ij}. \tag{10-17}$$

In this way, we can implement the Euclidean distance between weight vector $\boldsymbol{w}_i$ and input pattern $\boldsymbol{x}_k$ in the FCN architecture with Eq. (10-3) rewritten as

$$(d_{ki})^2 = \sum_{j=1}^{N} g_{ijk}^2, \tag{10-18}$$

where $g^2{}_{ijk}$ is as an input from the weight cell, w_{ij}, to the node cell i. The weights in the FCN architecture are changed in terms of the following functions:

$$w_{ij}^t = w_{ij}^{t-1} + \Delta w_{ij} \tag{10-19}$$

and

$$\Delta w_{ij} = B\sum_{k=1}^{P} S_{ki} g_{ijk} = B\sum_{k=1}^{P} AC_{ki}\mu_{ki} g_{ijk} \tag{10-20}$$

where w_{ij}^{t} and w_{ij}^{t-1} are the weights at time t and time $t-1$, respectively. Note that Eq. (10-20) is a variant of the fuzzy competitive learning update rule in Eq. (10-11). We can adjust two control parameters, A and B, equal or close to $1/\sigma^2$ and η, respectively.

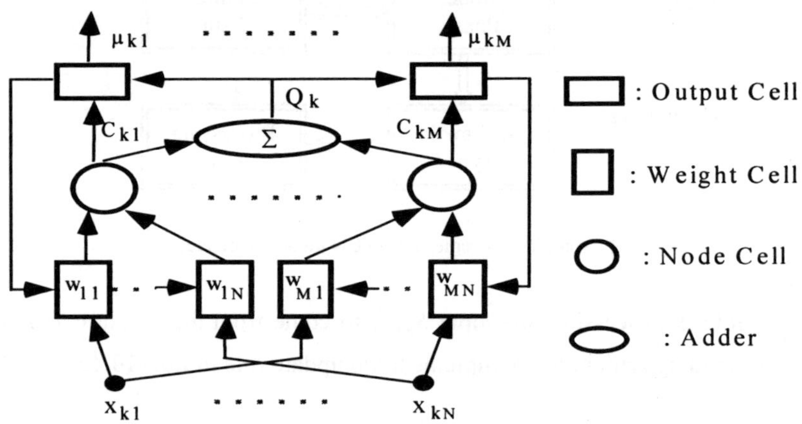

Fig. 10-4. Parallel FCN architecture with three processing cells plus adder.

10.3.3. *Performance Analysis*

Our goal is to implement FCN in VLSI. Using the current VLSI technology, the FCN architecture can be flexibly built by only three types of processing cells plus an adder. Each cell, which is built from some operators and memories, can be easily implemented using either a digital or a hybrid (combining digital with analog) approach.

The hardware complexity given by the design is usually considered as an important measure for VLSI implementation. Based on the FCN architecture, hardware complexity can be represented as

$$H_{FCNN} = M(NH_{weight} + H_{output} + H_{node}) + H_{\Sigma}, \tag{10-21}$$

where N and M are the dimensions of the input and output spaces, respectively; H_{Σ} is the complexity of the adder (Σ) in Fig. 10-4; H_{weight}, H_{output} *and* H_{node} are the complexities of three different cells, respectively.

Note that H_{FCN} is independent of P, the number of the input patterns, and H_{Σ} has a linear complexity in the number of connections of the nodes, M. Therefore the attached cost of direct competition in the FCN architecture, H_{Σ}, can be compared with the other competitive architectures, such as the MAXNET [14] with the connective complexity, M(M-1) / 2. This means that the connective cost of direct competition in the FCN is reduced by a factor of (M-1) / 2.

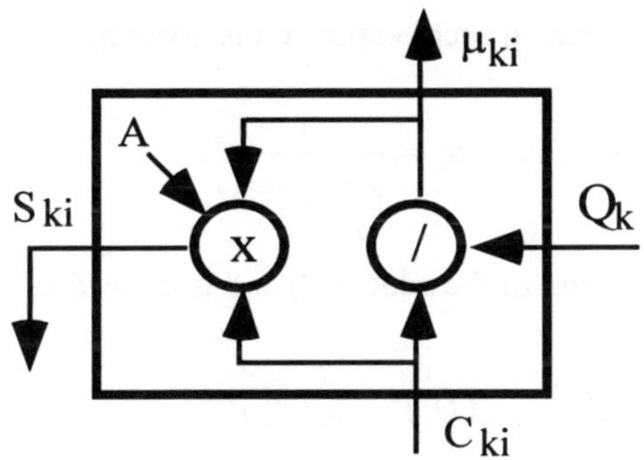

Fig. 10-5. FCN architecture: the output cell structure.

In order to analyze the effectiveness of the FCN architecture, we also take the architecture of fuzzy c-means (FCM) as our comparative target. The objective function for the FCM is given by [2-3]

$$J = \tfrac{1}{2}\sum_{i=1}^{M}\sum_{k=1}^{P}(\mu_{ki})^{\beta}(d_{ki})^{2} \tag{10-22}$$

where its membership function is

$$\mu_{ki} = \frac{1}{\sum_{l=1}^{M}(\frac{d_{ki}}{d_{kl}})^{2/(\beta-1)}} \tag{10-23}$$

Thus, the fuzzy competitive learning update rule to minimize Eq. (10-22) can be obtained as

$$\Delta_k w_{ij} = \eta\gamma_{\beta}(x_{kj} - w_{ij}) \tag{10-24}$$

where

$$\gamma_\beta = (\mu_{ki})^\beta [1 - \beta(\beta - 1)^{-1}(1 - \mu_{ki})] \tag{10-25}$$

with $\beta \in [1, \infty)$.

Like the FCN architecture in Fig. 10-4, the FCM can be implemented using the architecture shown in Fig. 10-7. It is evident that these two architectures have identical structure and the same number of building elements, but they do differ in the complexities of the three kinds of cells. The node cell for the FCM architecture is characterized by the Euclidean distance $(d_{ki})^2$ in Eq. (10-3) rather than the Gaussian nonlinearity C_{ki} in Eq. (10-2). The function characterizing each output cell is given by

$$\mu_{ki} = \frac{1}{(\frac{d_{ki}}{D_k})^{2/(\beta-1)}} \tag{10-26}$$

where $(D_k)^2$ is the output of the adder (Σ) and is defined as

$$(D_k)^2 = \sum_{i=1}^{M} (d_{ki})^2 \tag{10-27}$$

The updating rule for the weight cell is represented by Eq. (10-24). Except for the node cell, the other two cells in the FCM architecture are more difficult to implement in VLSI than the corresponding ones in the FCN architecture. This is due to the use of more complicated functions; especially in the weight cell where nearly five more operators including multiplier, subtractors, and power exponent, will be needed. However, the Gaussian function defined in the node cell in the FCN architecture can be replaced with two sigmoid functions which are easier to implement. One way is to use the difference of the sigmoid functions.

The definitions of the cells in the two architectures, discussed above, are summarized in Table 10-1. Here the number of the elements required is indicated in parentheses.

10.4. Systolic Array Design

10.4.1. *Mapping Strategies*

Systolic arrays (SAs) is regular interconnected arrays using a set of building elements, each performing some simple operation, where the data flow in a rhythmic fashion with only local interconnects between elements. It has already been shown that SAs provide a good medium to implement neural networks in VLSI [15-16]. Numerous researchers have proposed SAs for different neural network applications including signal / image processing [18], supercomputer [19], and content-addressable memory [20]. In this section, we present mapping strategies from FCN to SAs with implementation efficiency as our goal.

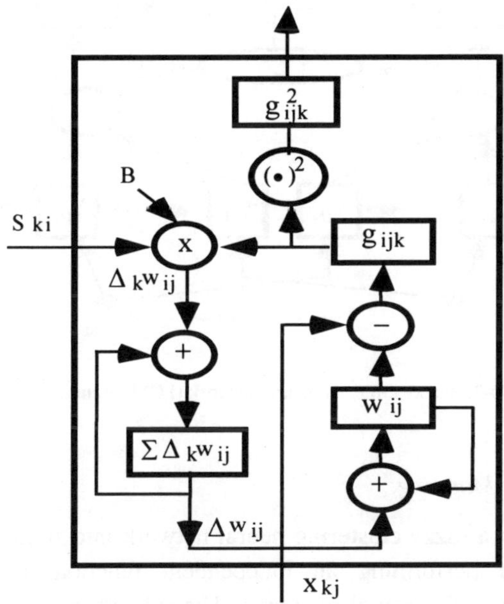

Fig. 10-6. FCN architecture: the weight cell structure.

As previously stated, the FCN described in Section 10.2 is specified by its network topology, node characteristics, and fuzzy competitive rule. The network topology defines how each node connects to other nodes. The node characteristics define the function which combines the various inputs and weights into a single quantity as well as the function which then maps this value to an output. The fuzzy competitive rule specifies an initial set of weights and indicates how weights should be adapted during use to accomplish the task at hand.

It is clear that the complexity of the FCN stems not from the complexity of its nodes but rather from the multitude of ways in which a large collection of these nodes can interact. Therefore, an important task is to build highly parallel, regular and modular SAs that are attractive for VLSI techniques. Mapping from the FCN architecture to SA implementation can be achieved by the following three types of mappings (See Fig. 10-8):

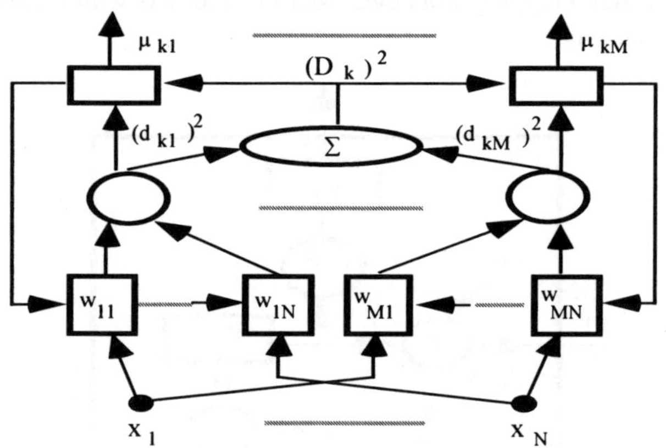

Fig. 10-7. Fuzzy c-means neural network (FCM) architecture.

Processing Mode Mapping:

Here we partition a fuzzy clustering neural network into some basic subnets, each capable of performing an independent function. Often a subnet represents a layer in the neural networks. The subnets are implemented by a corresponding SA, which are then cascaded according to the architectural definition.

Computing Property Mapping:

Each basic subnet function is reduced to a recursive form which is implemented by the corresponding pipeline matrix in terms of the systolic rules. In practice, this mapping transforms spatial parallelism to temporal parallelism.

Arithmetic Module Mapping:

A basic operation in recursive arithmetic is implemented by a building element. For example, a node can be divided into two parts: forming a weighted sum of N inputs and passing the result through a nonlinearity. The weighted sum can easily be integrated by a two-dimension recursive matrix using weight processing elements. To form the nonlinearity, a special element can be defined which may be cascaded with the recursive matrix as a bound node of its output.

10.4.2. *Typical FCN Subnets*

There have been a variety of topologies that FCNs use. They can be designed as different SA computing subnet structures. Some basic subnets in FCN and their typical SAs are defined as follows.

SA Computing Model:

A basic computing model for fuzzy clustering can be extracted from the self-organizing scheme presented by Kohonen [21]. The model, comprising a number of nodes connected in certain interconnection pattern, is defined as

$$d = f\left\{\sum_{j=1}^{N}(x_j - w_j)^2\right\}, \tag{10-28}$$

where w_j is the weight of the jth input x_j and f is a nonlinear function. The SA structure with 1×N processing elements (PEs), each defined as $Z = z + (x-w)^2$, plus a nonlinear function cell, is shown in Fig. 10-9. Here z is an intermediate result and the ring denotes a bound node.

Feed-Forward Subnet:

This subnet is often used as a computing layer, such as layer processing in multi-layer perceptrons or the calculation of membership function. A typical example is to use as a competitive layer in the learning vector quantization (LVQ) clustering algorithm [8]. This model can be expressed as

$$d_i = \sum_{j=1}^{N}(x_j - w_{ij})^2 \tag{10-29}$$

where $1 \leq i \leq M$. The subnet and its SA structure with M × N PEs are shown in Fig. 10-10. Two different functions of the PE are defined as x' = x and Z = z + $(x-w)^2$, respectively.

Feed-Back Subnet:

This class of subnets is frequently used in competitive learning. Typical example is the Maxnet competitive layer. The general expression for the subnet is

$$y_i(t+1) = \sum_{j=1}^{M} w_{ij} y_j(t), \tag{10-30}$$

where $1 \leq i, j \leq M$ and $y_j(t)$ is the output of node j at time t. The subnet and associated SA structure with M x M PEs are shown in Fig. 10-11, where each PE is defined as Y = z + yw.

10.4.3. *FCN Systolic Arrays*

Based on the given mapping strategies and basic computing subnets mentioned above, the FCN architecture obtained in Section 10.3 can be systematically implemented by the corresponding SAs (See Fig. 10-12), where two kinds of data flow paths, feedforward and feedback, are indicated in different lines. Each processing layer in the FCN architecture is achieved by the different SAs. The number of the PEs required for each SA is indicated in parentheses in Fig. 10-12. Since two simple 1-D SAs, output SA and node SA, can be easily built, we will concentrate here only on the discussion of 2-D SA.

One feedforward SA with M × N weight PEs and M node PEs is shown in Fig. 10-13, where the simple weight PE is defined in Fig. 10-10 (b). Arranging properly the input data flow, X, the corresponding output of the nodes, C_{ki}, can be obtained by this systolic array.

The other 2-D SA, i.e. the triangle SA with M adder PEs and $M[M-1]/2$ shifting registers, is shown in Fig. 10-14, where an adder PE is defined as: $S = r + s$ and r' = r. When the data flow, {C_{ki}}, which comes from the feedforward SA in Fig. 10-13, enters this array, each adder can accumulate its corresponding ΣC_{ki}. Note that after M steps both Q_k and $\{C_{ki}\}$, $1 \le i \le M$, $1 \le k \le P$, are available at the same time.

Table 10-1. Comparison of the progressing elements between FCN architecture and FCM architecture.

	FCN	*FCM*
Weight Cell *(M × N)*	$g_{ijk}^2 = (x_{kj} - w_{ij})^2$ $\Delta w_{ij} = \eta \Sigma S_i g_{ijk}$	$g_{ijk}^2 = (x_{kj} - w_{ij})^2$ $\Delta w_{ij} = \eta \sum (\mu_{ki})^{\beta} [1 - \beta(\beta-1)^{-1}(1-\mu_{ki})] g_{ijk}$
Node Cell *(M)*	$C_{ki} = \exp\left\{-\sum_{j=1}^{N} g_{ijk}^2 / 2\sigma^2\right\}$	$(d_{ki})^2 = \sum_{j=1}^{N} g_{ijk}^2$
Output Cell *(M)*	$\mu_{ki} = \frac{C_{ki}}{Q_k}$ $S_{ki} = \frac{1}{\sigma^2} C_{ki} \mu_{ki}$	$\mu_{ki} = \frac{1}{(\frac{d_{ki}}{D_k})^{2/(\beta-1)}}$
Adder (1)	$Q_k = \sum_{i=1}^{M} C_{ki}$	$(D_k)^2 = \sum_{i=1}^{M} (d_{ki})^2$

Considering to embed a feedback path, i.e., learning data flow, into the weight update arithmetic, a weight PE in systolic implementation is given in Fig. 10-15. It is mainly composed of four memory elements, including accumulation memory ($\Sigma \Delta_k w_{ij}$), weight memory (w_{ij}), difference memory (g_{ijk}) and its exponent memory (g_{ijk}^2), and four kinds of operators, i.e., adder, subtractor, multiplier and the unit that generates square of a number.

It is evident that when an input pattern, x_{kj}, passes through a weight PE, the outputs satisfy the given functions: $x'_{kj} = x_{kj}$ and $Z = z + g_{ijk}^2 = z + g_{ijk}^2 = z + (x_{kj} - w_{ij})^2$. Also the update rule defined in Eq. (10-16) is achieved as $\Delta_k w_{ij} = BS_{ki} g_{ijk}$, where B is the control parameter of the multiplier in the PE;

S_{ki} is an intermediate result which comes from the output PE in Fig. 10-5; and g_{ijk} is obtained in the previous phase and stored in the difference memory (g_{ijk}). We have used the update rule in batch mode, i.e., $\Delta w_{ij} = \sum_{k=1}^{P} \Delta_k w_{ij}$, which accumulates the changes Δ_k over all input patterns before updating the weight.

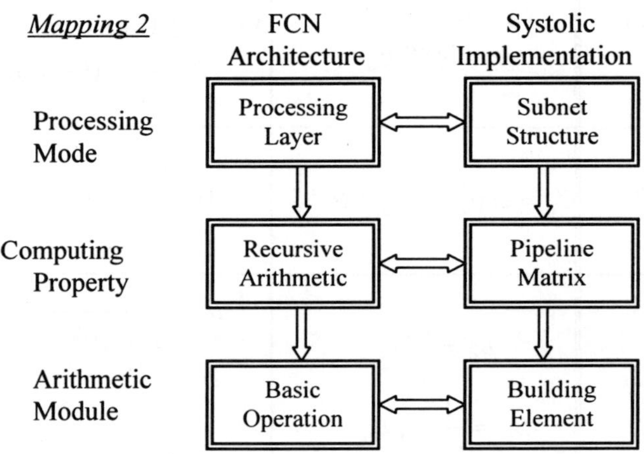

Fig. 10-8. Mapping strategies from FCN model to architecture.

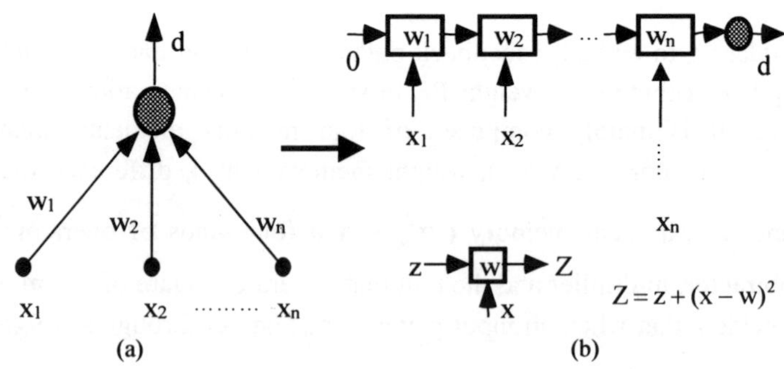

Fig. 10-9. (a) A basic FCN computing model. (b) the corresponding SA structure.

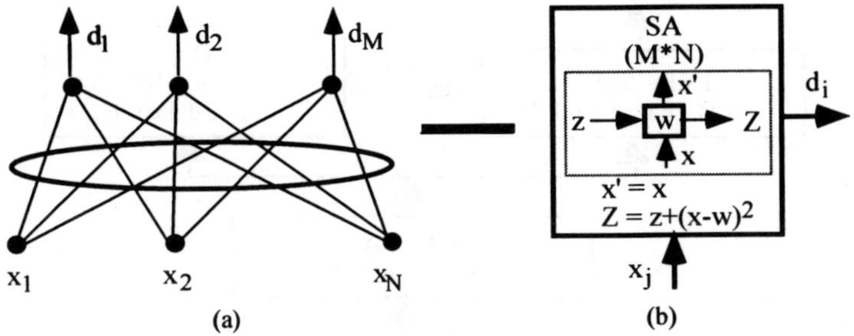

Fig. 10-10. (a) A typical feedforward subnet model. (b) The corresponding SA structure.

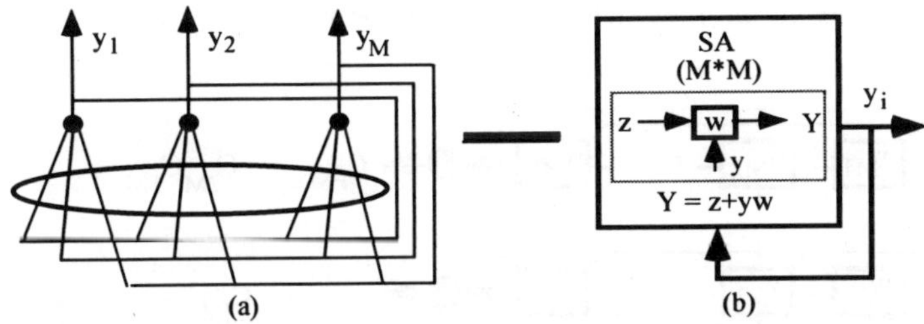

Fig. 10-11. (a) A typical feedback subnet model. (b) The corresponding SA structure.

10.5. Conclusions

In this chapter, we have presented a novel system design methodology for FCN. This methodology offers flexibility, programmability, and precision in computation, coupled with the advantages of large pipelined throughput and local interconnections. Two mapping strategies from FCN model to architecture and from architecture to SA implementation are described. The effectiveness of the methodology is illustrated by applying the design to an effective FCN model, developing the corresponding parallel architecture, comparing with FCM architecture, and building the SAs suitable for VLSI implementation. Although we have considered here the problems of clustering, one may use the said VLSI design methodology to other problem domains handled within a neuro-fuzzy framework.

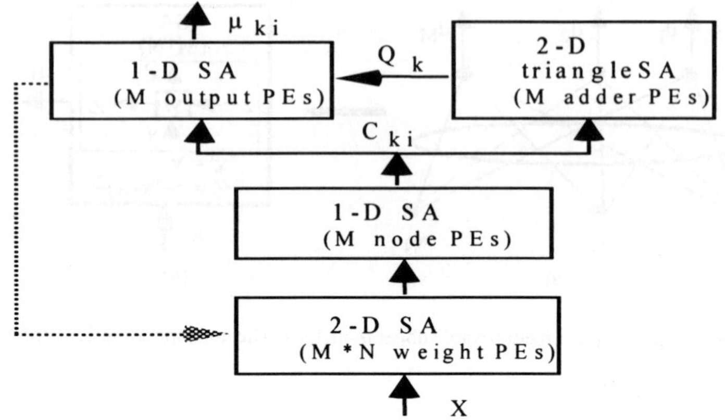

Fig. 10-12. The systolic array system implementation for FCN.

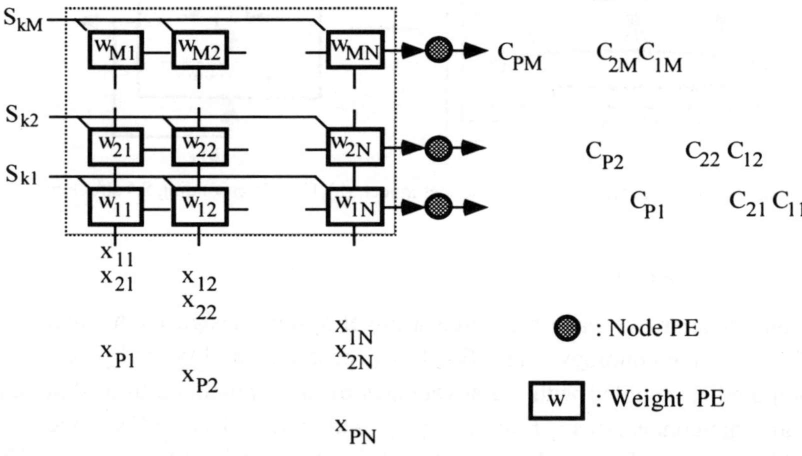

Fig. 10-13. Input / output data flow arrangement for the feedforward SA.

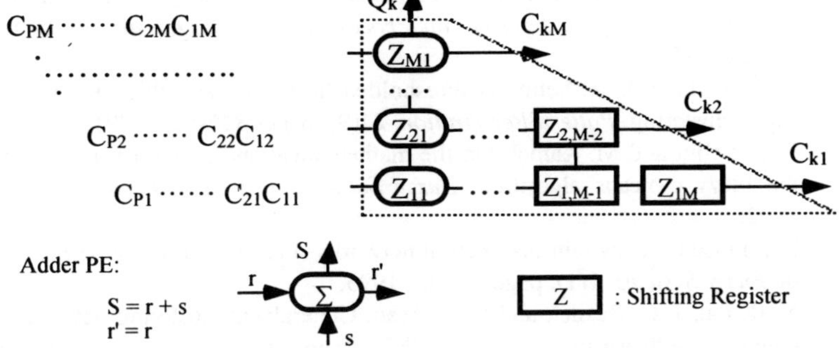

Fig. 10-14. The triangle SA with *M* adders and *(M[M-1])/2* registers.

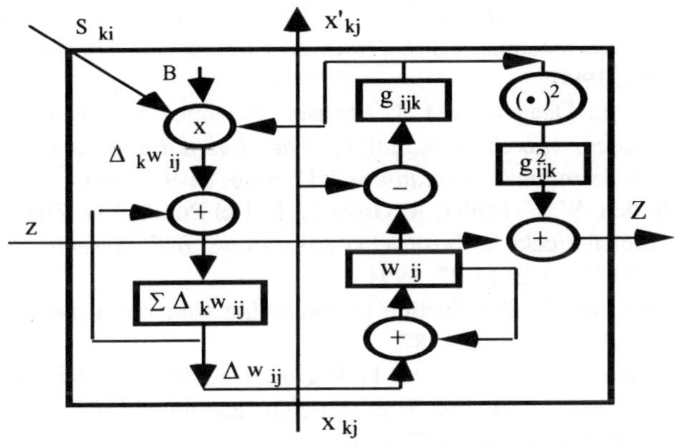

Fig. 10-15. Weight PE structure in systolic implementation.

References

[1] E. Ruspini. Numerical methods for fuzzy clustering. *Information Sciences,* 2, pages 319-350, 1970.

[2] J. C. Dunn. *A fuzzy relative of the ISODATA process and its use in detecting compact well separated clusters*. J. Cybernets, 3, pages 32-57, 1974.

[3] J. C. Bezdek. A convergence theorem for the fuzzy ISODATA clustering algorithms. *IEEE Trans. Pattern Anal. Machine Intelligence,* 2(1), pages 1-8, 1980.

[4] M. Kamel, B. Hadfield and M. Ismail. Fuzzy query processing using clustering techniques. *Information Processing & Management,* 26(2), pages 279-293, 1990.

[5] M. Kamel and S. Z. Selim. A thresholded fuzzy c-means algorithm for semi-fuzzy clustering. *Pattern Recognition,* 24(9), pages 825-833, 1991.

[6] S. Z. Selim and M. Kamel. On the mathematical and numerical properties of the fuzzy c-means algorithm. *Fuzzy Set and Systems,* 49, pages 181-191, 1992.

[7] P. Simpson. Fuzzy min-max neural networks--Part 2: Clustering. *IEEE Trans. on Fuzzy Systems,* 1(1), pages 32-45, 1993.

[8] N. R. Pal, J. C. Bezdek and E. C. Tsao. Generalized clustering networks and kohonen's self-organizing scheme. *IEEE Trans. Neural Networks,* 4(4), pages 549-557, 1993.

[9] S. Mitra and S. K. Pal. Self-organizing neural network as a fuzzy classifier. *IEEE Trans. on Systems, Man, and Cybernetics,* 24(3), pages 385-389, 1994.

[10] C. C. Jou. Fuzzy clustering using fuzzy competitive learning networks. in *Proc. IEEE IJCNN'92,* pages 714-719, 1992.

[11] D. Zhang, G. A. Jullien and W. C. Miller. A neural-like network approach to finite ring computations. *IEEE Trans. on Circuits and Systems,* 37(8), pages 1048-1052, 1990.

[12] D. Zhang, L. Deng and M. I. Elmasry. Pipelined architecture for neural-network-based speech recognition. *The International Journal: Neural, Parallel & Scientific Computations,* 2(1), pages 81-92, 1994.

[13] G. A. Jullien, W. C. Miller, R. Grondin, L. Del Pup and D. Zhang. Dynamic computational blocks for bit-level systolic arrays. *IEEE Journal of Solid-State Circuits,* 29(1), pages 14-22, 1994.

[14] R. P. Lipmann. An introduction to computing with neural nets. *IEEE ASSP Magazine,* 4, pages 4-22, 1987.

[15] C. Lehmann, M. Virredaz and F. Blayo. A generic systolic array building block for neural network with on-chip learning. *IEEE Trans. Neural Networks,* 4(3), pages 400-407, 1993.

[16] J. Chung, H. Yoon and S. R. Maeng, A systolic array exploiting the inherent parallelisms of artificial neural networks. *Microprocessing & Microprogramming,* 33(3), pages 145-159, 1992.

[17] D. Zhang. *Parallel VLSI Neural System Designs*, Springer-Verlag. Singapore, 1998.

[18] D. Zhang and M. I. Elmasry. Mapping neural networks onto systolic arrays. *The Journal: Neural, Parallel & Scientific Computations* 4(3), pages 341-352, 1996.

[19] K. G. Margaritis and D. J. Evans. Systolic implementation of neural networks for searching sets of properties. *Parallel Computing,* 18(3), pages 325-334, 1992.

[20] S. Y. Kung. *Digital Neural Networks,* Prentice Hall, 1993.

[21] T. Kohonen. *Self-Organization and Associative Memory,* Springer-Verlag. Berlin, 1989.

[22] D. Zhang, M. Kamel and M.I. Elmasry. Fuzzy clustering neural network (FCNN): Competitive learning and parallel architecture. *The Journal of Intelligent and Fuzzy Systems,* 2, pages 289-298, 1994.

CHAPTER 11

PIPELINED SYSTOLIC ARRAYS FOR TIME-DELAY NEURAL NETWORKS

D. Zhang

The Centre for Multimedia Signal Processing and Department of Computing
Hong Kong Polytechnic University, Kowloon, Hong Kong
Tel: (852) 2766 7271 Fax: (852) 2774 0842
E-mail: csdzhang@comp.polyu.edu.hk

S.K. Pal

Machine Intelligence Unit Indian Statistical Institute
203, B.T. Road, Calcutta 700035, India
E-mail: sankar@isical.ac.in

In this chapter, we describe a methodology for parallel time-delay window computing by considering the features and characteristics of such neural speech recognition systems. A model for time-delay window computing and its corresponding architecture definition are described in Section 11.2. Two kinds of processing stages used in pipelined architecture and their building elements are explained in Section 11.3. In Section 11.4, some mapping strategies from the window computing model into systolic array structures are defined. Three typical speech recognition applications and their performance analysis by parallel window computing are given in Sections 11.5 and 11.6, respectively. A brief conclusion is included in Section 11.7.

11.1. Introduction

Artificial neural networks (ANN), as processors of time-sequence patterns, have been successfully applied to several speaker-dependent speech recognition problems [1-14]. A variety of neural speech recognition algorithms has been developed. Numerous studies have demonstrated the effectiveness of multilayer systems with time-delay sequences as inputs to these systems [15-18]. Typical examples are: time-delay neural network

(TDNN) proposed by Waibel and Lang [19-21]; block-windowed neural network (BWNN) by Sawai [22]; and dynamic programming neural network (DNN) by Sakoe [23-24].

Some features used in these neural speech recognition systems are incorporation of time delays, temporal integration, or recurrent connections. Spectral inputs are applied to input nodes sequentially, one frame at a time, and their corresponding input matrix is formed [15-16]. Since only short time delays are used, these neural speech recognition systems can be integrated into real time speech recognizer. However, these systems concern, so far, mainly with algorithms; their behaviors and characteristics are primarily investigated by simulation on general purpose computers. The spatiotemporal computing parallelism inhered in such neural speech recognition systems is little explored; thereby restricting its application domain to real life problems.

11.2. Window Computing Model

11.2.1. *Definition and Notation*

Based on the neural systems with time-delay sequence input of feature parameters for speech recognition [15-18], we can develop a typical computing model composed of $p+2$ layers, which includes an input layer, p hidden layers, and an output layer. Both the input layer and the hidden layers are characterized by time-delay sequence input matrix of speech parameters, built by $m_s \times n_s$ ($s = 1, 2, \ldots, p+1$) memory elements, where $m_{p+1} = q$ is the number of pattern classes. The output layer consists in q units. The relation between node x_{ij} in Layer s and node y_{kl} in Layer $s+1$ can be defined as

$$y_{kl} = f(\sum\sum x_{ij} w^{kl}_{i-k+1,j-l+1} + \theta^{(kl)}) \qquad (11\text{-}1)$$

where $k \le i \le k + e_s - 1$ and $l \le j \le l + r_s - 1$; f is a sigmoid function; $w^{(kl)}$ and $\theta^{(kl)}$, $1 \le k \le m_{s+1}$ and $1 \le l \le n_{s+1}$, are referred to as weight value and bias value, respectively. Both values can be obtained from a small input submatrix (called "window"), where the size is $e_s \times r_s$ ($e_s \le m_s$ and $r_s \le n_s$) in Layer s, to the node y_{kl} in Layer $s+1$. This kind of time-delay window computing methodology is shown in Fig. 11.1. Obviously, there will be $m_{s+1} \times n_{s+1}$ windows formed by the input matrix ($m_s \times n_s$) in Layer s.

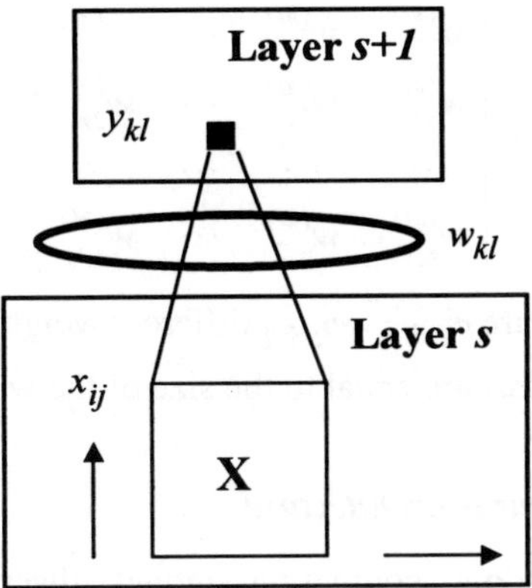

Fig. 11-1. Time-delay window computing model between Layer s and Layer s+1.

To implement such a time-delay window computing in Eq. (11-1), we can use only an input window built by $e_s \times r_s$ elements in Layer s. Instead of moving such a window to the whole input matrix, speech parameters in time-delay sequence are arranged to pass through the window in pipeline. Thus, the expression in Eq. (11-1) can be rewritten as

$$y_{kl} = f(\sum W_{kl} X^T) = f(\sum X W_{kl}^T) \tag{11-2}$$

where X is an input sub-matrix given by such a fixed window, which can be represented as

$$X = \begin{bmatrix} X_1 \\ X_2 \\ \cdots \\ X_{e_s} \end{bmatrix} = \begin{bmatrix} x_{11} & x_{12} & \cdots & x_{1r_s} \\ x_{21} & x_{22} & \cdots & x_{2r_s} \\ \cdots & \cdots & \cdots & \cdots \\ x_{e_s1} & x_{e_s2} & \cdots & x_{e_sr_s} \end{bmatrix} \tag{11-3}$$

and W_{kl} is the corresponding weight matrix from the window in Layer s to node y_{kl} in Layer $s+1$, i.e.,

$$W_{kl} = \begin{bmatrix} w_{11}^{(kl)} & w_{12}^{(kl)} & \cdots & w_{1r_s}^{(kl)} \\ w_{21}^{(kl)} & w_{22}^{(kl)} & \cdots & w_{2r_s}^{(kl)} \\ \cdots & \cdots & \cdots & \cdots \\ w_{e_s1}^{(kl)} & w_{e_s2}^{(kl)} & \cdots & w_{e_sr_s}^{(kl)} \end{bmatrix}. \qquad (11\text{-}4)$$

It is evident that there are $m_{s+1} \times n_{s+1}$ different weight matrices from Layer s to Layer $s+1$. Their sizes are equal to the size of the window in Layer s.

11.2.2. *Pipelined Neural Architecture*

The time-delay window computing model discussed above can be implemented by a pipelined neural architecture with $p+1$ processing stages, each with its own control sequence (See Fig. 11-2 (a)). In each processing stage, a fixed time-delay computing window is built as a connection to next stage. Loading an input submatrix, X, to the window in a pipeline mode and mapping the corresponding weight matrix, W_{kl} ($k = 1, 2, \ldots, m_{s+1}$; $l = 1, 2, \ldots, n_{s+1}$), the output result, y_{kl}, can be obtained. Since all time-delay computing windows in the pipelined neural architecture are capable of working at the same time, the potential parallelism inhered in such neural speech recognition systems can be well explored.

A basic time-delay neuron in the pipelined neural architecture is defined in Fig. 11-2 (b). The time-delay inputs, X_i ($i = 1, 2, \ldots, e_s$), are undelayed or delayed D_t ($= \Sigma D_j + \Delta$), where D is a delay unit and Δ is its increment ($j = 1, 2, \ldots, t-1$; $t = 1, 2, \ldots, r_s-1$). The time-delay speech inputs, X_i, will be multiplied by several weights, one for each delay and one for the undelayed input.

Note that two types of operations, namely, control flow and data flow, are used in this pipelined neural architecture. Master control unit not only gives each control sequence to the corresponding processing stage, but also arranges the time-delay speech input parameters of each frame as data flow input to the given computing window. Once the window is filled, the time-delay sequence input submatrix obtained is processed. Depending on the nature of the time-delay speech inputs, two different processing stages can be used in the architecture. These will be discussed in the next section.

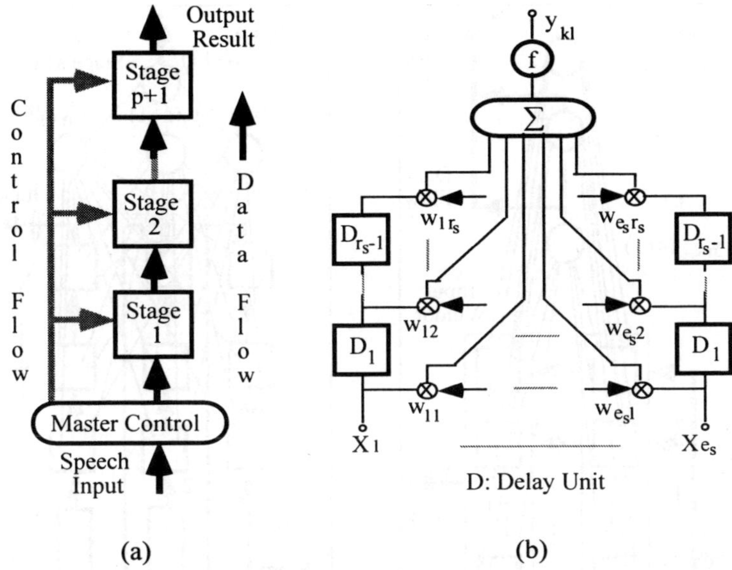

Fig. 11-2. (a) Pipelined neural speech recognition system with p+1 processing stages. (b) Time-delay neuron structure in the pipelined neural system.

11.3. Pipelined Architecture: Processing Stages

11.3.1. *Parallel Processing Stage*

In the pipelined neural system, a parallel processing stage can be defined in Fig. 11-3 (a), where a window built by $m_s \times r_s$ elements is utilized to receive the input data flow from its previous stage, and m_{s+1} neurons are used to send the output results to the next stage in parallel. In other words, the input data flow is passed through the window and transformed by this stage to generate the corresponding output data flow. The widths of both data flows are defined as m_s and m_{s+1}, respectively. A new feature parameter obtained by each neuron in Stage s can be represented as

$$y_i(s+1) = f(\sum W_i(s) X^T(s)) \tag{11-5}$$

where $s = 1, 2, ..., p+1$; $i = 1, 2, ..., m_{s+1}$; $W_i(s)$ is the *i-th* weight matrix and $X(s)$ is the input submatrix given by the window in Stage s.

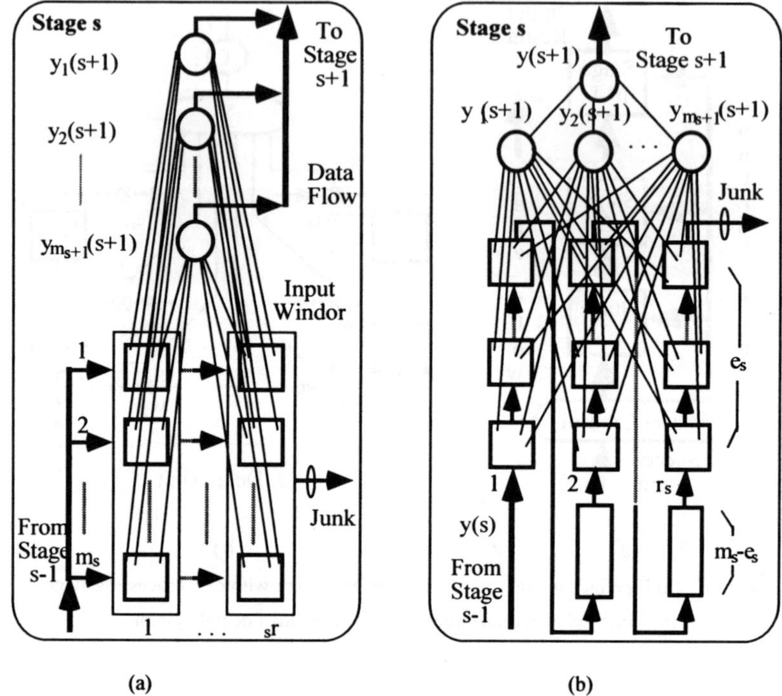

(a) (b)

Fig. 11-3. Two kinds of processing stages: (a) in parallel and (b) in serial.

11.3.2. *Serial Processing Stage*

In this processing stage, a pipe with single parameter width is made by a chain of serial shifting elements (see Fig. 11-3 (b)). A window structure is designed to implement the transformation between stages. The $r_s - 1$ line delays, each $m_s - e_s$ shifting elements, are built to receive a serial stream of parameters and to form the required window $(e_s \times r_s)$, which are input to m_{s+1} neurons. Thus, the total $(e_s \times r_s) + (m_s - e_s)(r_s - 1) = m_s (r_s - 1) + e_s$ elements are needed in window structure. The neurons associated with the window are defined in Eq. (11-5) with their common output

$$y(s+1) = OR\ (y_i\ (s+1)) \tag{11-6}$$

where $i = 1, 2, ..., m_{s+1}; s = 0, 1, ..., p+1$. Note that the output of each neuron can, in turn, be obtained as, either a single output, $y_i(s+1)$, or no output within a single clock interval. The output order in a cycle is: $y_1(s+1), \cdots, y_{m_{s+1}}(s+1), *_1, \cdots, *_{e_s-1}$, where $m_{s+1} + e_s - 1 = m_s$ and " * " indicates no output. There are a total of n_{s+1} $(= n_s - r_s + 1)$ processing cycles for each given speech input matrix, $m_s \times n_s$.

11.3.3. *Building Elements*

There are three kinds of building elements, including window, synapse and summing element, which are used in two different processing stages described before. A window element can be implemented by a regular shifting register and thus the following discussion will be focused on the other two building elements. Considering on-line back-propagation (BP) learning, which has successfully applied to the neural speech recognition systems, two processing phases, searching and learning, are defined in the building elements. They can be implemented by special feedforward and feedback paths, respectively.

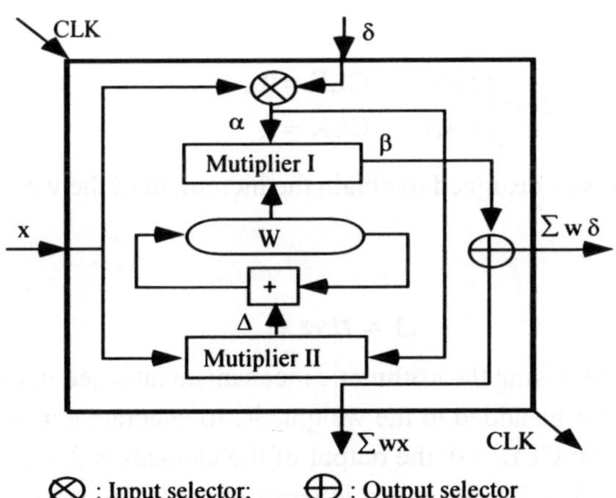

Fig. 11-4. Synapse building element structure.

Synapse Element:

A synapse element is used to store and change weight values. It is mainly composed of a weight memory (*W*), two multipliers (*I* and *II*) and two selectors (⊗ and ⊕), shown as in Fig. 11-4. A control clock, CLK, indicates the phase of the element. CLK = 0, means searching (or feedforward) phase and CLK = 1, means learning (or feedback) phase. In this element, there are two data inputs (x and δ) and two outputs (Σwx and $\Sigma w\delta$), where x / Σwx work in CLK = 0, and δ / $\Sigma w\delta$ in CLK = 1. Multiplier *I* can generate a common output,

$$\beta = \alpha W \tag{11-7}$$

where α is the output of the input parameter selector, ⊗, which is represented as

$$\alpha = \begin{cases} x & \text{CLK} = 0, \\ \delta & \text{CLK} = 1. \end{cases} \tag{11-8}$$

An output parameter selector, ⊕, can choose a correct output result of the element, i.e.,

$$\beta = \begin{cases} \Sigma wx & \text{CLK} = 0, \\ \Sigma w\delta & \text{CLK} = 1. \end{cases} \tag{11-9}$$

Multiplier *II* is only designed to obtain the increment of the weight value when CLK = 1,

$$\Delta = \eta\, \alpha x \tag{11-10}$$

where η is a gain. Using the arithmetic mechanism attached in the element, the increment, Δ, can be added to the weight, W, to generate a new weight value. In this way, when CLK = 0, the output of the element is Σwx; Otherwise, the output is $\Sigma w\delta$. Also, W is changed in terms of the following rule

$$W = W + \Delta = W + \eta\, \delta x. \tag{11-11}$$

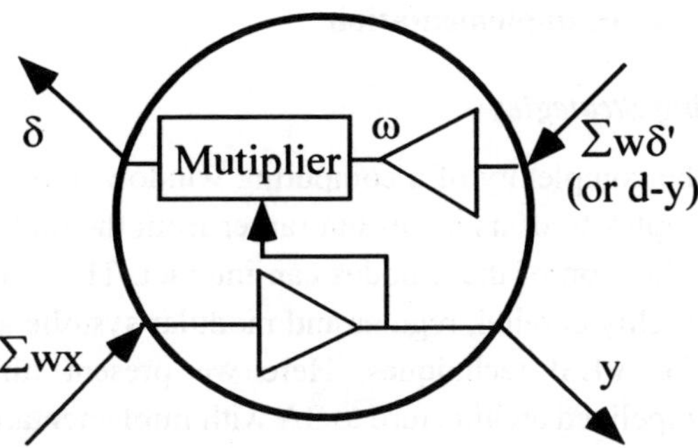

Fig. 11-5. Summing building element structure.

Summing Element:

This element is built to obtain two-direction accumulative results for both feedforward and feedback processing (See Fig. 11-5). It consists of two amplifiers and a multiplier. Two inputs (outputs), Σwx and $\Sigma w\delta'$ (δ and y), are from (to) the current stage and the next stage, respectively. Their input / output relations are $\Sigma wx\,/\,y$ and $\Sigma w\delta'\,/\,\delta$. When CLK = 0, the output of the element is represented as

$$y = f(\Sigma wx) \tag{11-12}$$

and when CLK = 1, the output is

$$\delta = y\,(1 - y)\;\omega = y\,(1 - y)\,f(\Sigma w\delta'). \tag{11-13}$$

Connection Network:

Three kinds of building elements can be easily implemented by the current VLSI technologies [13,25-26]. Using these simple building elements, a basic connection network in Stage s can be designed as in Fig. 11-6, where the size of the time-delay computing window is defined as $m_s \times r_s$. There are a total of $m_s \times r_s \times m_{s+1}$ synapse elements and m_{s+1} summing elements used in the network. Obviously, the whole pipelined neural system can be implemented by cascading such regular connection networks

11.4. Systolic Array Implementation

11.4.1. *Mapping Strategies*

It is clear that the complexity of a computing window implementation stems not from the complexity of its nodes but rather from the multitude of ways in which a large collection of these nodes can interact. Therefore, an important task is to build highly parallel, regular and modular systolic arrays (SAs) that are attractive for VLSI techniques. Here we present different mapping strategies from pipelined architecture to SA with implementation efficiency as our goal.

Processing Mode Mapping:

Here we partition a pipelined neural system into some basic processing stages with time-delay window, each capable of performing an independent function. Often a processing stage represents a layer in the neural networks. The processing stages are implemented using a corresponding SA, which are then cascaded.

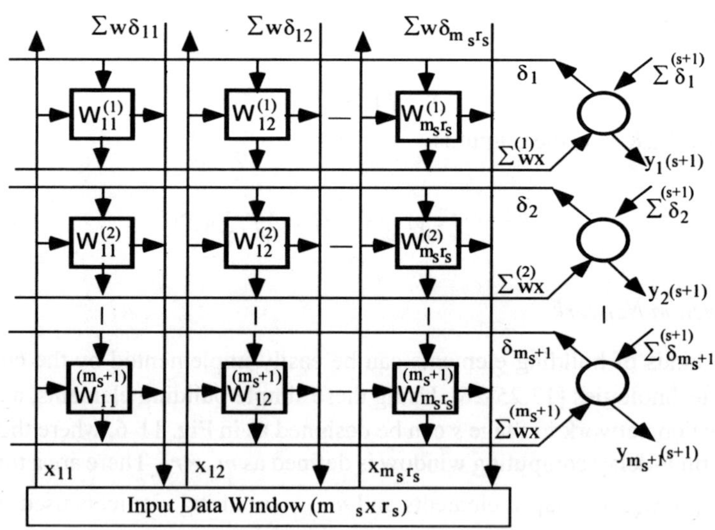

Fig. 11-6. Connection network with on-line learning in Stage s.

Computing Property Mapping:

Each processing stage function is reduced to a recursive form which is implemented by the corresponding pipeline matrix in terms of some systolic rules. In practice, this mapping changes parallelism in place to parallelism in time.

Arithmetic Module Mapping:

A basic operation in recursive arithmetic is implemented by a computing element. For example, a node is divided into two parts: forming a weighted sum of N inputs and passing the result through a nonlinearity. The weighted sum can easily be integrated by a two-dimensional (2-D) recursive matrix. To form the nonlinearity, a special element is defined which may be cascaded with the recursive matrix as a bound node of its output.

11.4.2. *SA Structures: Computing Cell*

Using the aforesaid mapping strategies two kinds of processing stages, in parallel and in serial, as obtained in Section 11.3 can be systematically implemented by the corresponding SAs (see Fig. 11-7 and 11-8). In both arrays, the line delays built by shifting elements are used to receive a data stream of parameters and to construct the window required. There are a total of $\{m_s (m_s -1)\} / 2$ and $\{(2m_s -2e_s - r_s) (r_s -1)\} / 2$ shifting elements in parallel processing SA and in serial processing SA, respectively. The adder arrays are built as the accumulators to compress the output results of the computing window. Obviously, some regular shifting registers and adders can implement the line delays and adder arrays.

Computing cells, defined in the both SAs, can be properly arranged to form each computing window in parallel or in serial. However, all of these computing cells have an identical structure with special feedforward and feedback paths. They are mainly composed of weight memory (W_i) $(i = 1, 2, ..., m_{s+1})$, adder (⊕) and multiplier (⊗). Three data inputs, x, $\{z_i\}$ and $\{\delta_i\}$, and their outputs, x', $\{Z_i\}$ and $\{\Omega_i\}$, are defined in the computing cell, where $\{z_i\}$ and $\{Z_i\}$ are used for CLK = 0; $\{\delta_i\}$ and $\{\Omega_i\}$ for CLK = 1. Note that each input (output) is transmitted by m_{s+1} data except x (x'). When CLK = 0, the outputs of the cell in feedforward path are defined as: $Z = x\, w + z$ and $x' = x$;

otherwise, the output as $\Omega = w\delta$. At the same time, W is changed in terms of the rule in Eq. (11-11).

It is evident that the SAs shown in Fig. 11-7 and 11-8 are regular interconnected arrays using a set of computing cells, each performing some simple window computing, where the data flows in a rhythmic fashion with only local interconnects between cells. They can provide a good medium to implement the pipelined neural system in VLSI.

11.5. Speech Recognition Applications

In this section, we provide the results of our study on three types of neural systems using the parallel time-delay window computing in the pipelined neural architecture. The neural systems selected are motivated by speech recognition applications and they have been widely used [15-24].

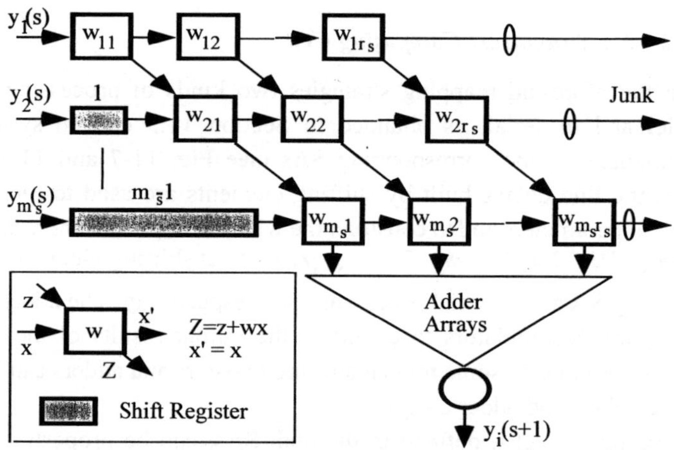

Fig. 11-7. Parallel data flow window computation.

11.5.1. *Time-Delay Neural Network (TDNN)*

TDNN is a neural system that can take into account the "dynamic nature of speech". It is used to represent temporal relationships between successive acoustic frames, while providing some invariance under time translation [19]. It has been demonstrated that the TDNN computing can provide excellent discrimination ability among speech sounds. Speech recognition performance obtained by using the TDNN has often exceeded that of many conventional approaches [20-21].

The basic TDNN system is composed of an input layer, two hidden layers and an output layer [19]. Except the output layer, each layer has an $m_s \times n_s$ ($s = 1, 2, 3$) matrix of memory elements, where $n_2 = n_1 - r_1 + 1$, $n_3 = n_2 - r_2 + 1$ and $m_3 = q$. The relation between the input layer and the 1st hidden layer

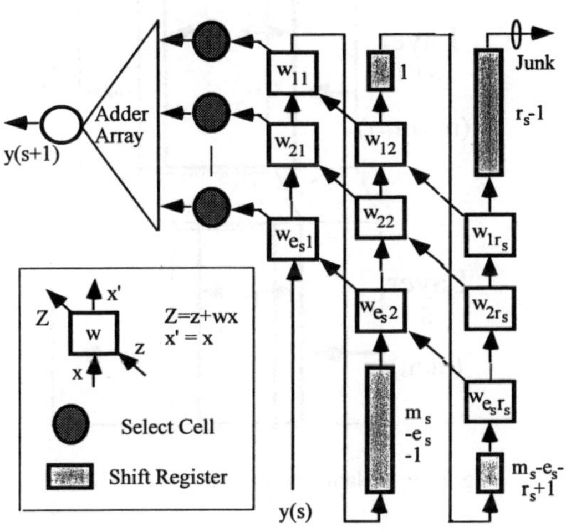

Fig. 11-8. Serial data flow window computation.

(and also between 1st and 2nd hidden layers) (See Fig. 11-9.) is represented as

$$y_{kl} = f(\Sigma\Sigma x_{ij}\, w^{(k)}{}_{i,j-l+1} + \theta^{(k)}) \tag{11-14}$$

where $1 \le i \le m_s$ and $1 \le j \le l + r_s + 1$. The TDNN computing can be implemented by the pipelined system with three parallel processing stages. Except the last stage without the data window, each input parameter matrix ($m_s \times n_s$) in the first two stages is pipelined to pass through its window ($m_s \times r_s$), $s = 1, 2$. When the window is filled by successive data flow, m_{s+1} new values of the parameters can be, in parallel, obtained and simultaneously fed into the window in the next stage. It is evident that there are m_{s+1} different weight matrices and n_{s+1} input data windows from Stage s to Stage $s+1$, and their sizes are equal to the size of the window in Stage s, i.e., $m_s \times r_s$. Using this parallel window computing to implement the TDNN, only $m_s \times r_s$ window elements, instead of $m_s \times n_s$ (generally, $r_s << n_s$) elements, are needed in Stage s.

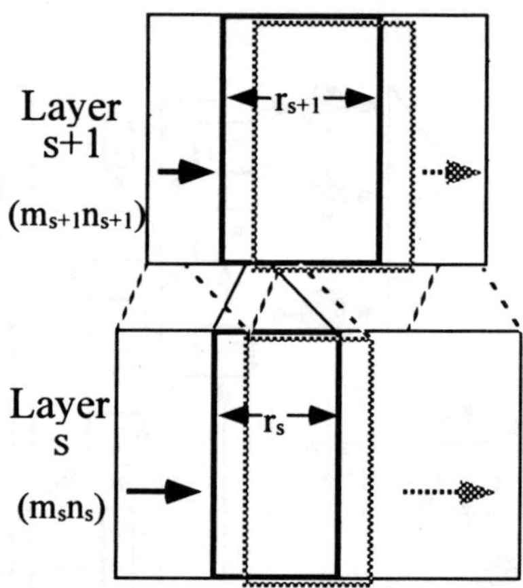

Fig. 11-9. Relation between layers for TDNN.

11.5.2. *Block-Windowed Neural Network (BWNN)*

BWNN is based on windowing each layer of the neural network with overlaped local time-frequency windows. This neural system makes it possible to capture global features from the upper layers as well as precise local features from the lower layers. It is proved to be robust for speech sound variations in both frequency- and time-domains among different speakers [22].

The BWNN system is composed of an input layer, three hidden layers and an output layer [22]. Except the output layer, each layer has a $m_s \times n_s$ ($s = 1, 2, 3, 4$) matrix of memory elements and their relation between layers satisfies

$$\begin{cases} m_{s+1} = m_s - e_s + 1 \\ n_{s+1} = n_s - r_s + 1 \end{cases} \tag{11-15}$$

where $m_4 = q$ and $e_s = r_s$, i.e., the length and the width of the submatrix in Layer s are the same (see Fig. 11-10). It is clear that the TDNN structure is a special example of the BWNN if $e_s = m_s$.

The use of the pipelined neural system to implement the BWNN involves four serial processing stages. Like the TDNN implementation, the last stage is the output stage without the data window. Each input matrix in the other stages can form an $m_s \times n_s$ ($s = 1, 2, 3$) pipeline with the width of a single parameter and passes through its window ($e_s \times r_s$), parameter by parameter. An output result obtained from Stage s within a single clock interval is sent to the window in Stage $s+1$ without any delay. This means that only an input window built by $e_s \times r_s$ shifting elements and some line delays by $(m_s - e_s)(r_s - 1)$ shifting elements, i.e., the total $m_s(r_s - 1) + e_s$ window elements, instead of $m_s \times n_s$ elements (in general, $e_s << m_s$ and $r_s << n_s$), are needed in Stage s.

11.5.3. *Dynamic Programming Neural Network (DNN)*

DNN is proposed on the integration of multilayer neural network and dynamic programming based matching. Researchers have used DNN extensively in speaker-independent word recognition, and proved that it has excellent time normalization ability, flexible learning facility, expandability to continuous speech recognition, and high tolerance to the spectral pattern variation [23].

The DNN can be implemented by the pipelined neural system with three processing stages (see Fig. 11.11). An input pattern, $x_1, ..., x_i, ..., x_I$, is defined as a warping function $i = i(j)$ between input pattern time i and window element j, where $j = 1, 2, ..., J$. Without an input matrix with $I \times J$ memory elements [23], a window is built by $2 \times J$ window elements and J neurons are used in the first stage. When the input patterns, x_i and x_{i-1}, pass through the window, the corresponding output for each neuron can be represented as

$$y_{ij} = f(w_{j0} x_i + w_{j1} x_{i-1}) \tag{11-16}$$

where w_{j0} and w_{j1} are weighs from two window elements to neuron j.

In the second stage, a window with $J \times 1$ window elements is used to receive y_{ij} ($j = 1, 2, ..., J$), in parallel. Each neuron in the stage is used as a multiplier, i.e.,

$$q_{ij} = f(w_j y_{ij}) = w_j y_{ij} \tag{11-17}$$

The third stage is built by a serial processing structure. Its input data, q_{ij}, is arranged in a pipeline mode of a single parameter like $q_{iJ} \rightarrow \cdots \rightarrow q_{ij} \rightarrow \cdots \rightarrow q_{i1}$. In other words, the parallel outputs from the previous stage will be changed as the serial inputs to this stage. It is composed of a four-element window, two *J-1* line delays and a processing element (PE), shown in Fig. 11-12 (a). The PE is designed by the standard dynamic programming algorithm [24]. Its initial condition is set at $g_{11} = q_{11}$, implemented by the external control. Then, the data is processed with

$$g_{ij} = q_{ij} + max(g_{i,j-1}, g_{i-1,j-1}, g_{i-2,j-1}) \tag{11-18}$$

The PE shown in Fig. 11-12 (b) implements this maximization problem. The PE consists of a tri-comparator subnet for extracting the maximum of three analog inputs [31] and an adder. Given an input parameter, q_{ij}, an output of the PE, g_{ij}, can be obtained and fed into the window to generate the following new values. This process is continued until the total cumulating value, $Z = g(I, J)$, is reached. Such a process is represented in Fig. 11-13.

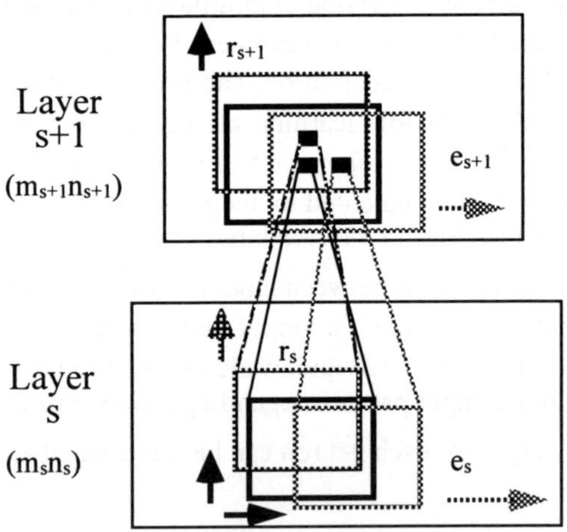

Fig. 11-10. Relation between layers for BWNN.

11.6. Structure Analysis

For a given neural system, both the structure design and access time needed to solve the problem are two most important performance measures [13,25-30].

In this section, we will analyze these measures for our pipelined neural architecture, where parallel processing stage defined in Fig. 11-3 (a) and serial processing stage in Fig. 11-3 (b) are referred to as Type 1 and Type 2, respectively. The way of selecting the property parameters for parallel time-delay window computing is also discussed in this section.

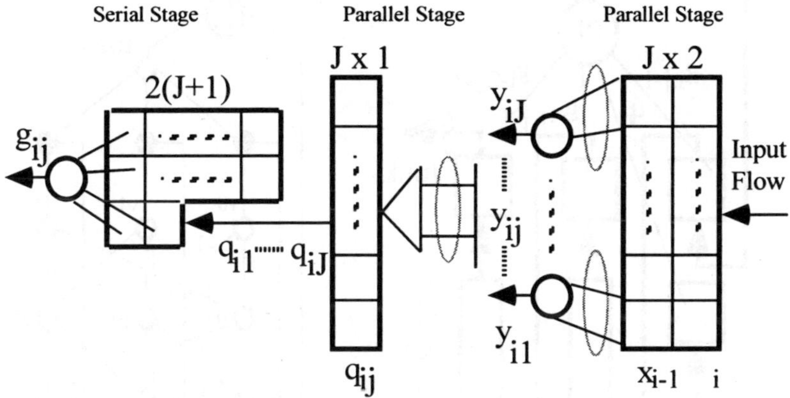

Fig. 11-11. Pipelined neural architecture for DNN implementation.

11.6.1. *Structure Complexity*

We choose a typical TDNN computing for comparison with our methodology. In Section V, it has been shown that the parallel time-delay window computing can implement TDNN and greatly reduce the memory elements in each layer of the neural networks to a small number of window elements in the processing stage. This is because only a limited window is connected to its next stage and the parameters shifted out from the window are discarded. Since the speech feature parameters are applied to each layer sequentially one frame at a time, this reduction of memory elements is feasible.

Note that both memory element used in traditional TDNN computing and window element in parallel window computing have the same hardware complexity because they are based on a regular register. In this way, we can perform the traditional TDNN computing by using the three kinds of building elements given in Section III. According to the basic TDNN definition [19-20], it is assumed for the traditional computing that in Layer *s* ($s = 1, 2, ...,$

$p+1$), the number of window elements is taken as $m_s \times n_s$, the number of synapse elements as $m_s \times n_s \times m_{s+1}$, and the number of summing elements as $m_{s+1} \times n_s + 1$.

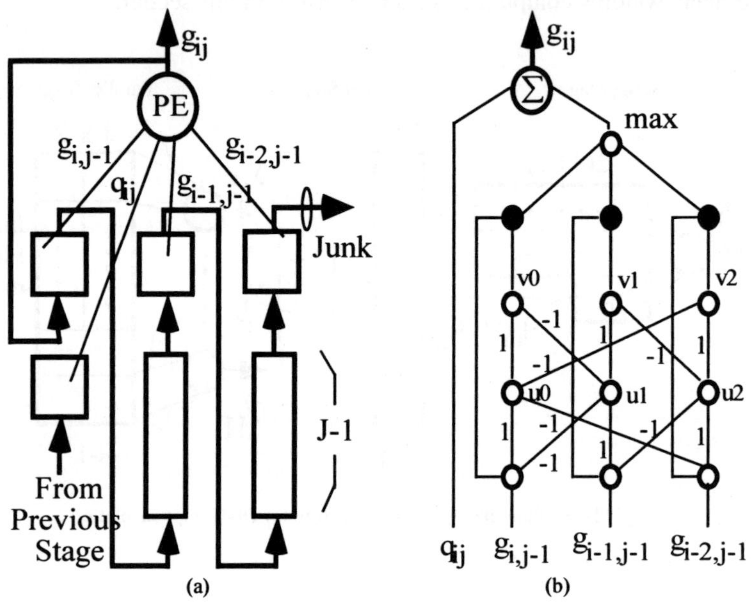

Fig. 11-12. (a) Serial processing stage for DNN and (b) its PE structure.

In the parallel window computing, the numbers of window elements used in Stage s for Type 1 and Type 2 (see Section 11.3) have been given as $m_s \times r_s$ and $m_s\,(r_s - 1) + e_s$, respectively. Their measures for window elements can be defined as follows:

$$D_{Type\ 1(s)} = \frac{m_s \times r_s}{m_s \times n_s} = \frac{r_s}{n_s} \tag{11-19}$$

And

$$D_{Type\ 2(s)} = \frac{m_s\,(r_s - 1) + e_s}{m_s \times n_s} \tag{11-20}$$

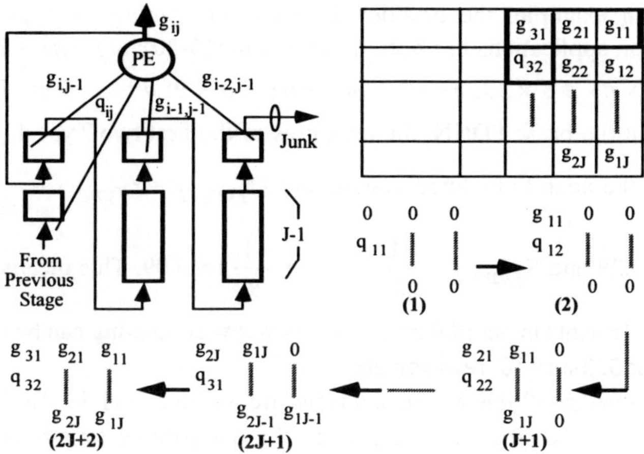

Fig. 11-13. The window process for DNN implementation.

Similarly, the numbers of both synapse and summing elements used in Stage *s* for Type 1 and Type 2 can be obtained from Section III. Thus, the measures for synapse element are

$$Y_{Type\ 1(s)} = \frac{m_s \times r_s \times m_{s+1}}{m_s \times n_s \times m_{s+1}} = \frac{r_s}{n_s} \tag{11-21}$$

and

$$Y_{Type\ 2(s)} = \frac{e_s \times n_s \times m_{s+1}}{m_s \times n_s \times m_{s+1}} = \frac{e_s \times r_s}{m_s \times n_s} \tag{11-22}$$

Both the numbers of summing elements used in Stage *s* for Type 1 and Type 2 are m_{s+1}. Hence these two kinds of processing stages have the same measure, i.e.,

$$S_{Type\ 1(s)} = S_{Type\ 2(s)} = \frac{m_{s+1}}{m_{s+1}\, x\, n_{s+1}} = \frac{1}{n_{s+1}}\ . \tag{11-23}$$

It is evident that the measures of the entire system for three kinds of building elements are their mean of over each Stage *s*.

As an example, the traditional TDNN computing for typical speech recognition applications has been described in [19-20]: $m_1 \times n_1 = 16 \times 15$, $r_1 = 3$; $m_2 \times n_2 = 8 \times 13$, $r_2 = 5$; $m_3 \times n_3 = 3 \times 9$; $q = 3$. Then, in order to implement the basic TDNN, the measures of the first two Type 1 processing stages in the neural pipelined system are: $D_{Type\ 1} = Y_{Type\ 1} = \frac{1}{2}(\frac{3}{15} + \frac{5}{13}) \approx 0.29$ and $S_{Type\ 1} = \frac{1}{2}(\frac{1}{13} + \frac{1}{9}) \approx 0.09$. This means that three building elements in parallel time-delay window computing can be reduced by a factor of 3, 3 and 10, respectively.

The results of the above analysis are summarized in Table 11-1. It indicates that the structure complexity for our parallel time-delay window computing is much less than that of the traditional TDNN computing.

11.6.2. *Throughput Rate*

The neural speech recognition systems are well suited to pipelining because of their multilayer networks as processors of time-delay sequence patterns. In the pipelined system embedding parallelism or concurrency, the throughput rate can be fixed and it does not vary with the size of the problem grows, i.e.,

$$T = O(1) \tag{11-24}$$

Hence, a high throughput rate can be maintained in such pipelined neural systems, where the clock of the master control element is selected from the longest time delay among processing stages.

11.6.3. *Window Parameter*

Computing window in the pipelined neural system is not only an important component, but also an obvious feature which differs from other neural systems. The window size has a direct relation to the properties of the pipelined system, such as the number of window elements, Ω, and the computing time, Ψ. The smaller the window, the fewer is the number of window elements, and the longer is the computing time required.

In Type 1, these two trade-off properties for Stage s are

$$\begin{cases} \Omega_{Type\,1} = r_s \times m_s, \\ \Psi_{Type\,1} = n_s - r_s + 1. \end{cases} \tag{11-25}$$

We define their product as

$$E(r_s) = \Omega_{Type\,1} \times \Psi_{Type\,1} = (r_s \times m_s)(n_s - r_s + 1). \tag{11-26}$$

To maximize the $E(r_s)$ function, take derivative with respect to window size r_s, i.e.,

$$E'(r_s) = m_s n_s - 2m_s r_s + m_s\,. \tag{11-27}$$

Let $E'(r_s) = 0$. The optimal size of window for Type 1 (See Fig. 11.14(a)) can be then selected as

$$r_s = \frac{n_s + 1}{2}\,. \tag{11-28}$$

Note that this optimal window size is not a function of the length of the window, m_s. In the same way, the two trade-off properties in Type 2 can be written as:

$$\begin{cases} \Omega_{Type\,2} = (r_s - 1)\, m_s + r_s \\ \Psi_{Type\,2} = m_s n_s - (r_s - 1)\, m_s - r_s \end{cases} \tag{11-29}$$

where $r_s = e_s$, i.e., a square window is used. The size of the window can be selected directly from the relation $\Omega_{Type\,2} = \Psi_{Type\,2}$ (see Fig. 11.14(b)), which leads to

$$(r_s - 1)\, m_s + r_s = m_s n_s - (r_s - 1)\, m_s - r_s. \tag{11-30}$$

Hence, the choice of the window size for Type 2 is

$$r_s = \frac{m_s(n_s + 2)}{2(m_s + 1)}\,. \tag{11-31}$$

11.7. Conclusions

In this chapter, a novel parallel structure for time-delay neural networks are used in speech recognition applications is presented. The effectiveness of the design has been illustrated by extracting a window computing model from the time-delay neural systems, developing the corresponding pipelined architecture with parallel or serial processing stages and comparing its performance with the traditional TDNN computing. Applying this parallel window to a typical time-delay neural network, it has been shown that the methodology can greatly reduce the structure complexity while maintaining a high throughput rate.

Table 11-1. Comparison between traditional TDNN computing and parallel window computing in Stage *s*.

		Traditional TDNN Computing	*Parallel Window Computing*	
			Type 1	*Type 2*
Window Element	*Element #*	$m_S \times n_S$	$m_S \times r_S$	$m_S (r_S - 1) + e_S$
	Complexity	*1*	$\frac{r_S}{n_S}$	$\frac{m_s (r_s - 1) + e_s}{m_s \times n_s}$
Synapse Element	*Element #*	$m_S \times n_S \times m_{S+1}$	$m_S \times r_S \times m_{S+1}$	$e_S \times r_S \times m_{S+1}$
	Complexity	*1*	$\frac{r_S}{n_S}$	$\frac{e_s \times r_s}{m_s \times n_s}$
Summing Element	*Element #*	$m_{S+1} \times n_{S+1}$	m_{S+1}	m_{S+1}
	Complexity	*1*	$\frac{1}{n_{s+1}}$	$\frac{1}{n_{s+1}}$

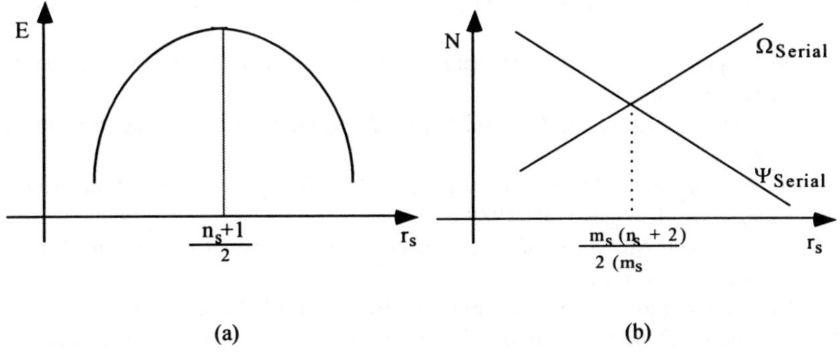

Fig. 11-14. (a) Window parameter selection for Type 1.
(b) Window parameter selection for Type 2.

References

[1] S. Furui. Current technology for speech recognition, *IEEE Press, Piscataway*, N.J., 1995.

[2] N. Morgan and H. Bourlard. Neural networks for statistical recognition of continuous speech. *Proc. IEEE*, 83(5), pages 742-770, 1995.

[3] K. Chen, D. Xie and H. Chi. A modified HME architecture for text-dependent speaker identification. *IEEE Trans. on Neural Networks*, 7(5), pages 1309-1314, 1996.

[4] H. P. Campbell. Speaker recognition: Tutorial. *Proc. IEEE*, 85(9), pages 1437-1463, 1997.

[5] K. Chen, et al.. Speaker identification using time-delay HMEs. *International Journal of Neural Systems*. 7(1), pages 29-43, 1996.

[6] G. Doddington. Speaker recognition – identifying people by their voice. *Proc. IEEE*, 73(11), pages 1651-1664, 1986.

[7] D. O'Shaugenessy. Speaker recognition. *IEEE ASSP Magazine*, 3, pages 4-17, 1986.

[8] H. Bourlard and C. J. Wellekens. Links between markov models and multilayer perceptrons. *IEEE Trans. Pattern Analysis and Machine Intelligence*, 12(12), pages 1167-1178, 1990.

[9] T. Matsui and S. Furui. Speaker recognition technology. *NTT Review*, 7(2), pages 42-48, 1995.

[10] D. P. Morgan and C. L. Scofield. *Neural networks and speech processing*. Kluwer Academic Publishers, Boston, 1991.

[11] L. M. Fu. *Neural networks in computer intelligence*. McGraw-Hill, Inc., New York, 1994.

[12] C. Bishop. *Neural networks for pattern recognition*, Oxford University Press, Oxford, 1995.

[13] M. I. Elmasry. *VLSI artificial neural networks engineering.* Kluwer Academic Publishers, Boston, 1994.

[14] D. Zhang. *Parallel VLSI Neural System Designs*, Springer-Verlag, Singapore, 1998.

[15] R. P. Lippmann. Review of neural networks for speech recognition. *Neural Computing*, 1, pages 1-38, 1989.

[16] S. Furui and M. M. *Sondhi. Advances in speech signal processing.* Marcel Dekker Inc., N.Y., 1992.

[17] M. Koerner (ed.). *Speech recognition*, Prentice Hall, Austin, 1996.

[18] J. Junqua. *Robustness in automatic speech recognition: fundamentals and applications.* Kluwer Academic Publishers, Boston, 1995.

[19] A. Waibel, T. Hanazawa, G. Hinton, K. Shikano and K. J. Lang. Phoneme recognition using time-delay neural networks. *IEEE Trans on ASSP*, 37(3), pages 328-339, 1989.

[20] A. Waibel, Modolar construction of time-delay neural networks for speech recognition, *Neural Computing*, 1, pages 38-46, 1989.

[21] K. J. Lang and A. Waibel. A time-delay neural networks architecture for isolated word recognition. *Neural Networks*, 3, pages 23-43, 1990.

[22] H. Sawai. Frequency-time-shift-invariant time-delay neural networks for robust continuous speech recognition. *Proceeding of ICASSP*-91, S-2(1), pages. 45-48, 1991.

[23] H. Sakoe, R. Isotani, K. Yoshida, K. Iso and T. Watanabe. Speaker-independenct word recognition using dynamic programming neural networks. *Speech Recognition*, pages 439-442, 1989.

[24] H. Sakoe and S. Chiba. Dynamic programming algorithm optimization for spoken word recognition. *IEEE Trans. on Acoustics, Speech, and Signal Processing,* 26(1), pages 43-39, 1978.

[25] D. Zhang and M. I. Elmasry. VLSI compressor design for digital neural network implementation. *IEEE Transactions on Very Large Scale Integration (VLSI) Systems*, 5(4), pages 230-233, 1997.

[26] B. A. White and M. I. Elmasry. The digi-neocognitron: a digital neocognitron neural network model for VLSI, *IEEE Transactions on Neural Networks*, 3(1), pages 73-85, 1992.

[27] J. B. Burr. "Digital neural network *implementation", in Neural Networks, Concepts, Applications, and Implementations,* Prentice Hall, Englewood Cliffs, pages 237-285, 1991.

[28] S. Y. Kung. *Digital neural networks*, Prentice Hall, Englewood Cliffs, 1993.

[29] F. Y. Shih and J. Moh. Implementing morphological operations using programmable neural networks. *Pattern Recognition*, 25(1), pages 89-99, 1992.

[30] Y. Takefuji. *Neural network parallel computing*, Kluwer Academic Publishers, Boston, 1994.

[31] F. Y. Shih and J. Moh. Implementing morphological operations using programmable neural networks. *Pattern Recognition*, 25(1), pages 89-99, 1992.

CHAPTER 12

AN INTEGRATED INTELLIGENT CLASSIFICATION ENGINE (I^2CE) FOR BIOSIGNAL ENGINEERING

A.N. Kastania

Department of Informatics,
Athens University of Economics and Business,
10434 Athens, Hellas, Greece
E-mail: ank@aueb.gr

M.P. Bekakos

Department of Electrical and Computer Engineering
School of Engineering, Democritus University of Thrace,
67100 Xanthi, Hellas, Greece
E-mail: mpb@aueb.gr

The aim of this chapter is to provide a review of different approaches to biosignal classification using adaptive logic networks (ALNs), compare their performance on a challenging medical data set, and draw conclusions on their applicability to realistic biomedical engineering problems. In the proposed neural classification perspective, the engineering disciplines of pattern recognition, modeling and prediction are supplemented by the machine learning challenge of reproducing intelligence itself. Herein, various adaptive logic networks architectures were built and evaluated based on the relational approach for ALNs and focused on the development and control of an integrated intelligent classification engine (I^2CE) for biosignal engineering. The fundamental theme of representing knowledge in hardware neural networks was investigated with relation to constructing ALN based classifiers. The rules for fast and efficient design of intelligent engines based on ALN classifiers are also discussed. Finally, a new neurochip design strategy is introduced which combines neural network approaches with the machine learning objective of imitating human intelligence.

12.1. Introduction

Three main historical strands of research can be identified when the ultimate goal is to solve a classification problem: ***statistical***, ***machine learning*** and ***neural network***. Many pattern recognition and classification systems have been developed for electrophysiologic signals in research and clinical systems using the neural network perspective [6,12,13,14,15,16].

A challenging biosignal engineering problem [8,11,20,21] is the classification of patients with essential hypertension [9,19,23] disease into prognostic classes of physiologic and pathologic behavior on the basis of 24h biosignal Holter measurements [5,7].

The integrated intelligent classification engine (I^2CE) is defined to be a system constructed with neural networks data mining techniques capable of recognizing and classifying specific diagnostic output states. The basic entity of the I^2CE system is a classification function which is feasible to be produced using an adaptive logic network (ALN) [1]. This function is capable to determine whether or not biosignal input and diagnostic output quantities are in a certain relationship to each other.

In the ALN approach used herein, I^2CE learned the linear pieces functions [1,2] from the available empirical data during training. Each ALN extracted after simultaneous training and testing for a specific number of epochs, is a type of feedforward multilayer perceptron [14] that uses linear threshold units in a first hidden layer, and logic gates AND and OR in other hidden layers and in the output layer. The units form a tree with a single boolean output.

In this work, the rules for designing intelligent engines based on ALN classifiers are investigated. Furthermore, the global behavior of the I^2CE engine is exploited using modeling techniques.

12.2. The Concept of a Neural Classifier

Any context of decision or forecast on the basis of currently available information is covered by the term of ***classification***. A ***classification procedure*** is a formal method for repeatedly making such judgements in new situations.

Generally, neural classifiers [10] consist of layers of interconnected nodes, each node producing a non-linear function of its input. The input to a node may come from other nodes or directly from the input data. Also, some nodes are identified with the output of the network. The complete network therefore, represents a very complex set of interdependencies which may incorporate any degree of nonlinearity, allowing very general functions to be modeled and perform the desired classification task. Neural networks can produce very good predictions, but they are neither easy to use nor easy to understand. The difficulties with ease of use stem mainly from the extensive data preparation required to get good results from a neural network model. The results are difficult to understand because a neural network is a complex nonlinear model that does not produce rules.

Training a neural network is the process of setting the weights on the inputs of each of the units in such a way that the network best approximates the underlying function, or expressing it in data mining terms, does the best job of predicting the target variable. A distinction is often made, in pattern recognition [22], between ***supervised*** and ***unsupervised*** learning. The former describes the case where the training data are accompanied by labels indicating the class of events that the measurements represent, or more generally, a desired response to the measurements. The well known supervised neural network architectures are the Perceptron and Multi Layer Perceptron, the Cascade Correlation learning architecture, and Radial Basis Function Networks. Unsupervised learning refers to the case where measurements are not accompanied by class labels. Networks exist which can model the structure of samples in the measurement, or attribute space, usually in terms of a probability density function, or by representing the data in terms of cluster centres and widths. Such models include Gaussian mixture models and Kohonen networks.

Adaptive logic networks (ALNs) use an evolutionary mechanism to grow neural networks to have success as classifiers. An ALN is an artificial neural system which uses only simple logical functions for building a decision tree architecture during training [1]. The knowledge is stored in the network architecture and in the functions of the system nodes. ALNs use hard limiters instead of sigmoids, and the hidden layers above the first one are all logic gates AND or OR [2]. During training, the ALN applies least squares fitting using many linear pieces at once; after training a decision tree is created that partitions the input space into blocks, in each of which the function is represented by a simple expression containing a few linear pieces connected by MAXIMUM and MINIMUM operations of Armstrong, *et. al* [3]. As the tree in the ALN learns which branches are not providing any additional information, it evolves by deleting those branches. ALNs have been proved ideal for supervised classification [4,18].

12.3. Biosignal Quality Issues

In this section, the biosignal acquisition methods used to assure data quality which is a fundamental requirement in data mining processes are summarized. Two different research designs were adopted and compared. The first approach involved the extraction of efficient ALN classifiers for each essential hypertension waveform parameter (i.e. 24-hour systolic arterial pressure (SAP), 24-hour diastolic arterial pressure (DAP) and 24-hour heart rate (HR)) on biosignals quantified every 15 minutes. In the second approach all essential hypertension parameters were used for the extraction of a combined ALN classifier on biosignals quantified every 30 minutes. The same group of patients was used for both the experimental designs and patient records containing missing values were excluded from the study due to the constraint that neural networks cannot deal with missing values.

12.3.1. *Patients and Ambulatory Monitoring*

Two hundred and thirty patients with essential hypertension were studied. Mean age $\pm$ SE of these patients was 52.82 $\pm$ 12.05 years. All patients had normal renal function tests and normal plasma renin activity. None had evidence of other disease that might had affected the myocardial structure. Study entry criteria included an eight-to-twelve year history of arterial hypertension. After discontinuing all drugs for at least two weeks, the patients underwent ambulatory blood pressure in a typical workday. Blood pressure and heart rate were recorded every fiftheen minutes. Therefore, during a 24-hour period, up to 96 blood pressure and 96 heart rate values were obtained. A Spacelabs 90202 ABP monitor (Spacelabs Inc., Redmond, Wash.) was used for 24-hour ambulatory monitoring. The accuracy of the monitoring system was compared to mercury sphygmomanometer readings both before each patient left the hospital and on the following day when he returned.

12.3.2. *Echocardiographic Measurements*

All echocardiograms were obtained by a skilled sonographer utilizing a Sigma-1C echocardiograph (Kontron Instruments Inc., Everett, Mass.) and a 3.5 MHz transducer. Two-dimensional guided M-mode echocardiograms at the level of the chordae tendineae were recorded. Numbers were randomly assigned to all echocardiograms, blinding all patient identification and time sequence. Each echocardiogram was read by two investigators. Four to six cycles of septal and posterior wall thicknesses and left ventricular internal diastolic and systolic dimensions were measured by each reader, using the American Society of Echocardiography (ASE) convention of measurement. The ASE convention marks diastole as the onset of the QRS complex and measures wall thickness and chamber dimensions from one leading edge to another. The mean measurements for each echocardiogram were recorded and the average of the two investigators was used for all calculations.

12.3.3. *Clinical Classification Criteria*

Patients were clinically characterized as ***pathologic*** in cases that the available echocardiographic indices expressed per body surface area satisfied at least one of the following criteria: (i) $PW_D > 1.1$, or (ii) $IVS_D > 1.1$, or (iii) $LV_{mass} > 110$. In all other cases patients were characterized as ***physiologic*** (Table 12-1).

Table 12-1. Echocardiographic left ventricular hypertrophy indices.

IVS_D	end-Diastolic IntraVentricular Septum thickness
PW_D	left ventricular Posterior Wall Diastolic thickness
LV_{mass}	Left Ventricular mass

12.4. Classification via Adaptive Logic Networks (ALNs)

Various *Adaptive Logic Networks* (ALNs) were trained for a different number of epochs to produce the classification functions together with the errors and the percentages of the classification accuracy. The basic components of an ALN are: (i) ***the Linear Threshold Units*** (LTUs), which are the leafs of the decision tree architecture and (ii) ***other nodes*** which are AND and OR operators with two or more inputs [1]. Since the ALN is trained, several linearform equations are produced.

In all experiments the training strategy adopted was: (i) tree growable starting with one linear function (LF), (ii) min RMSE: 0.001, (iii) learning rate: 0.2, (iv) jitter off and (v) confidence interval (for error):99.90. The sinusoidal ALN classifiers are presented on the basis of tree growing behavior and the elapsed time to achieve training for different predefined number of epochs scenarios. Time starting point was set to 13:00 for all input patterns.

12.4.1. *Extraction of Sinusoidal ALN Classifiers*

In this approach, the 96 different values of each essential hypertension waveform parameter per patient;***measured at 15 min intervals within the 24-hour period***, form the input variables (features) for sinusoidal ALN classifier design (Tables 12-2, 12-3, 12-4).

The ALN architectures were evaluated using an unseen data set consisting of physiologic and pathologic patients.

12.4.2. *The Combined Biosignal Engineering Approach*

In this approach the training patterns of pathologic and physiologic patients were consisted of 24-hour systolic arterial pressure (SAP), diastolic arterial pressure (DAP) and heart rate (HR) values ***measured every 30 minutes.***

Each linear piece L_i in the ALN architecture under exploitation accepts the input features (equation 12-1) of the essential hypertension waveform parameters, where the input variables represent the systolic arterial pressure input pattern (i.e. $x_1, .., x_{48}$), the diastolic arterial pressure input pattern (i.e. $x_{49}, .., x_{96}$), the heart rate pattern (i.e. $x_{97}, .., x_{144}$) and also the clinical characterization as pathologic or physiologic (i.e. x_{145}).

$$ALN - input - variables = (x_1, x_2, ..x_{144}, x_{145}) \tag{12-1}$$

Various ALN architectures were built on the basis of the combined patterns, to learn functions from the empirical data presented to them dur-

Table 12-2. Experimental systolic arterial pressure ALN architectures.

Epochs	LFs	Elapsed Time
50	11/14	mins: 00, secs: 02
100	28/33	mins: 00, secs: 04
200	48/53	mins: 00, secs: 08
300	49/55	mins: 00, secs: 12
400	53/56	mins: 00, secs: 17
500	53/56	mins: 00, secs: 21
600	50/56	mins: 00, secs: 26
700	51/56	mins: 00, secs: 30
800	52/56	mins: 00, secs: 35
900	53/57	mins: 00, secs: 39
1000	54/57	mins: 00, secs: 44
2000	60/67	mins: 01, secs: 33
3000	59/67	mins: 02, secs: 26
4000	57/67	mins: 03, secs: 11
5000	51/67	mins: 07, secs: 25
6000	47/67	mins: 04, secs: 36
7000	48/68	mins: 05, secs: 39
8000	44/68	mins: 06, secs: 29
9000	44/68	mins: 07, secs: 20
10000	49/68	mins: 07, secs: 48

Table 12-3. Experimental diastolic arterial pressure ALN architectures.

Epochs	LFs	Elapsed Time
50	10/11	mins: 00, secs: 01
100	25/28	mins: 00, secs: 03
200	56/61	mins: 00, secs: 08
300	86/91	mins: 00, secs: 12
400	81/93	mins: 00, secs: 19
500	86/93	mins: 00, secs: 25
600	90/93	mins: 00, secs: 30
700	87/93	mins: 00, secs: 37
800	86/93	mins: 01, secs: 10
900	88/93	mins: 00, secs: 48
1000	88/93	mins: 00, secs: 54
2000	88/93	mins: 01, secs: 56
3000	85/93	mins: 03, secs: 07
4000	76/93	mins: 08, secs: 41
5000	74/93	mins: 20, secs: 12
6000	68/93	mins: 25, secs: 57
7000	63/93	mins: 31, secs: 35
8000	58/93	mins: 33, secs: 50
9000	60/94	mins: 34, secs: 49
10000	58/94	mins: 34, secs: 18

Table 12-4. Experimental heart rate ALN architectures.

Epochs	LFs	Elapsed Time
50	16/17	mins: 00, secs: 02
100	35/37	mins: 00, secs: 04
200	54/56	mins: 00, secs: 08
300	51/57	mins: 00, secs: 13
400	54/57	mins: 00, secs: 18
500	54/57	mins: 00, secs: 21
600	54/57	mins: 00, secs: 27
700	52/57	mins: 00, secs: 31
800	55/57	mins: 00, secs: 36
900	52/57	mins: 00, secs: 40
1000	51/58	mins: 00, secs: 45
2000	53/78	mins: 01, secs: 50
3000	57/85	mins: 04, secs: 21
4000	56/95	mins: 12, secs: 07
5000	52/106	mins: 20, secs: 33
6000	53/109	mins: 26, secs: 49
7000	51/109	mins: 31, secs: 50
8000	42/109	mins: 34, secs: 48
9000	49/109	mins: 35, secs: 49
10000	48/109	mins: 34, secs: 49

ing training and testing for a specific number of epochs (Table 12-5).

12.4.3. *Tree Growing and Speed Comparisons*

The experimental results of the previous subsections (Tables 12-2, 12-3, 12-4, 12-5) indicate that the simplest ALN architectures are produced in the combined engineering approach. This finding is supported by the small number of linear functions (LFs) extracted in each experiment conducted (Table 12-5). Since these units form a tree with a single boolean output, the ALN architectures of the combined biosignal engineering approach exhibit less complexity than the others (those extracted in the exploitation of the ALN architectures expressing the independent effect of each essential hypertension sinusoidal component; SAP,DAP,HR).

Furthermore, in the neural network training scientific literature, it is generally accepted that large number of input features make it more difficult for the network to find patterns and can result in long training phases that never converge to a good solution. The combined biosignal

Table 12-5. Experimental combined biosignal engineering ALN architectures.

Epochs	LFs	Elapsed Time
50	11/12	mins: 00, secs: 03
100	21/22	mins: 00, secs: 05
200	25/25	mins: 00, secs: 09
300	25/25	mins: 00, secs: 15
400	25/25	mins: 00, secs: 19
500	25/25	mins: 00, secs: 23
600	25/25	mins: 00, secs: 28
700	25/25	mins: 00, secs: 34
800	25/25	mins: 00, secs: 37
900	25/25	mins: 00, secs: 44
1000	23/25	mins: 00, secs: 47
2000	25/25	mins: 01, secs: 36
3000	24/25	mins: 02, secs: 25
4000	24/25	mins: 03, secs: 15
5000	24/25	mins: 04, secs: 04
6000	21/25	mins: 04, secs: 48
7000	20/25	mins: 05, secs: 46
8000	17/25	mins: 06, secs: 31
9000	18/25	mins: 07, secs: 26
10000	19/25	mins: 08, secs: 13

engineering approach exploited herein, extracts classification results at extremely highest speed compared to the single component approach (Tables 12-2, 12-3, 12-4, 12-5). This finding indicates that in the combined approach, ALNs perform at high speed although having more input features.

12.5. Classifier Quality Assurance

The quality metrics to accept a neural classifier as reliable and efficient are the maximization of the classification accuracy in unseen data sets and the minimization of the errors produced in training, test and validation phases. These metrics were applied in the design of the Integrated Intelligent Classification Engine (I^2CE) and resulted to mathematical models both for the error and the classification accuracy prediction.

12.5.1. *Methodology for Behavioral Modeling*

Curve-fitting process for each state of the I^2CE engine (i.e. sinusoidal or combined) consists of three stages: (a) linear fit, (b) non-linear fit and (c) determination of best curve fit.

In the stage of the linear fit, the progress in the fitting of the up to 3304 linear equations [17] was examined and the linear equations were fitted by a two-pass direct method that was proved very fast. The maximum number of terms in an equation was set to be between 1 and 11 and the term significance between 2 and 15. The Gauss-Jordan, LUDecomp, and SVDecomp options were used selectively by the linear curve-fitting algorithm for all equations, both to solve for the coefficients and for the errors/confidence intervals. In further detail the estimation of the Coefficients and Errors/Confidence Intervals varied with the equations as follows: (a) Coefficients by direct matrix solution, Errors/Confidence Intervals by LU Decomposition (b) Coefficients by Gaussian Elimination, Errors/Confidence Intervals by LU Decomposition and (c) Coefficients by SVD (Singular Value Decomposition), Errors/Confidence Intervals by SVD. In the stage of the nonlinear fit there were used 14 possible non-linear equations [17]. The maximum numbers of iterations were set from 2 to 999. The converge threshold option sets the criteria for stopping the iterative fitting and could range from 3 to 15. A value of 6 was set to represent stopping criteria whereby the R^2 coefficient was unchanged within 10^{-6} for five iterations.

In order to obtain the best curve fit, the curve-fitted equations were sorted selectively by the R^2 coefficient of determination, the degree-of-freedom adjusted R^2, the overall fit standard error (Root MSE), or by the F-statistic. The R^2 options and standard error sorts were used for approximating functions whereas the F-statistic was used where a parametric model was desired.

12.5.2. *Classification Accuracy Behavior*

The classification accuracy of the experimental ALN architectures is governed by mathematical models derived using the proposed behavioral modeling methodology.

12.5.2.1. *Behavioral Models of Sinusoidal ALN Classifiers*

Classification accuracy models of the ALN classifiers as a function of the number of epochs are presented for systolic arterial pressure in Fig.

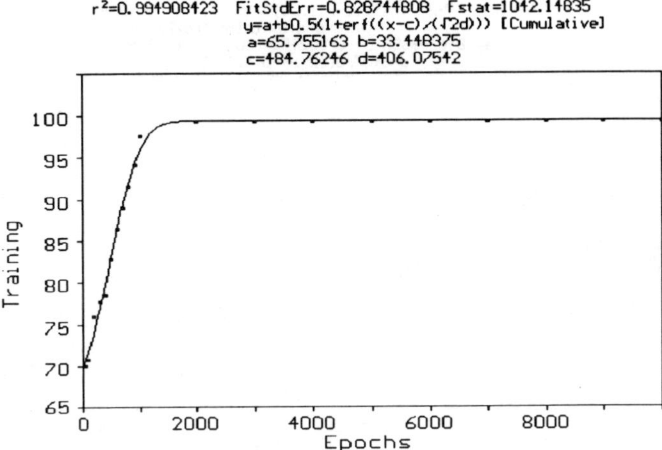

Fig. 12-1. Behavioral model of systolic arterial pressure ALNs classification accuracy for the training set.

12-1, 12-2, 12-3, for diastolic arterial pressure in Fig. 12-4, 12-5, 12-6 and for heart rate in Fig. 12-7, 12-8, 12-9.

12.5.2.2. *Behavioral Models of the Combined ALN Classifiers*

The classification accuracy models [16] for the combined biosignal engineering approach are presented in Fig. 12-10, 12-11, 12-12.

12.5.3. *Error Behavior*

The ALN engines were also found to terminate operation with different train and test error values for the different number of epochs scenarios used for experimentation. Errors are presented in Tables 12-6, 12-7, 12-8, 12-9 together with the classification accuracy percentages achieved during training and testing the architectures for the predefined numbers of epochs.

Validation results of the ALN architectures with an unseen dataset containing pathologic and physiologic essential hypertension patterns are presented in Fig. 12-13.

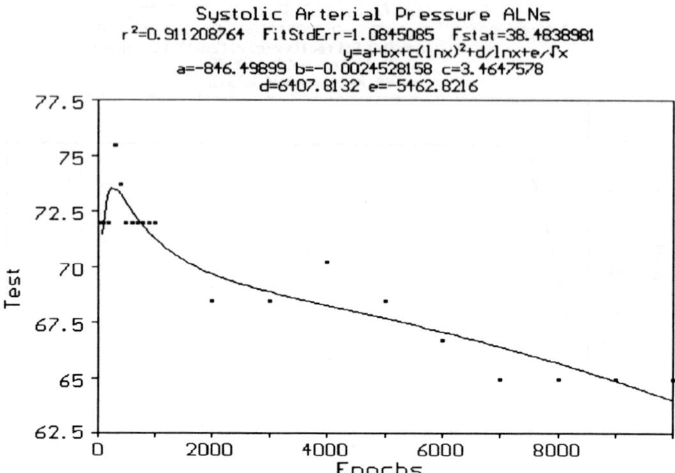

Fig. 12-2. Behavioral model of systolic arterial pressure ALNs classification accuracy for the test set.

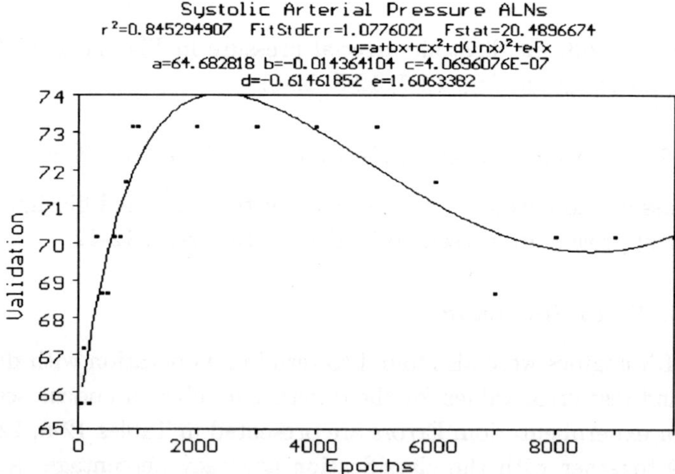

Fig. 12-3. Behavioral model of systolic arterial pressure ALNs classification accuracy for the validation set.

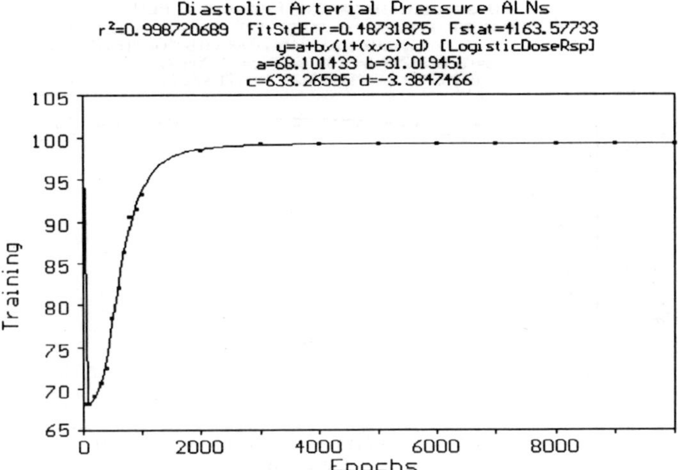

Fig. 12-4. Behavioral model of diastolic arterial pressure ALNs classification accuracy for the training set.

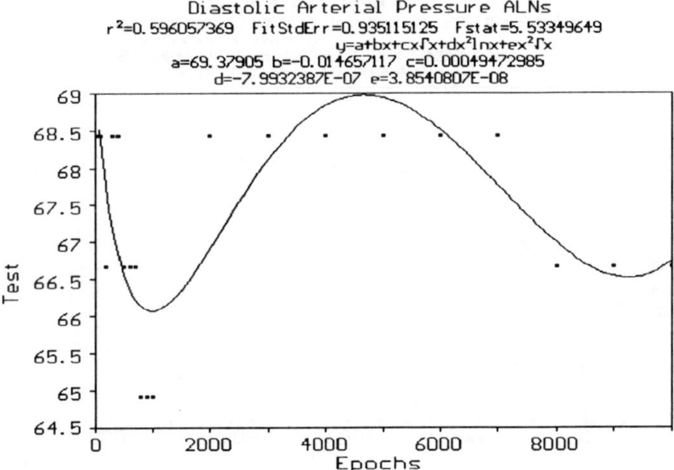

Fig. 12-5. Behavioral model of diastolic arterial pressure ALNs classification accuracy for the test set.

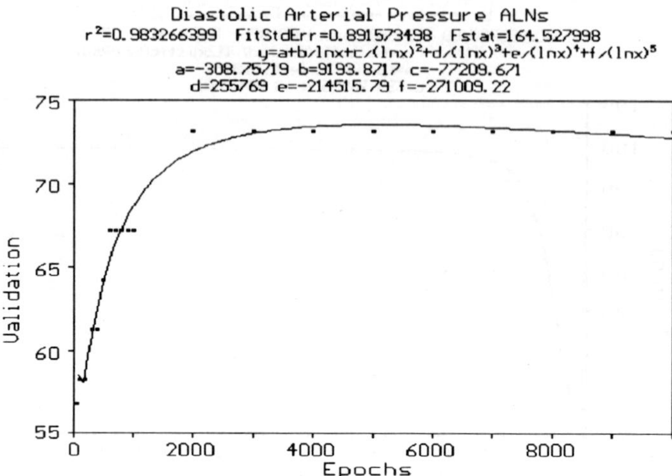

Fig. 12-6. Behavioral model of diastolic arterial pressure ALNs classification accuracy for the validation set.

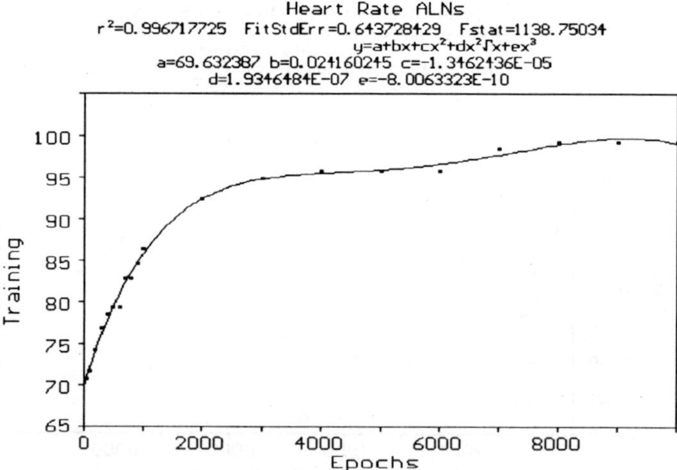

Fig. 12-7. Behavioral model of heart rate ALNs classification accuracy for the training set.

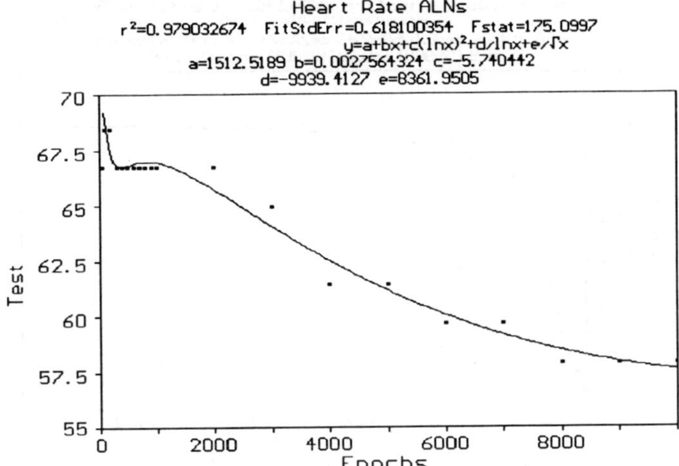

Fig. 12-8. Behavioral model of heart rate ALNs classification accuracy for the test set.

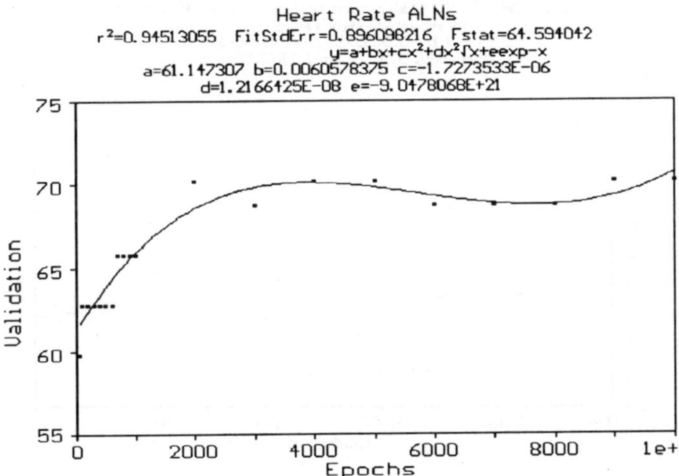

Fig. 12-9. Behavioral model of heart rate ALNs classification accuracy for the validation set.

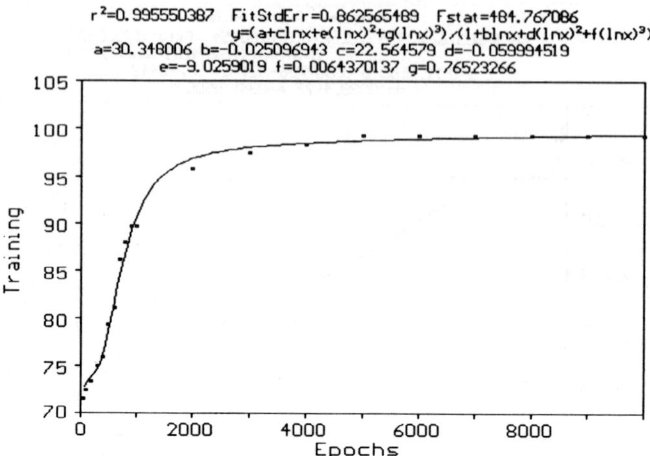

Fig. 12-10. Behavioral model of the combined biosignal engineering ALNs classification accuracy for the training set.

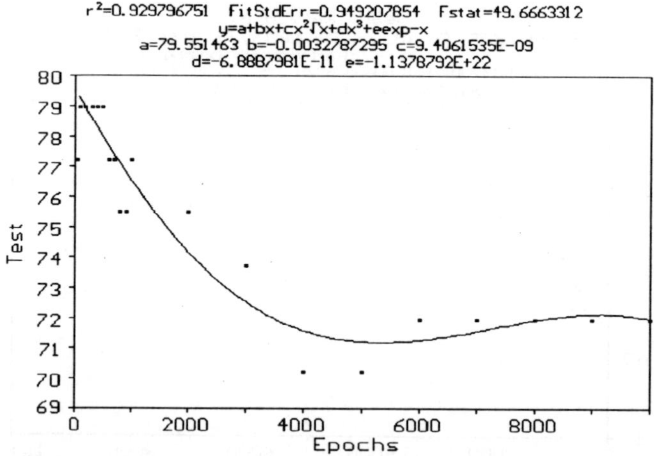

Fig. 12-11. Behavioral model of the combined biosignal engineering ALNs classification accuracy for the test set.

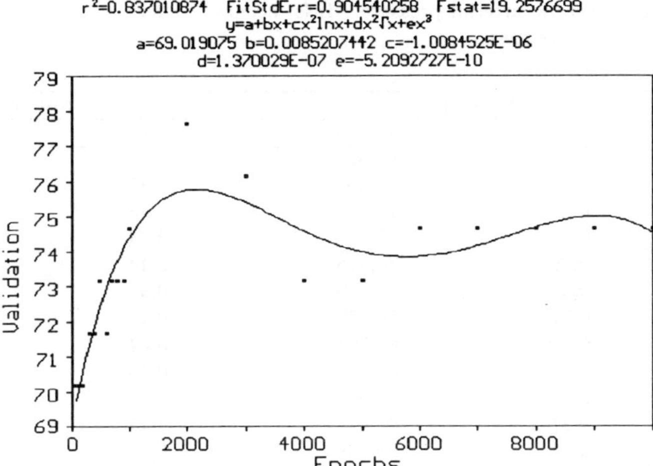

Fig. 12-12. Behavioral model of the combined biosignal engineering ALNs classification accuracy for the validation set.

12.6. Conclusions and Outlook

In this chapter, various adaptive logic networks were built and evaluated based on the relational approach for ALNs. Research focus was on the development and control of an integrated intelligent classification engine (I^2CE) for biosignal engineering.

The proposed research design strategies revealed the power of the adaptive logic networks as classifiers in biosignal engineering investigations. Apart from the classification quality issues, tree growing behavior and execution speed were also exploited on a challenging medical data set.

Experimental results from essential hypertension biosignals indicate that the design of the (I^2CE) engine should be based on the combined biosignal engineering approach. The combined ALNs, extracted herein, appart from their high classification accuracy provide also high speed of execution and simplicity for digital design of the ALN classification function.

Future work will be focused on the extensive validation of the combined biosignal engineering ALN architectures aiming at the identification of the combined ALN architecture that detects with 100% classi-

Table 12-6. Systolic arterial pressure ALNs; accuracy refers to the percentage of correctly classified patterns.

Epochs	Training set		Test set	
	error	accuracy(%)	error	accuracy(%)
50	0,44	69,828	0,43	71,930
100	0,43	70,690	0,43	71,930
200	0,40	75,862	0,43	71,930
300	0,38	77,586	0,44	75,439
400	0,36	78,448	0,44	73,684
500	0,34	82,759	0,44	71,930
600	0,32	86,207	0,44	71,930
700	0,31	88,793	0,44	71,930
800	0,29	91,379	0,45	71,930
900	0,28	93,966	0,45	71,930
1000	0,27	97,414	0,45	71,930
2000	0,18	99,138	0,47	68,421
3000	0,13	99,138	0,48	68,421
4000	0,11	99,138	0,48	70,175
5000	0,09	99,138	0,49	68,421
6000	0,08	99,138	0,49	66,667
7000	0,08	99,138	0,49	64,912
8000	0,07	99,138	0,49	64,912
9000	0,07	99,138	0,49	64,912
10000	0,07	99,138	0,50	64,912

fication accuracy the hypertensives without hypertension (i.e. without target organ damage prognostic prediction). This architecture will permit the digital implementation of the (I^2CE) engine as a neurochip to be embedded in biomedical engineering devices for 24-hour blood pressure and heart rate monitoring.

References

[1] W.W. Armstrong and M.M. Thomas. Adaptive logic networks, Sec. C.1.8 in the Handbook of neural computation, E. Fieseler and R. Beale (eds.). Institute of physics publishing and Oxford university press, USA, (1996).

[2] W.W. Armstrong and G. Godbout. Properties of binary trees of flexible elements useful in pattern recognition. IEEE International Conf. on Cybernetics and Society, San Francisco, USA. IEEE Cat. No.75, CHO 997-7 SMC., pp.447-449, (1975).

[3] W.W. Armstrong, R.B. Stein, A. Kostov, M. Thomas, P. Baudin, P. Gervais, D. Popovic. Application of adaptive logic networks and dy-

Table 12-7. Diastolic arterial pressure ALNs; accuracy refers to the percentage of correctly classified patterns.

Epochs	Training set		Test set	
	error	accuracy(%)	error	accuracy(%)
50	0,46	68,103	0,46	68,421
100	0,45	68,103	0,46	68,421
200	0,42	68,966	0,46	66,667
300	0,41	70,690	0,46	68,421
400	0,39	72,414	0,46	68,421
500	0,37	78,448	0,46	66,667
600	0,36	81,897	0,46	66,667
700	0,35	86,207	0,46	66,667
800	0,33	90,517	0,46	64,912
900	0,32	91,379	0,46	64,912
1000	0,31	93,103	0,46	64,912
2000	0,23	98,276	0,47	68,421
3000	0,17	99,138	0,47	68,421
4000	0,14	99,138	0,48	68,421
5000	0,12	99,138	0,49	68,421
6000	0,11	99,138	0,49	68,421
7000	0,10	99,138	0,49	68,421
8000	0,09	99,138	0,49	66,667
9000	0,08	99,138	0,50	66,667
10000	0,08	99,138	0,50	66,667

namics to study and control of human movement. *Second International Symposium on 3D Analysis of Human Movement*, Poitiers, June 30-July 3. pp.81-84, (1993).

[4] W. Armstrong, C. Chu and M. Thomas. Feasibility of using Adaptive Logic Networks to Predict Compressor Unit Failure. in *Proceedings Battelle Pacific Northwest Laboratories Workshop on Environmental and Energy Applications of Neural Networks*, Richland WA, USA, March 30-31, (1995).

[5] K. Benson and A.J. Hartz, A comparison of observational studies and randomized, controlled trials. *N Engl J Med*, **342**, pp.1878-6, (2000).

[6] F. De Carli. Neural networks for pattern recognition and classification in the analysis of electrophysiologic signals. in *Proceedings of Neural Networks in Biomedicine*, Singapore, World Scientific, pp.287-302, (1994).

[7] J. Concato, N. Shah and R.I. Horwitz. Randomized, controlled trials, observational studies, and the hierarchy of research designs. *N Engl J Med*, **342**, pp.1887-92, (2000).

[8] F. Davidoff, B. Haynes, D. Sackett and R. Smith. Evidence based medicine. *BMJ*, **310**, pp.1085-6, (1995).

Table 12-8. Heart rate ALNs; accuracy refers to the percentage of correctly classified patterns.

Epochs	Training set		Test set	
	error	accuracy(%)	error	accuracy(%)
50	0,45	70,690	0,48	66,667
100	0,44	71,552	0,48	68,421
200	0,42	74,138	0,48	68,421
300	0,41	76,724	0,48	66,667
400	0,39	78,448	0,48	66,667
500	0,38	79,310	0,48	66,667
600	0,37	79,310	0,48	66,667
700	0,36	82,759	0,48	66,667
800	0,35	82,759	0,48	66,667
900	0,34	84,483	0,48	66,667
1000	0,33	86,207	0,49	66,667
2000	0,26	92,241	0,49	66,667
3000	0,22	94,828	0,50	64,912
4000	0,19	95,690	0,50	61,404
5000	0,17	95,690	0,51	61,404
6000	0,16	95,690	0,51	59,649
7000	0,14	98,276	0,51	59,649
8000	0,13	99,138	0,51	57,895
9000	0,12	99,138	0,51	57,895
10000	0,11	99,138	0,51	57,895

[9] Guidelines Subcommittee, World Health Organization-International Society of Hypertension guidelines for the management of hypertension. *J Hypertens*, **17**, pp.151-83, (1999).

[10] J. Hertz, A. Krogh and R. Palmer. Introduction to the Theory of Neural Computation. Addison-Wesley, (1991).

[11] H. Hiatt and L. Goldman L. Making medicine more scientific. *Nature*, **371**:100, 1994.

[12] A.N. Kastania. Unsupervised categorization of multivalued input patterns: an application to biosignal clustering. *Proceedings of Fourth Hellenic-European Conference on Computer Mathematics and its Applications*,(ed. E.A. Lipitakis), LEA Press, Greece, pp.444-450, (1998).

[13] A.N. Kastania and M.P. Bekakos. Mathematical networks for classification:the biosignal pattern matching paradigm. *Proceedings of the Fourth Hellenic-European Conference on Computer Mathematics and its Applications*, (ed. E.A. Lipitakis), LEA Press, Greece, pp.346-354, (1998).

[14] A.N. Kastania and M.P. Bekakos. Neurochip architectural design for pattern recognition and classification. *Applications of High Performance Computing in Engineering VI*,(eds. M. Ingber, H. Power, C.A.

Table 12-9. Combined biosignal engineering ALNs; accuracy refers to the percentage of correctly classified patterns.

Epochs	Training set		Test set	
	error	accuracy(%)	error	accuracy(%)
50	0,45	71,552	0,43	77,193
100	0,43	72,414	0,43	78,947
200	0,41	73,276	0,43	78,947
300	0,40	75,000	0,43	78,947
400	0,38	75,862	0,43	78,947
500	0,37	79,310	0,43	78,947
600	0,36	81,034	0,44	77,193
700	0,35	86,207	0,44	77,193
800	0,34	87,931	0,44	75,439
900	0,33	89,655	0,44	75,439
1000	0,32	89,655	0,45	77,193
2000	0,25	95,690	0,47	75,439
3000	0,20	97,414	0,49	73,684
4000	0,17	98,276	0,50	70,175
5000	0,15	99,138	0,50	70,175
6000	0,13	99,138	0,51	71,930
7000	0,12	99,138	0,51	71,930
8000	0,11	99,138	0,51	71,930
9000	0,10	99,138	0,51	71,930
10000	0,09	99,138	0,52	71,930

Brebbia), WIT Press., pp.259-270, (2000).

[15] A.N. Kastania and M.P. Bekakos. NN-modelling for non linear functions approximations. *Applications of High Performance Computing in Engineering VI*, (eds. M. Ingber, H. Power, C.A. Brebbia), WIT Press, pp.249-258, (2000).

[16] A.N. Kastania and M.P. Bekakos. Behavioral modeling of the Integrated Intelligent Classification Engine. *Proceedings of the 2001 International Conference on Parallel and Distributed Processing Techniques and Applications* (PDPTA'2001), June 25-28, Las Vegas, Nevada, USA, CSREA Press, pp.1780-1786, (2001).

[17] A.N. Kastania and M.P. Bekakos. A statistical approach to curve-fitting exploitation of biomedical waveforms. *Computer Simulations in Biomedicine*, H. Power and R.T. Hart (eds.), Computational Mechanics Pubs., pp.521-530, (1995).

[18] A. Kostov, D.B. Popovic, R.B. Stein, and W.W. Armstrong. EMG patterns learning by Adaptive Logic Network. *Proc. of the 15th Annual International Conference of the IEEE Engineering in Medicine and Biology Society*, San Diego, USA, pp.1135-1136, (1993).

[19] G. Perera. Hypertensive vascular disease:description and natural his-

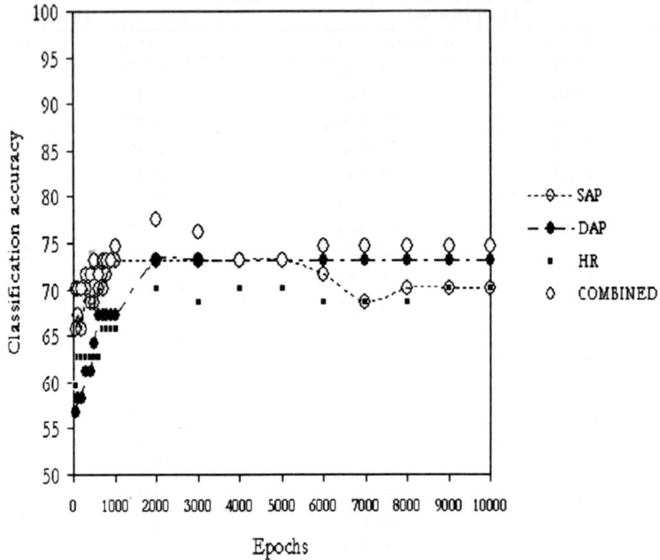

Fig. 12-13. Classification accuracy comparisons of the ALN architectures.

tory. *J Chronic Dis* ,**1**, pp.33-42, (1955).

[20] Rosenberg W and Donald A. Evidence based medicine:an approach to clinical problem-solving. *BMJ* , **310**, pp.1122-6, (1995).

[21] D.L. Sackett, W.C. Rosenberg , J.A. Muir Gray , R.B. Haynes and W.S. Richardson. Evidence-based medicine:what it is and what it isn't. *BMJ*, **312**:71, (1996).

[22] R.J. Schalkoff. Pattern Recognition:Statistical,Structural and Neural Approaches. Wiley, Singapore, (1992).

[23] J.D. Swales. Evidence-based medicine and hypertension. *J Hypertens*, **17**, pp.1511-6, (1999).

CHAPTER 13

MULTIPLIERLESS DESIGNS FOR ARTIFICIAL NEURAL NETWORKS

H.K. Kwan*

Department of Electrical and Computer Engineering
University of Windsor, 401 Sunset Avenue, Windsor, Ontario, Canada N9B 3P4
E-mail: kwan1@uwindsor.ca

**Visiting Professor*
Department of Electrical and Electronic Engineering
The University of Hong Kong, Pokfulam Road, Hong Kong

Optimal architectures for designing multiplierless multilayer feedforward neural networks suitable for various combinations of discrete/continuous input to discrete/continuous output mappings are introduced. For illustration, the details of one design for discrete input to discrete output mapping and one design for continuous input to continuous output mapping are presented. For discrete input to discrete output mapping, the network has continuous weights in the first layer and one-powers-of-two weights at the second and higher layers. The simplified sigmoid activation functions are used at the hidden neurons and the step activation functions are used at the output neurons. Consequently, no multiplier is needed for discrete input to discrete output mapping. For continuous input to continuous output mapping, the network adopts the simplified sigmoid activation functions at the output neuron.s, continuous weights in the second and higher layers, 3-level activation functions (of ±1 and 0 outputs, Quantized Neurons with S=0) at the hidden neurons, and one-powers-of-two weights in the input layer. Consequently, no multiplier is required for continuous input to continuous output mapping. Results show that each of these multiplierless multilayer neural networks can retain nearly identical recall performance of the corresponding continuous-weight network, while having reduced hardware cost and increased computational speed in digital implementation.

13.1. Introduction

Multilayer feedforward neural networks (MFNNs) are widely used for building artificial neural network systems. A multilayer feedforward neural network (MFNN) has a parallel structure that involves massive computational operations. In an MFNN, the basic operations involve taking the sum of weighted inputs, and then computing the activation through a nonlinear function. Nonlinear activation function computation and multiplication are the two most expensive items in the digital implementation of a MFNN.

The most popular nonlinear activation function that has been used in MFNNs is the sigmoid activation function (SAF). In [1], it has been shown that a SAF can be implemented by a simplified sigmoid activation function (SSAF) with one binary multiplication, one binary shift, one binary addition, and one comparator. This provides a simple way for digital implementation of the SAF especially when the sum of weighted inputs (as one of the binary multiplication inputs) can be adequately represented in a short binary word length format.

Assuming that M and N are, respectively, the word lengths in bits for representing the weight and the activation function output. A MFNN contains many multiplications and each M-bit x N-bit multiplication has a computational complexity proportional to MxN. The two inputs that are to be multiplied by a multiplier inside a MFNN are a M-bit weight and a N-bit activation function output. To simplify the MxN multiplication, we can reduce the word length requirement of either the weight or the activation function output by adopting a reduced word length representation format. This idea leads to two approaches for designing digital multiplierless multilayer feedforward neural networks (MMFNNs) such that multiplications can be eliminated. Before describing further, we consider the one-powers-of-two (OPOT) digital format, in which a value is represented in one of the following allowed discrete values $[\pm 1, \pm 2^{-1}, ..., \pm 2^{-S}, 0]$. This OPOT format is especially suitable for reduced word length digital representation because a multiplication can now be reduced to a binary shifting operation that can be directly implemented by a shift register.

The first approach to eliminate multiplications in a MFNN is to represent each weight in an OPOT format such that no multiplication is needed. We shall call this as the OPOT approach. OPOT weight designs have been presented in [2-5] for discrete (represented in binary [0,1] or bipolar [-1,+1]) input to discrete output mapping application. In [2-3], OPOT weights are used to quantize continuous weights, OPOT weights and slopes of activation functions are then adjusted adaptively to reduce the sum of squared output errors to a specified limit for discrete input to discrete output mapping. In [4-5], in addition to OPOT weights, the simplified sigmoid activation functions (SSAFs) are used at the outputs of hidden and output neurons, OPOT weights and biases are then adjusted to compensate the weights' quantization errors. Adapting biases [4-5] is a much simpler implementation strategy than that of adapting slopes of sigmoid activation functions (SAFs) [2-3].

The second approach to eliminate multiplications in a MFNN is to represent each activation function by a quantized activation function with OPOT output. We shall call this OPOT quantized activation function and its associated neuron as a quantized neuron (QN) [6-7]. We shall call this as the QN approach. In so doing, a multiplier, once again, is now reduced to a simple shift register. In [6-7], a method for designing MMFNNs using the quantized neurons (QNs) at the hidden layer and continuous weights for discrete input to discrete output mapping has been presented.

In real-world applications, the input and output formats can be continuous (represented in fixed-point or floating-point) or discrete (represented in binary or bipolar). By utilizing the properties of the given input to output format of an application, digital arithmetic can be designed in an optimal manner so as to reduce the cost of hardware implementation as well as to increase the speed of computation. So far, the previous two approaches have been focused on multiplierless designs that may not yet fully utilized the properties of some given input to out formats. In this chapter, we are going to describe a third approach which makes use of all possible combinations of OPOT weights, QNs, and a given input to output format to arrive at an optimal architecture for any discrete/continuous input and discrete/continuous output mapping application. We shall call this approach as the Mixed OPOT or Mixed QN approach. Individual design details of the third approach can be found in [8-11]. In [8], an optimal design for MMFNNs tailored to discrete input to discrete output type of applications has been advanced and shall be described in Section 13.5. Instead of adopting continuous weights in the first layer and OPOT weights in the second layer as in [8], a study on the adoption of uniformly quantized sum-of-powers-of-two (SOPT) weights in both layers and more choices of uniformly quantized SOPT weights are given in the first layer is presented in [9]. In [10-11], an optimal MMFNN architecture is introduced for continuous input to continuous output mapping and the details shall be described in Section 13.6. In this chapter, our focus is on optimal MMFNN architectures, in which one of the two inputs to be multiplied is represented in OPOT format, that make use of a specified input to output mapping format.

There are a number of other design techniques for reducing the hardware complexity in the digital implementation of artificial neural networks. A design of recurrent neural networks with [-1, 0, +1] weights, which is a special case of OPOT weights with S=0 (see 13-16), for word sequence learning is described in [12-13]. A multiplierless one-layer feedforward neural network for maximum or minimum computation is described in [14]. In [15], weight values of multilayer perceptrons are restricted to powers-of-two or sum of powers-of-two, thus saving chip area and computation time. The design was applied to pattern recognition problems with a binary input to binary output format. In [16-17], integer weights in the set [-3, -2, -1, 0, 1, 2, 3] are used in multilayer feedforward neural networks for applications to XOR, encoder/decoder, and MONK type of binary input to binary output problems. The results obtained in [17] also suggest that, in many cases, limited weight resolution can be offset by an increase in the size of its hidden layer. In [18-19], hard-limiter activation functions, integer weights, and integer thresholds are used to facilitate the actual hardware implementation of multilayer binary neural networks. The design was applied to the circular region, a 4-bit odd-parity function, and a 7-bit random function type of binary input to binary output problems. In [20], the VLSI implementation of a multilayer neural network with ternary activation functions and limited integer weights is described, and multiplexing is used to solve I/O limitation. The design was applied to a handwritten digital recognition problem. In [21], multiplierless designs for the Hopfield network and the multi-layer perceptron are presented. In the designs, one-to-four shifts and zero-to-three additions type of operations are used to replace a multiplication. In addition, sign-digit representation of weights is adopted which leads to a minimal number of non-zero digitals needed for the operations. The multiplierless Hopfield network design was applied to estimate motion between successive frames of digital video, and the multiplierless multilayer perceptron design was applied to a binary input to binary output mapping problem.

The organization of this chapter is as follows: The backpropagation-based adaptive algorithms for the weights and for the biases of a MFNN are summarized in Section 13.2. The equations defining the SSAF and the three-level activation function (3-LAF) are given in Section 13.3. Section 13.4 summarizes the optimal MMFNN architectures for various combination of discrete/continuous input to discrete/continuous output mappings. An optimal

MMFNN architecture for discrete input to discrete output mapping is presented in Section 13.5. In Section 13.5, we also include the design strategy, the design algorithm, the simulation results, and the hardware comparison of the design. In Section 13.6, the design of an optimal MMFNN architecture for continuous input to continuous output mapping is presented. Similarly, in Section 13.6, we also include the design strategy, the design algorithm, the simulation results, and the hardware comparison of the design. Finally, a conclusion is given in Section 13.7.

13.2. Weight and Bias Adaptation

13.2.1. *Weight Adaptation*

In an MFNN, the activation of the jth neuron at the hth layer during the presentation of the kth input pattern $\mathbf{X}_k = [x_{1k}, x_{2k}, x_{3k}, \ldots, x_{N_0k}]$ can be computed as:

$$y_{jk}^{[h]} = F[\sum_{i=1}^{N_{h-1}} w_{ij}^{[h]} \; y_{ik}^{[h-1]} + b_j^{[h]}] \tag{13-1}$$

for j=1 to N_h, h=1 to H, and k=1 to K. $y_{ik}^{[h-1]}$ is the output of a neuron i at the layer h-1; $w_{ij}^{[h]}$ is the weight between a neuron i at layer h-1 and a neuron j at layer h; $b_j^{[h]}$ is the bias of a neuron j at layer h; N_h is the number of neurons at layer h; F[.] represents the activation function; and

$$y_{ik}^{[0]} = x_{ik} \;\; for\; i = 1\; to\; N_0 \tag{13-2}$$

The sum of the squared output errors (SSE) related to the input pattern k is defined as:

$$e_k = \sum_{j=1}^{N_H} (t_{jk} - y_{jk}^{[H]})^2 \tag{13-3}$$

where t_{jk} represents the jth element of the target pattern k and h=H refers to the output layer. Based on gradient descent, the change in weights [22-23] can be expressed as:

$$\Delta_k w_{ij}^{[h]} = -\varepsilon \; \partial e_k / \partial w_{ij}^{[h]} \tag{13-4}$$

where ε is a learning rate parameter for weights. After some derivations, we obtain

$$\Delta_k w_{ij}^{[h]} = \varepsilon\, \delta_{jk}^{[h]}\, y_{ik}^{[h-1]} \tag{13-5}$$

where

$$\delta_{jk}^{[h]} = F'[z_{jk}^{[h]}] \sum_{i=1}^{N_{h+1}} \delta_{ik}^{[h+1]}\, w_{ji}^{[h+1]} \tag{13-6}$$

for h<H, and

$$\delta_{jk}^{[H]} = 2(t_{jk} - y_{jk}^{[H]})F\,[z_{jk}^{[H]}] \tag{13-7}$$

Here F'[z] represents the partial of F[z] with respect to z.

13.2.2. *Bias Adaptation*

Similarly, based on gradient descent, the bias of a neuron can also be adapted [22-23] as

$$\Delta_k b_j^{[h]} = -\,\varepsilon_b\ \partial e_k / \partial b_j^{[h]} \tag{13-8}$$

where ε_b is a step size for bias adaptation. We can show that

$$\Delta_k b_j^{[h]} = \varepsilon_b\, \delta_{jk}^{[h]} \tag{13-9}$$

Both weights and biases are updated at the end of each epoch.

13.3. Activation Functions

13.3.1. *Simplified Sigmoid Activation Function*

A type of activation function commonly used in MFNNs is the SAF in which its bipolar version is of the following form

$$G(z) = g(1 - e^{-\alpha z})/(1 + e^{-\alpha z}) \tag{13-10}$$

where g is a scaling factor. The above expression involves an infinite exponential series and traditionally it can be implemented by a look-up-table

(LUT). Recently, a simplified version of the above sigmoid activation function is described in [1] with an approximate order of complexity of one binary multiplication, which provides an attractive method for digital implementation of a SAF. This SSAF, saturates to ±1 at and beyond points ±L, can be expressed as

$$G_s(z)=\begin{cases} 1 & L\leq z \\ H_s(z) & -L<z<L \\ -1 & z\leq -L \end{cases} \tag{13-11}$$

where

$$H_s(z)=\begin{cases} z(\beta-\theta z) & 0\leq z\leq L \\ z(\beta+\theta z) & -L\leq z<0 \end{cases} \tag{13-12}$$

and $\Theta=1/L^2$, $\beta=2/L$, and L is an arbitrary positive number determining the saturation points. If an OPOT value of L is used, the above defined activation function $G_s(z)$ can be implemented by one binary multiplication, one binary shift, one binary addition, and one comparator.

To prevent the learning process from being stuck at either saturation region, a small positive value σ is assigned to the derivative of $G_s(z)$ at and beyond ±L, hence

$$G'_s(z)=\begin{cases} \sigma & L\leq z \\ H'_s(z) & -L<z<L \\ \sigma & z\leq -L \end{cases} \tag{13-13}$$

where $G'_s(z)$ and $H'_s(z)$ represent, respectively, the partial derivatives of $G_s(z)$ and $H_s(z)$ with respect to z.

13.3.2. *Three-level Activation Function*

The 3-LAF applied at the output of each hidden neuron is defined as

$$G_3(z)=\begin{cases}1 & t_3\leq z\\ 0 & -t_3<z<t_3\\ 1 & z\leq -t_3\end{cases} \qquad (13\text{-}14)$$

where t_3 (=0.33) is the threshold value. Before training, the three intersections of $G_3(z)$ with the scaled bipolar SAF (at g=1.1 in (13-10)) were found. The derivatives of $G_3(z)$ at these three intersections of z were used, respectively, as the approximate derivatives of $G_3(z)$ in the three regions of (13-14) during training.

13.4. Optimal Architectures for MMFNN Designs

As briefly introduced in Section 13.1, there are a total of three approaches to arrive at optimal architectures for MMFNN designs. The first approach is the OPOT approach that adopts only OPOT weights (W_{OPOT}) and SSAFs for continuous input to continuous output mapping. The second approach is the QN approach that adopts QNs (except StepAFs at the output layer) and continuous weights (W_C) for discrete input to discrete output mapping.

The third approach is the Mixed approach, which can be applied to various combinations of discrete/continuous input to discrete/continuous output mappings. The Mixed approach can be subdivided into the Mixed OPOT approach and the Mixed QN approach. In the Mixed OPOT approach, W_C have to be used as weights in the input layer for discrete input, and StepAFs have to be used at the output layer for discrete output. In the Mixed QN approach, OPOT weights (W_{OPOT}) have to be used in the input layer for continuous input, and SSAFs have to be used in the output layer for continuous output.

For illustration, these optimal architectures are summarized in Tables 13-1 to 13-4. The dotted line '… … …' at the middle of each of the Tables 13-1 to 13-4 represents that the same hidden (weights and neurons) architecture can be inserted for additional hidden layers.

Table 13-1. Optimal MMFNN architectures based on OPOT and QN approaches.

	OPOT Approach	QN Approach
Output format	Continuous	*Discrete*
Output layer neurons	SSAFs	*StepAFs*
Output layer weights	WOPOT	WC
(H-1)th Hidden layer neurons	SSAFs	QNs
(H-1)th Hidden layer weights	WOPOT	WC
...	...	...
2nd Hidden layer neurons	SSAFs	QNs
2nd Hidden layer weights	WOPOT	WC
Input layer neurons	SSAFs	QNs
Input layer weights	WOPOT	WC
Input format	Continuous	Discrete

13.5. Mixed OPOT Optimal MMFNN Architecture for Discrete Input to Discrete Output Mapping Applications [8]

13.5.1. *Design Strategy*

We shall illustrate the design strategy for the case of a 2-layer feedforward neural network (2FNN) for an application on numerals 0 to 9 recognition problem with bipolar (±1) input to bipolar (±1) output mapping format (see Fig. 13-2). For bipolar inputs, continuous-valued weights can be used in the first layer to allow maximum flexibility for adaptation while without requiring any multiplication. To reduce the implementation cost of bipolar SAFs, bipolar SSAFs are used at the output of all the hidden neurons. The outputs of the hidden layer are represented in a continuous format. In order to eliminate multiplications, OPOT weights are used in the second layer such that a multiplication is reduced to just shifting of binary bits. To facilitate a smooth adaptation in learning, bipolar SAFs are used at the outputs of output neurons for learning. After learning, bipolar step activation functions (StepAFs) are used at the outputs of the output neurons to reduce the hardware complexity. Based on the given bipolar input to bipolar output format, the above design strategy optimizes the arithmetic operations such that no multiplication is required in the hardware implementation. Moreover, the inexpensive bipolar SSAFs and the simple bipolar StepAFs are used to replace the expensive bipolar SAFs at the outputs of the respective hidden and output layers. The concept of the design can be extended and generalized to other discrete/continuous input to discrete/continuous output formats, as well as to any M-layer feedforward neural networks with M greater than 2.

Table 13-2. Optimal MMFNN architectures based on Mixed OPOT and Mixed QN approaches.

	Mixed OPOT	Mixed QN
Output format	*Discrete*	*Continuous*
Output layer neurons	*StepAFs*	*SSAFs*
Output layer weights	WOPOT	WC
(H-1)th Hidden layer neurons	SSAFs	QNs
(H-1)th Hidden layer weights	WOPOT	WC
...	...	...
2nd Hidden layer neurons	SSAFs	QNs
2nd Hidden layer weights	WOPOT	WC
Input layer neurons	SSAFs	QNs
Input layer weights	*WC*	*WOPOT*
Input format	*Discrete*	*Continuous*

13.5.2. *Design Algorithm*

The objective of the algorithm is to design an optimal multiplierless multilayer feedforward neural network (MMFNN) suitable for bipolar input to bipolar output mapping. The detailed design algorithm is given as follows:

Step 1: Prepare a set of random weights and zero biases, with bipolar SAFs at the output layer and bipolar SSAFs (with a OPOT L) at the hidden layer.

Step 2: Starting with the latest weights and zero biases, adaptively train the weights of the network using the backpropagation algorithm without adjusting biases until the total sum of squared output errors (TSSE) of all patterns K is

$$\sum_{k=1}^{K} e_k < E \tag{13-15}$$

where E is a prespecified error level. The network obtained at this point is denoted as Network 1.

Table 13-3. Optimal MMFNN architectures based on Mixed OPOT and Mixed QN approaches for continuous input to discrete output mapping.

	Mixed OPOT	Mixed QN
Output format	*Discrete*	*Discrete*
Output layer neurons	*StepAFs*	*StepAFs*
Output layer weights	WOPOT	WC
(H-1)th Hidden layer neurons	SSAFs	QNs
(H-1)th Hidden layer weights	WOPOT	WC
...	...	...
2nd Hidden layer neurons	SSAFs	QNs
2nd Hidden layer weights	WOPOT	WC
Input layer neurons	SSAFs	QNs
Input layer weights	WOPOT	*WOPOT*
Input format	Continuous	*Continuous*

Step 3: Find the maximum absolute value wmax[2] among the weights in the output layer and normalize these weights by wmax[2].

Step 4: Adjust the parameter α of the sigmoid activation functions applied at the output neurons as αwmax[2].

Step 5: Quantize all the weights in the output layer to their nearest OPOT values chosen from the following set:

$$\{\ \pm 1, \pm 2^{-1}, \pm 2^{-2}, \ldots, \pm 2^{-S}, 0\ \} \tag{13-16}$$

where S determines the number of quantization levels. The quantization curve when S=4 is depicted in Fig. 13-1.

Step 6: Calculate the TSSE. If (13-15) is not satisfied, proceed to step 7; otherwise, go to step 8.

Step 7: Adapt all continuous weights of the first layer and the biases of neurons at both layers using the backpropagation algorithm until either (13-15) is satisfied or convergence is reached in which no further improvement in TSSE can be obtained.

Step 8: Find the maximum absolute value wmax[1] among the weights in the first layer, and set wmax[1]=2p[1], where 2p[1] is the smallest OPOT value greater than or equal to wmax[1].

Step 9: Normalize the weights in the first layer by 2p[1] and set parameters ß and Θ of the SSAFs applied at hidden neurons as ß=2p[1]ß and Θ=22p[1]Θ, respectively, such that they remain in OPOT format.

Step 10: Replace the SAFs at the output layer by the StepAFs. Then stop and denote the network obtained here as Network 2.

Table 13-4. Optimal MMFNN architectures based on Mixed OPOT and Mixed QN approaches for discrete input to continuous output mapping.

	Mixed OPOT	Mixed QN
Output format	Continuous	*Continuous*
Output layer neurons	SSAFs	*SSAFs*
Output layer weights	WOPOT	WC
$(H-1)^{th}$ Hidden layer neurons	SSAFs	QNs
$(H-1)^{th}$ Hidden layer weights	WOPOT	WC
…	…	…
2nd Hidden layer neurons	SSAFs	QNs
2nd Hidden layer weights	WOPOT	WC
Input layer neurons	SSAFs	QNs
Input layer weights	*WC*	WC
Input format	*Discrete*	Discrete

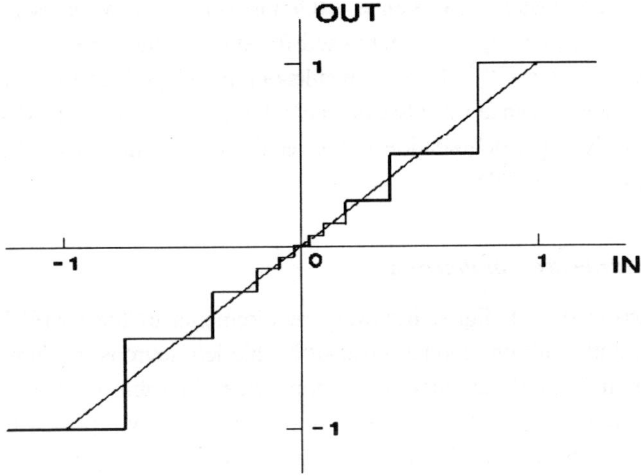

Fig. 13-1. Weight quantization curve (S=4).

13.5.3. *Simulation Results*

Simulations have been conducted to verify the proposed design algorithm. The input patterns used in training were 10 numerals as given in Appendix 13-1, each represented by 10x10 bipolar binary codes. Corresponding desired output patterns were 4-bit bipolar binary codes given below each input pattern. The network had 100 inputs, 4 outputs, and one hidden layer with various numbers of neurons. After training, 100 noisy versions of each of the 10 input patterns, totally 1000, were presented to test the recall accuracy of the network obtained. A noisy pattern was constructed by inverting randomly a specified percentage (it was 5% in this simulation) of the elements of the original pattern. The recall accuracy was obtained by taking the average over all 1000 testing patterns. Simulation results are summarized in Tables 13-5 and 13-6. All data given in these tables were averages of five designs, starting with different initial random weights uniformly distributed in [-0.1, +0.1]. For the purpose of comparison, the results of the corresponding continuous-weight MFNN (CMFNN), which had the same topology but continuous weights and SAFs at both layers, were also obtained and included in these tables. The total number of epochs under the MMFNN is the sum of epochs required to obtain both Network 1 and Network 2.

In order to get faster convergence, ±0.9 instead of ±1 were used for target patterns. The remaining parameters used for simulations were: ε=0.01, ε_b=0.1, E=0.01, α=2, L=2, $\sigma=10^{-2}$. Based on results obtained in Tables 13-5 and 13-6, we can see that convergence was always reached in the training of the MMFNN and there was only slight degradation in the recall performance of the MMFNN as compared to the CMFNN.

13.5.4. *Hardware Comparison*

A comparison of the digital hardware requirements of the CMFNN and the MMFNN, both with one hidden layer of N_1 hidden neurons, N_0 inputs, and N_2 outputs (with StepAFS at outputs), is summarized in Table 13-7. The hardware savings include '$N_1(N_0+N_2)$' 2-input multipliers are now replaced by 'N_1xN_2' OPOT shift registers, and 'N_1+N_2' SAFs are now replaced by N_1 SSAFs and N_2 StepAFs.

Table 13-5. Convergence speed of discrete input to discrete output mapping (in number of epochs).

Number of Hidden Neurons	CMFNN	MMFNN		
		S=2	S=4	S=8
10	202.8	721.2	629.0	696.4
20	98.0	182.6	172.2	172.2
40	78.2	155.8	141.8	141.2
60	71.4	141.2	146	146.8
80	67.4	120.2	126.6	126.6
100	56.0	99.2	99.8	100.6

13.6. Mixed QN Optimal MMFNN Architecture for Continuous Input to Continuous Output Mapping Applications [10-11]

13.6.1. *Design Strategy*

We shall illustrate the design strategy for the case of a 2FNN for an application on continuous input to continuous output mapping. For continuous inputs, OPOT weights are used in the first layer so that no multiplication is required. To reduce the implementation cost of bipolar SAFs, 3-level activation functions (3-LAFs) (of ±1 and 0 outputs, a special case of QNs with S=0) are used at the output of all the hidden neurons. The outputs of the hidden layer are represented in the 3-level discrete format. In order to allow a maximum flexibility for learning adaptation, continuous weights can be used in the output layer while without requiring any multiplication. Bipolar SSAFs are used at the outputs of output neurons. Based on the given continuous input to continuous output format, the above design strategy optimizes the arithmetic operations such that no expensive multiplication is required in the hardware implementation. The simple 3-LAFs and the inexpensive bipolar SSAFs are used to replace the expensive bipolar SAFs at the outputs of the respective hidden and output layers. The concept of the design can be extended and generalized to other discrete/continuous input to discrete/continuous output formats, as well as to any M-layer feedforward neural networks with M greater than 2.

Table 13-6. Recall performance of discrete input to discrete output mapping (in percentage correctness).

Number of Hidden Neurons	CMFNN	MMFNN		
		S=2	S=4	S=8
10	99.50	99.43	99.08	98.96
20	99.72	99.60	99.68	99.66
40	99.56	99.46	99.56	99.58
60	99.54	99.56	99.60	99.60
80	99.62	99.58	99.60	99.60
100	99.62	99.46	99.58	99.58

13.6.2. *Design Algorithm*

Step 1: Prepare a set of random weights and zero biases.

Step 2: Starting with the latest weights and zero biases, train the network using the backpropagation algorithm with the SSAFs at the output neurons and the 3-LAFs at the hidden neurons. Update the weights without adjusting biases until the TSSE becomes less than a prespecified error level E. The obtained network is denoted as Network 1.

Step 3: Find the maximum absolute value wmax among all the weights in the first layer and normalize these weights by wmax.

Step 4: Scale the threshold value t3 of all the 3-LAFs applied to hidden neurons as t3/wmax.

Step 5: Quantize those normalized weights in the first layer to their nearest OPOT values from the set of {±1, ±2-1, ..., ±2-S, 0}.

Step 6: Calculate the TSSE. If TSSE<E, stop and denote the network obtained here as Network 2; otherwise, proceed to Step 7.

Step 7: Re-adapt all continuous weights in the second layer using the backpropagation algorithm.

Step 8: Adapt the biases of neurons at both layers.

Step 9: Go to Step 6.

Table 13-7. Hardware comparison of discrete input to discrete output mapping.

Components	CMFNN	MMFNN
N_0-Input Adder	N_1	0
N_1- Input Adder	N_2	0
(N_0+1)- Input Adder	0	N_1
(N_1+1)- Input Adder	0	N_2
2- Input Multiplier	$N_1(N_0+N_2)$	0
OPOT Shift Register	0	$N_1 x N_2$
SAF	N_1+N_2	0
SSAF	0	N_1
StepAF	0	N_2

13.6.3. *Simulation Results*

Simulation results are summarized in Tables 13-8 and 13-9. Two normalized orthogonal continuous real vector sets, one as input pattern set $\mathbf{X}=[\mathbf{X}_1, \mathbf{X}_2, \mathbf{X}_3, \ldots, \mathbf{X}_{10}]$ and the other as target set $\mathbf{Y}=[\mathbf{Y}_1, \mathbf{Y}_2, \mathbf{Y}_3, \ldots, \mathbf{Y}_{10}]$, were used for training and recall. Each vector set consists of 10 vectors and each vector consists of 25 continuous real elements, which are generated by using a method described in [24]. These ten continuous input to continuous output pattern-pairs are plotted in Appendix 13-3. The network was used as a pattern associator, which had 25 inputs, 25 outputs, and one hidden layer with a variable number of neurons. After training, 100 noisy versions of each of the 10 input vectors were presented to the network to test the recall performance. The noisy vectors were constructed by adding random noise within the interval of ±R to each element of each input vector. R represents a percentage of the maximum element value among all the 10 input vectors. In the simulations presented here, R was 10% or 20%. The output vector was identified based on its cross correlations with all ideal output vectors. The ideal output vector with maximum cross correlation was selected as the recall vector. For comparison, the simulation results of the corresponding CMFNN, which had the same topology but continuous weights and bipolar SAFs at both layers, were also obtained. The values summarized in Tables 13-8 and 13-9 represent the average of five designs, starting with different initial random weights uniformly distributed within ±0.1. The learning rate parameter for weights was ε=0.01, the step size for bias adjustment was ε_b=0.01, and other parameters used were δ=0.01, α=2, D=2, M=4, and $E_0=10^{-6}$. It can be seen that the proposed MFNNs with SSAFS, 3-LAFs, and OPOT weights have a similar recall performance as the original MFNNs with SAFs and continuous weights at a cost of additional training epochs.

All simulations were written in Fortran language and run using a 486-33MHz PC: (a) when the number of hidden neurons was 10, the average training times of the CMFNN, the Network 1, and the Network 2 were, respectively, 0.211, 0.215, and 0.172 CPU seconds per epoch, and the average recall times of 1000 noisy input vectors of the CMFNN and the MMFNN were respectively, 5.54 and 3.35 CPU seconds; and (b) when the number of hidden neurons was 30, the corresponding average training times were, respectively, 0.431, 0.479, and 0.334 CPU seconds per epoch, and the corresponding average recall times were respectively, 12.20 and 5.98 CPU seconds.

13.6.4. *Hardware Comparison*

A comparison of the digital hardware requirements of the CMFNN and the MMFNN, both with one hidden layer of M hidden neurons, N inputs, N output neurons, is summarized in Table 13-10. The most expensive hardware item is multipliers in which the CMFNN requires a total number of 2xNxM whereas the MMFNN requires only NxM OPOT shift registers. Assuming b-bit implementation (where one multiplier is approximately equivalent to b adders), the number of hidden neurons in the MMFNN can be up to about '2b' times that of the CMFNN for a similar level of hardware cost. The hardware cost of the 'N' SSAF [1] and the 'M' 3-LAF is expected to be lower than that of 'N+M' SAF. In Table 13-10, we have not yet taken into account the hardware savings of: (a) all the zero connections in both layers of the MMFNN, and (b) storing only the 'NxM' OPOT weights in the MMFNN over that of the 'NxM' continuous weights in the CMFNN in their first layers.

Table 13-8. Convergence speed of continuous input to continuous output mapping (in number of epochs).

Number of Hidden Neurons	CMFNN	MMFNN	
		Network 1	Network 2
10	587.0	771.6	2690.0
15	380.2	427.8	313.4
20	339.8	256.6	211.4
25	221.6	124.8	99.2
30	174.8	117.2	70.0

Table 13-9. Recall performance of continuous input to continuous output mapping (in percentage correctness).

Number of Hidden Neurons	10% Random Noise		20% Random Noise	
	CMFNN	MMFNN	CMFNN	MMFNN
10	100.0	97.48	100.0	94.28
15	100.0	99.96	100.0	99.68
20	100.0	100.0	100.0	99.84
25	100.0	100.0	100.0	100.0
30	100.0	100.0	100.0	100.0

13.7. Concluding Remarks

Optimal MMFNN architectures suitable for discrete/continuous input to discrete/continuous output mappings, as illustrated by the above two cases of multiplierless 2-layer feedforward neural networks, have been presented. The design strategies utilize a combination of OPOT shift operations to replace multiplications, SSAFs to replace SAFs, as well as the specified input to output format to arrive at optimal design architectures. A designed MMFNN can retain nearly identical recall performance of the corresponding MFNN with continuous weights and SAFs, while having increased computational speed in applications and reduced cost in digital hardware implementation. In general, the same principle of optimal MMFNN architectures illustrated can also be applied to other types of artificial neural networks. Some systolic implementation methods for neural networks can be found in [25-27].

Table 13-10. Hardware comparison of continuous input to continuous output mapping.

Components	CMFNN	MMFNN
N_0-Input Adder	N_1	0
N_1- Input Adder	N_2	0
(N_0+1)- Input Adder	0	N_1
(N_1+1)- Input Adder	0	N_2
2-Input Multiplier	$N_1(N_0+N_2)$	0
OPOT Shift Register	0	$N_0 x N_1$
SAF	N_1+N_2	0
SSAF	0	N_2
3-LAF	0	N_1

References

[1] H. K. Kwan. Simple sigmoid-like activation function suitable for digital hardware implementation. *Electronics Letters*, 28(15), pages 1379-1380, July 16, 1992.

[2] C. Z. Tang and H. K. Kwan. Multilayer feedforward neural networks with single powers-of-two weights. *IEEE Transactions on Signal Processing*, 41(8), pages 2724-2727, Aug. 1993.

[3] C. Z. Tang and H. K. Kwan. Feedforward neural networks with powers-of-two weights. *Proceedings of International Joint Conference on Neural Networks*, Beijing, China, II, pages 93-98, Nov. 3-6, 1992.

[4] H. K. Kwan and C. Z. Tang. Designing multilayer feedforward neural networks using simplified sigmoid activation functions and one-powers-of-two weights. *Electronics Letters*, 28(25), pages 2343-2344, Dec. 3, 1992.

[5] H. K. Kwan and C. Z. Tang. Multilayer feedforward neural networks with powers-of-two weights and piecewise activation functions. *Proceedings of Canadian Conference on Electrical and Computer Engineering*, Toronto, Canada, II, pages. TM6.14.1-4, Sept. 13-16, 1992.

[6] C. Z. Tang and H. K. Kwan. Digital implementation of neural networks with quantized neurons. *Proceedings of IEEE International Symposium on Circuits and Systems*, Hong Kong, 1, pages 649-652, June 9-12, 1997.

[7] C. Z. Tang and H. K. Kwan. Feedforward neural networks without multipliers. *Proceedings of International Conference on Signal Processing Applications and Technology*, Boston, Massachusetts, U.S.A., 2, pages. 1194-1201, Nov. 2-5, 1992.

[8] H. K. Kwan and C. Z. Tang. Multiplierless multilayer feedforward neural networks. *Proceedings of 36th Midwest Symposium on Circuits and Systems*, Detroit, Michigan, U.S.A., 2, pages 1085-1088, August 15-18, 1993.

[9] H. K. Kwan and C. Z. Tang. A multilayer feedforward neural network model for digital hardware implementation. *Proceedings of IEEE International Symposium on Circuits and Systems*, London, U.K., 6, pages 343-346, May 30-June 2, 1994.

[10] H. K. Kwan and C. Z. Tang. Multiplierless multilayer feedforward neural network design suitable for continuous input-output mapping. *Electronics Letters*, 29(14), pages 1259-1260, July 8, 1993.

[11] H. K. Kwan and C. Z. Tang. A design method for multilayer feedforward neural networks for simple hardware implementation. *Proceedings of IEEE International Symposium on Circuits and Systems*, Chicago, Illinois, U.S.A., 4, pages 2363-2366, May 3-6, 1993.

[12] H. K. Kwan and J. Yan. Recurrent neural network design for temporal sequence learning. in Special Section S6 - *Neural Networks and Applications, Proceedings of 43rd Midwest Symposium on Circuits and Systems*, Michigan, Aug. 8-11, 2000.

[13] H. K. Kwan and J. Yan. Second-order recurrent neural network for word sequence learning. *Proceedings of International Symposium on Intelligent Multimedia, Video & Speech Processing*, Hong Kong, pages 405-408, May 2-4, 2001.

[14] H. K. Kwan. One-layer feedforward neural network for fast maximum/minimum determination. *Electronics Letters*, 28(17), pages 1583-1584, 20th Aug. 1992.

[15] M. Marchesi, G. Orlandi, F. Piazza, and A. Uncini. Fast neural networks without multipliers. *IEEE Transactions on Neural Networks*, 4(1), pages 53-62, Jan. 1993.

[16] A. H. Khan and E. L. Hines. Integer-weight neural nets. *Electronics Letters*, 30(15), pages 1237-1238, July 21, 1994.

[17] A. H. Khan and R. G. Wilson. Integer-weight approximation of continuous-weight multilayer feedforward nets. *Proceedings of IEEE International Conference on Neural Networks,* 1, pages 392-397, 1996.

[18] J. H. Kim and S.-K. Park. The geometrical learning of binary neural networks. *IEEE Transactions on Neural Networks*, 6(1), pages 237-247, Jan. 1995.

[19] J. H. Kim, S.-K. Park, Y. Han, H. Oh, and M. S. Han. Efficient VLSI implementation of a 3-layer threshold network. *Proceedings of IEEE International Conference on Neural Networks*, 2, pages. 888-893, 1997.

[20] B. G. Hoskins, M. R. Haskard, and G. R. Curkowicz. A VLSI implementation of multi-layer neural network with ternary activation functions and limited integer weights. *Proceedings of 20th International Conference on Microelectronics (MIEL'95),* 2, NIS, Serbia, pages. 843-846, Sept. 12-14, 1995.

[21] M. Cavanagh and V. K. Jain. Multiplierless neural networks for application to digital video. *Proceedings of IEEE International Conference on Acoustics, Speech, and Signal Processing*, 6, pages 3414-3417, 1996.

[22] C. Z. Tang and H. K. Kwan. Convergence and generalization properties of multilayer feedforward neural networks. *Proceedings of IEEE International Symposium on Circuits and Systems*, San Diego, California, U.S.A., 1, pages 65-68, May 10-13, 1992.

[23] C. Z. Tang and H. K. Kwan. Parameter effects on convergence speed and generalization capability of backpropagation algorithm. *International Journal of Electronics*, 74(1), pages 35-46, Jan. 1993.

[24] H. K. Kwan, Z. Wang, and J. J. Soltis. Method for generating partially correlated vector sets for neural network simulations. *International Journal of Electronics*, 74(4), pages 523-528, April 1993.

[25] H. K. Kwan. Systolic architectures for Hopfield network, BAM, and multi-layer feed-forward network. *Proceedings of the IEEE International Symposium on Circuits and Systems*, Portland, Oregon, U.S.A., 2, pages 790-793, May 9-11, 1989.

[26] H. K. Kwan and P. C. Tsang. Systolic implementation of multilayer feed-forward neural network with back-propagation learning scheme. *Proceedings of International Joint Conference on Neural Networks,* Washington, D.C., II, pages 155-158, Jan. 15-19, 1990.

[27] H. K. Kwan and P. C. Tsang. Systolic implementation of counterpropagation networks. *Proceedings of IEEE International Conference on Acoustics, Speech, and Signal Processing*, Albuquerque, New Mexico, U.S.A, 2, pages 953-956, April 3-6, 1990.

Appendix 13-1

Ten discrete input to discrete output pattern-pairs:

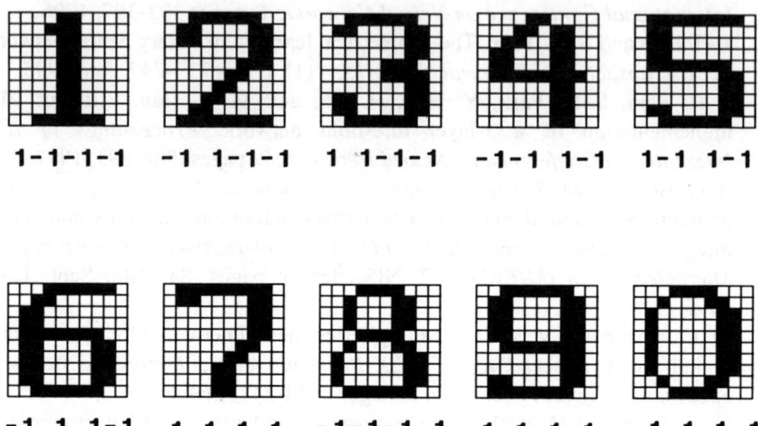

Appendix 13-2

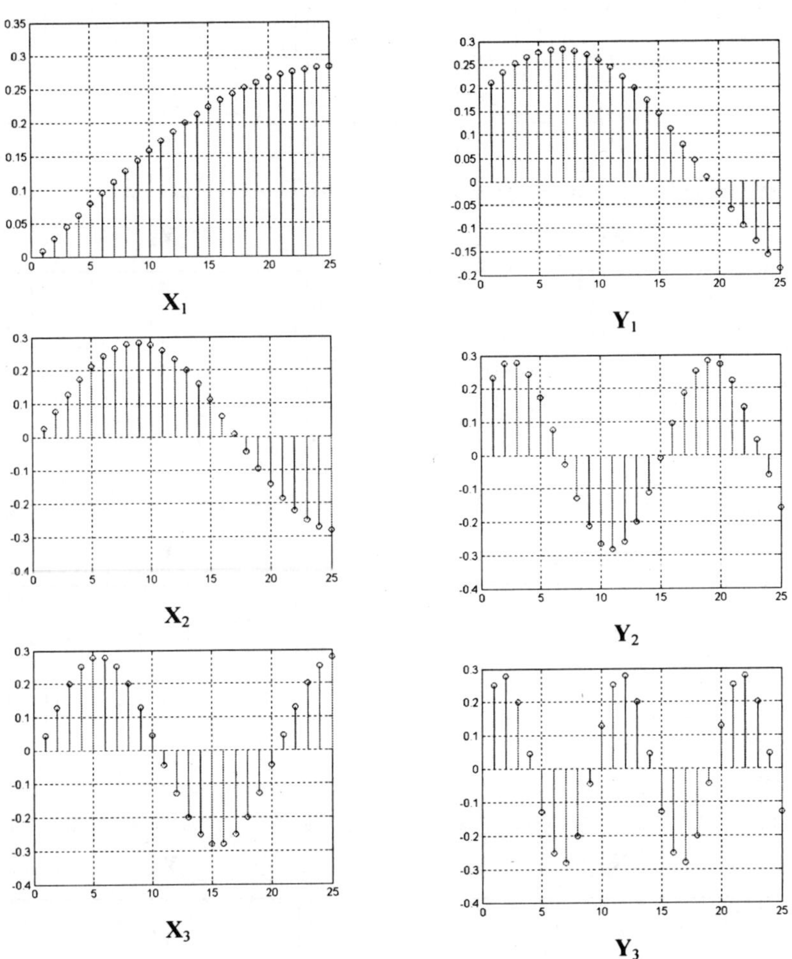

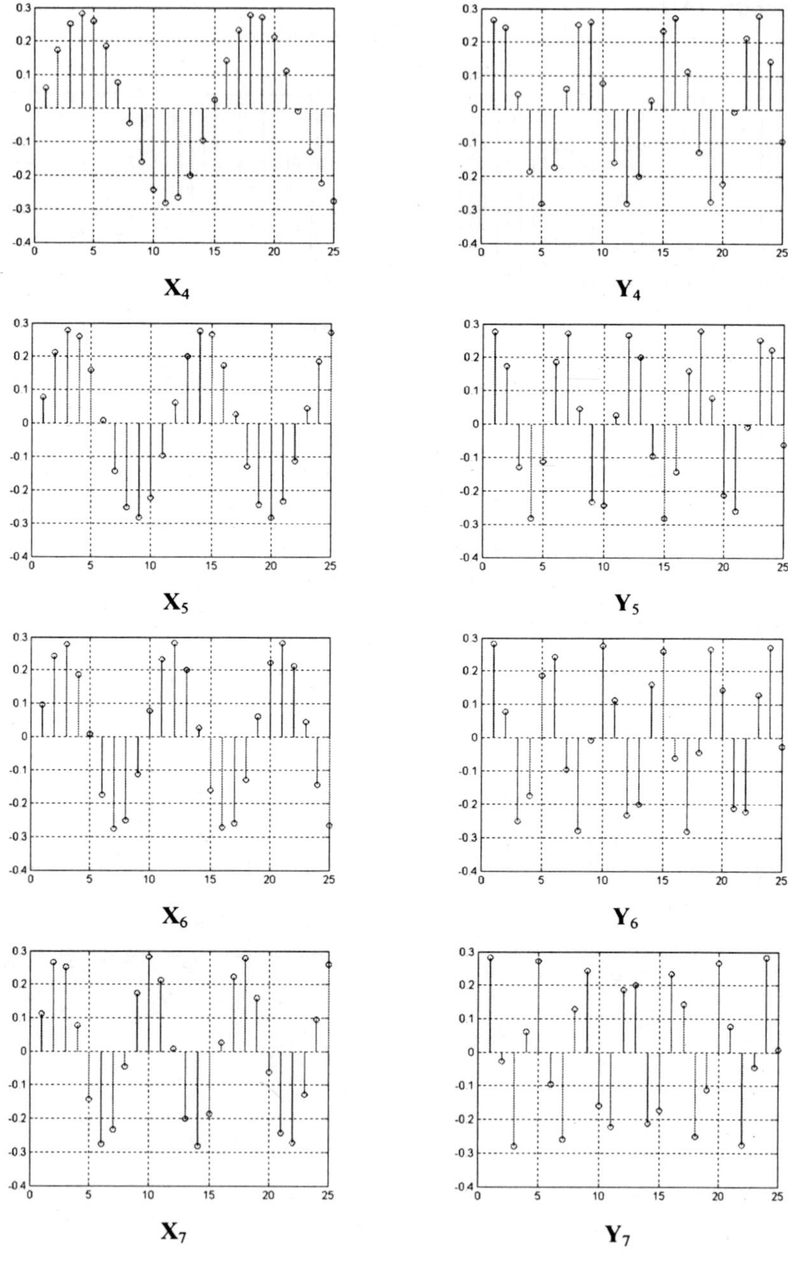
$\mathbf{X}_4$
$\mathbf{Y}_4$
$\mathbf{X}_5$
$\mathbf{Y}_5$
$\mathbf{X}_6$
$\mathbf{Y}_6$
$\mathbf{X}_7$
$\mathbf{Y}_7$

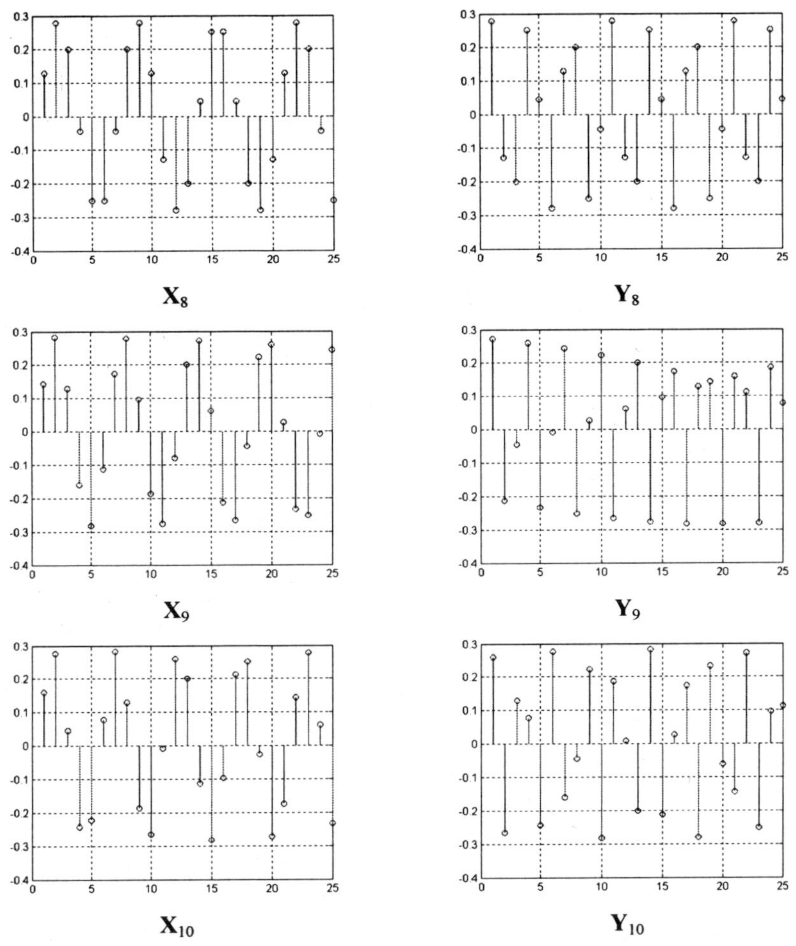

$\mathbf{X}_8$ $\mathbf{Y}_8$

$\mathbf{X}_9$ $\mathbf{Y}_9$

$\mathbf{X}_{10}$ $\mathbf{Y}_{10}$

CHAPTER 14

A VLSI SYSTEM FOR INTELLIGENT DECISION MAKING IN REAL-TIME

N. Ranganathan

Department of Computer Science and Engineering
Center for Microelectronics Research
University of South Florida, Tampa, FL 33620, USA
E-mail: ranganat@csee.usf.edu

M.I. Patel

Department of Computer Science and Engineering
Center for Microelectronics Research
University of South Florida, Tampa, FL 33620, USA

In this chapter we describe the design of a VLSI system for intelligent decision making in real time. The system architecture is an integration of a learning system (an Artificial Neural Network), and a rule based fuzzy expert system implemented as linear systolic arrays.

14.1. Introduction

Intelligent decision making is an important task in artificial intelligence, with applications in fields such as character recognition, medical diagnosis, navigational guidance, stock market prediction, and financial analysis, just to mention a few. The design of systems that adaptively learn in a dynamically changing environment and make decisions on-the-fly is a challenging task. Since such systems are computationally intensive, the development of hardware architectures for intelligent decision making is of research interest. This chapter will discuss the design of a real-time intelligent decision making system (RAPID) that integrates the capabilities of an artificial neural network and a fuzzy expert system on a VLSI chip. The goal is to design a system that is implementable in VLSI in order to obtain high speed and real-time response.

An Artificial neural network (ANN) is a network of highly interconnected processing elements, a computing system, inspired by the physiology of the human brain. ANNs possess an adaptive feature which allows each cell within the network to modify its state in response to experience. The neural network can then learn or self modify. Often, ANNs have been used to mimic expert systems. Expert systems are commonly used in decision making. An expert system is an intelligent database stored in a computer, built from knowledge extracted with the help of an expert in a given field. The database is accessed by a series of rules governing the decision making process. Learning systems such as ANNs have the ability to reveal new relationships between concepts and hypotheses by analyzing examples of solved cases. A learning system can be trained without the need for an expert. Although the learning systems represent an attractive approach, learning from sample experience alone is inadequate, if it excludes the context within which problem solving is being carried out. Consequently, there is a need to combine the learning abilities of an ANN and the domain specific knowledge of an expert system. Various implementations of expert systems can be found in the literature. Several system architectures for implementing neural networks and a few schemes for implementing expert systems exist in the literature. The problem of integrating a neural network and a fuzzy expert system in hardware is discussed here.

14.2. Why VLSI for Intelligent Decision Making?

VLSI technology has progressed rapidly in the past two decades. High packaging densities, low gate delays, and fabrication cost, powerful CAD design automation tools, and reliable and fault-tolerant design strategies are some of the advantages of present VLSI technology. The attributes of parallelism, concurrency, pipelining, modularity and regularity have become standard features of special purpose hardware designs.

In general purpose computers, high level code undergoes multiple levels of translation before being executed. Each instruction thus requires several microinstructions to be executed, resulting in processing times of many clock cycles. In VLSI architectures, most of the basic operations can be completed within a single clock cycle, resulting in faster execution rates. Many algorithms exhibit inherent parallelism which can be exploited to decrease processing times. Parallel machines are extremely expensive and occupy significant amounts of space. On the other hand, VLSI solutions are compact, cost effective, and off-load a major portion of the computation from the host, freeing it for other operations.

The compute-intensive nature of both neural networks and expert systems make their software implementations inadequate in meeting the demands of applications that need real-time response. This is true especially in systems that have to adapt to constantly changing parameters within the environment. By exploiting the pipelining and concurrency possible in neural networks as well as in rule-based expert systems, an intelligent decision making system can be implemented using special purpose hardware for achieving response times of the order of nanoseconds.

14.3. Background

Hardware designs for neural networks can be divided into two classes: (i) general-purpose neurocomputers and (ii) special-purpose neurocomputers. Most general-purpose neurocomputers are based on either commercial co-processor boards or parallel processor arrays. Commercial co-processor boards consist of floating point or signal processing accelerator boards with large amounts of memory. Some examples are: (i) CONE [4,45], (ii) NNETS [38], (iii) NETSIM [38], (iv) RAP [34], (v) Lneuro 1.0 [33], (vi) neuroTRAM [10], (vii) HERCULES [23], (viii) Kotilainen [35], (ix) the WAP [26], and (iix) Spert [21]. There have been several commercial neural network chips developed in the industry. Such chips include (i) Micro device's MD12202, (ii) Neuralogix NLX420, NLX 110, and NLX 230, (iii) INOVA N640009, (iv) 1000 NAP10C (neurocomputer array processor), (v) IBM's Z1SC036 (zero instruction set computer) chip 13, and (vi) Nestor and Intel's Ni100015, just to mention a few. The second type of general purpose neural computer design is based on parallel arrays. The system consists of simple cellular arrays of PEs arranged to form a regular topology. Many of the later implementations of neurocomputers used parallel arrays. These architectures include (i) Kung [41], (ii) Zubair [30], (iii) Ouali [20], (iv) MLP [26], (v) Distantel [12], (vi) Distante2 [12], (vii) CNAPS [7], (viii) Synapse [46], (ix) GENES IV [32], (x) JNC [47] and (xi) SPIN-L [39]. The literature is rich on neural network hardware and thus the set of works cited above is not complete.

The general purpose architectures have been classified as those using (i) hardware accelerators or co-processors, and (ii) parallel arrays. The co-processor based architectures use the characteristics of the processor to emulate the neuron functions. However, a large number of co-processors are required to represent a neural system and its interconnection. Consequently, these designs consume large amounts of memory both in the neural and the interconnection line representations. The parallel array based architectures use simple PEs consisting of basic neural operations. High performance is achieved through the distribution of work through the parallel array. Although parallel arrays achieve high performance, the selection of array dimensionality is critical. The 1-dimensional array is the simplest in communication, control, and data flow. The 2-dimensional array requires complex algorithms for ANN mapping, large number of I/O ports, and complex control and communication.

Many types of machines have been used to implement expert systems. The types of systems include RISCs (Reduced Instruction Set Computers) [8], LISP machines [1], Personal Sequential Machines [15], data flow machines, multiprocessors and others. Additionally, expert system controllers exist where special purpose expert systems are used to control a particular process. Based on pre-defined rules, an example process sends inputs to the controller. The expert system uses its rule base to calculate (infer) a decision based on the process inputs. Controllers are used in a variety of applications from automobile control [31,2,25,40], flight control [17,24], to temperature control [5].

14.4. RAPID System Architecture

RAPID is a real-time intelligent decision making system that computes decisions within a dynamically changing application environment. The architecture consists of an artificial neural network and a fuzzy rule based expert system. The architectural blocks and the data flow for the proposed RAPID system are shown in Fig. 14-1. Sensors are used to detect the surrounding environmental conditions in this model. The sensors send crisp data inputs to the artificial neural network. The fuzzification unit assigns fuzzy labels to the outputs of the ANN. These labels indicate the degree to which each crisp value is a member of a domain.

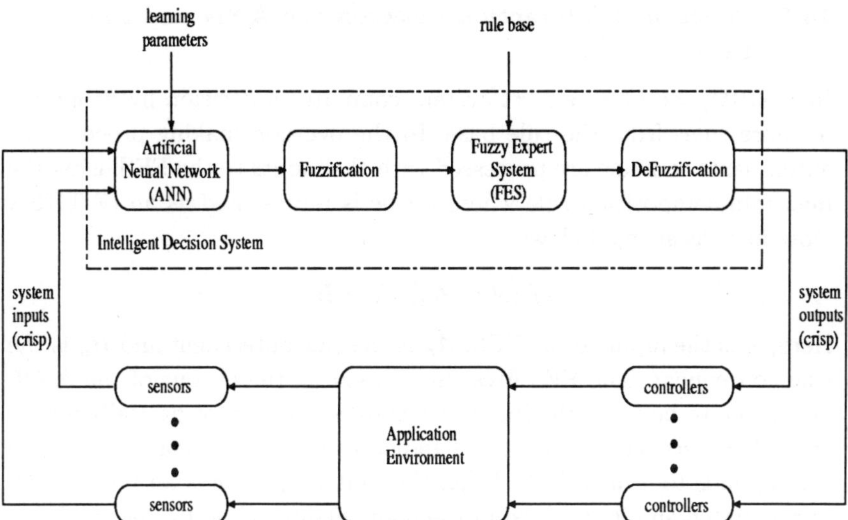

Fig. 14-1. RAPID system.

Then, the fuzzy expert system fires the rules based on these fuzzy values. The defuzzification unit converts the computed decisions into crisp values that are used to control the environment within an application. The details of such an integrated system can be found in [37].

The integrated intelligent decision making system requires an *artificial neural network* model that can handle a wide range of applications and is simple in terms of implementation and training. Based on the existing research [36,11,43,18], the backpropagation model has been found to be the most suitable choice. The ANN model used in this work is a fully connected, single hidden layer backpropagation network. The model is used in many applications ranging from speech synthesis to loan application scoring.

The *fuzzification unit* receives crisp data from the neural network and maps the data onto fuzzy sets within an input universe of discourse.

14.5. Functional Interaction between the ANN and the FES

In RAPID, the fuzzy expert system computes a decision by firing one or more rules from the rule base. In the decision making process, the antecedents of a rule are processed with the inputs to the FES using the max min composition rule where a rule is represented as an IF-THEN statement as shown below:

$$if\ (A = A_n)\ then\ B_n$$

Here, A is the input to the FES, A_n is the rule antecedent and B_n is the rule consequent. The FES fires rules based on the output of the ANN. For a rule to be fired, the input A has to completely or partially match the rule antecedent. A value of 1 indicates a complete match while a 0 corresponds to a mismatch. In partial matches, there is an imprecision in the match and is reflected in the membership value. In the rule base, the response of each rule is weighted according to the confidence or degree of membership of its inputs. When there are multiple antecedents (antecedent A^j indexed by j), a rule (i) is fired based on the weighted combinations of the antecedents and the inputs to the FES as shown in the equations below:

$$\mu_{B_i'}(v) = max_{u \in U}\ min(\mu_{B_i}, \bigcap_{j=1}^{r} min(\mu_{A_i^j}(u), \mu_{A^j(u)})) \quad (14\text{-}1)$$

$$\mu_{B_n'}(v) = max(\mu_{B_0'}, \dots \mu_{B_i'} \dots, \mu_{B_{n-1}'}) \quad (14\text{-}2)$$

Here, r is the number of antecedents of A_i^j in rule i indexed by j, A^j is the jth input to the rule i and μ_{B_i} is the membership function of the rule consequent. The operator '$\bigcap$' indicates the min operation of r elements, $\mu_{B_i'}$ is the decision of rule i with r antecedents and $\mu_{B_n'}(v)$ is the combined decision of n rules.

In RAPID, the ANN provides the inputs to the fuzzy expert system. Therefore, the above equation becomes

$$\mu_{B'}(v) = max_{u \in U}\ min(\mu_B(v), min(\mu_{y_j}(u), \mu_{A_n}(u))) \quad (14\text{-}3)$$

The computed FES decision, $\mu_{B'}(v)$, is shown to be biased by the ANN output which is assigned a fuzzy label $\mu_{y_j}(u)$. Since the ANN outputs are obtained from the different weighted connection lines, y_j is the bias value that depends on the decision surface after learning. Thus, the rule firing sequence is dynamically determined by the values of y_j obtained from the knowledge in the ANN.

As an example of the integrated system, the adaptive traffic light application is shown in Fig. 14-2.

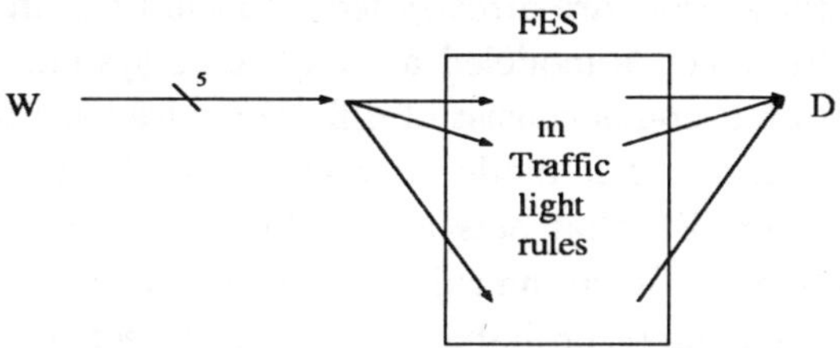

(a) Static traffic light system

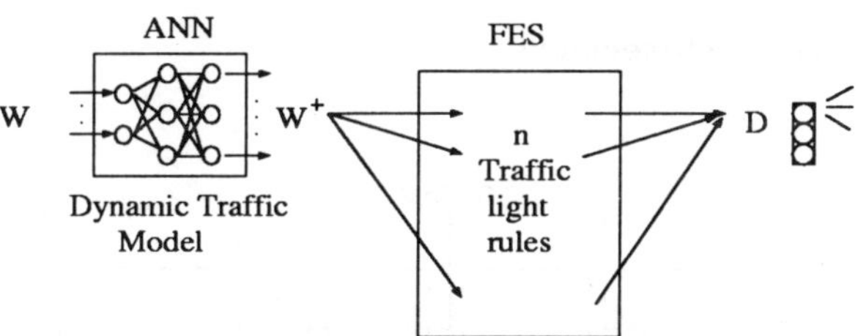

(b) Adaptive traffic light system

$$W = \begin{bmatrix} HS \\ CS \\ SD \\ VD \\ RG \end{bmatrix}$$

example rule: if (HS and CS and SD and VD and RG) then (CT and GT and GText)

$$D = \begin{bmatrix} CT \\ GT \\ GText \end{bmatrix}$$

Fig. 14-2. Two models for traffic light control.

The environment provides an input weight vector **W** that represents the current traffic conditions. A vector consists of the highest saturation (HS) and the cross saturation (CS) values, the saturation difference (SD), the volume difference (VD) and the required green time extension (RG). The weight vector W^+ represents the predicted next traffic conditions provided by the ANN. This is used to fire dynamically the selected rules from the rule base. The output vector **D** represents the new traffic light cycle values. It consists of values for the cycle time (CT), the green time (GT) and the green time extension (GText) for adjacent intersections. The rules used for decision making consist of five antecedents and three consequents in this example.

In Fig. 14-2-a, the traffic environment is modeled as a static system using a FES consisting of m rules. For the static system, the input values from the traffic sensors are directly fed into the FES. In Fig. 14-2-b, the traffic environment is modeled as a dynamic system using an ANN and a FES. Both systems consist of different rules used to adjust the traffic light cycle times. The model uses an ANN which provides a traffic prediction vector W^+ that acts as weights or bias for selecting rules in the FES rule base. Since the rule base in the integrated system is smaller ($n < m$), it is more probable that the right set of rules will be fired for cycle adjustment.

14.6. Hardware Architecture

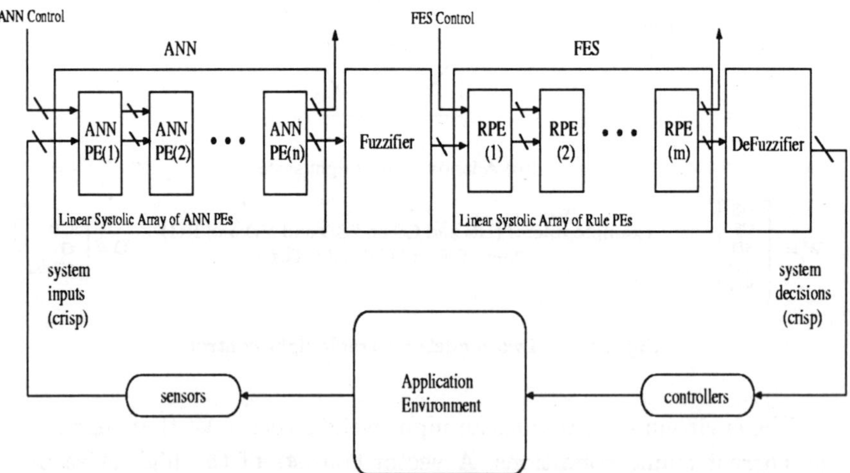

Fig. 14-3. RAPID hardware architecture.

The RAPID hardware architecture is designed such that the input data from the application environment flows through the architecture in a linear fashion as shown in Fig. 14-3. The sensors within the environment send crisp data vectors into the shift registers. These values are then transmitted to the artificial neural network. The ANN is composed of a linear array of processing elements (PEs) and the fuzzy expert system is a linear array of rule processing elements (RPEs). The system computes a new decision every clock cycle. The outputs of the ANN are assigned fuzzy labels that are stored in a RAM and the computed decisions are defuzzified into crisp values using a multiplier/adder tree. These crisp values are used as input signals to control devices which manage the application environment.

The input and the output signals to the RAPID system are divided into two sets: (i) control and (ii) data. The control signals are used to initialize and change the execution of the artificial neural network and the fuzzy expert system. The data signals represent (i) the fuzzy input and output values, and (ii) the crisp input and output values. The control signals for the neural network carry information about (i) the neural topology, (ii) the control registers that store the interconnection and node ID values, (iii) memory addressing, (iv) the ANN execution steps, (v) the inter-node connection settings and (vi) the ANN operational modes. The control for the fuzzy expert system involves down-loading the rule base and setting the antecedent and consequent counters. The output of the RAPID system is a 4-bit crisp decision sent to the application environment.

The size and the width of the fuzzy subsets (membership functions) are important considerations in the design of the inferencing unit. The FES uses fuzzy values in the decision making process. Since the fuzzy expert system uses fuzzy membership values, the bus width between the rule storage and the inferencing unit is dependent on the width of the membership values. As the width decreases, the precision in the output of the FES monotonically decreases. In the RAPID system, the range of values a membership function can have is limited. This limitation allows less communication signals to be used between the rule base and the inferencing mechanism. The membership function is implemented as a universe of discourse of 16 sets, where each set is discretized into 16 grades of membership as shown in Fig. 14-4. A value of 0 corresponds to non-membership and a value of 15 corresponds to a complete membership while the intermediate values indicate partial membership. After defuzzification the crisp values are 4-bits long ranging from 0 to 15.

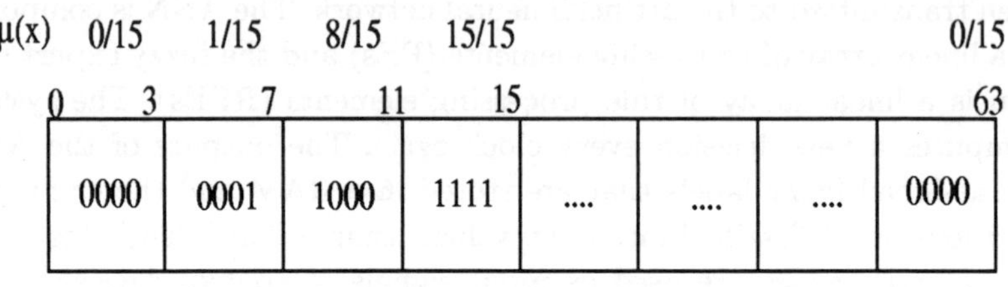

Fig. 14-4. Memory representation of a membership function.

14.6.1. *Fuzzification Unit*

The fuzzification unit receives 4-bit values generated by the neural network, and converts them into fuzzy values with a membership function $\mu(x)$. The architecture is shown in Fig. 14-5. The fuzzification unit consists of (i) a 64 x 64 bit RAM, (ii) an ANN output sequencer, (iii) a floating point decimal to integer converter and (iv) four 16 : 1 bit multiplexors. The membership functions are stored in a RAM. The RAM address is formed by concatenating the ANN output and the 2-bits from the ANN output sequencer. Since the ANN output is a floating point decimal between 0 and 1 and is used for memory addressing, it is converted into an integer value between 0 and 15. To convert the floating point decimal into an integer, a 4-bit shift register is used that shifts the decimal point 4-bit positions to the right. The resulting value is an approximation to the corresponding integer value. Since an ANN can have several outputs each corresponding to various linguistic variables and memberships, an ANN output sequencer is used to coordinate the appropriate linguistic variables stored in the RAM with each ANN output.

The 64 x 64 bit RAM can hold four membership functions each having a maximum of 14 linguistic variables. The RAM is partitioned into four sections where each section has 16 membership entries. The sec-

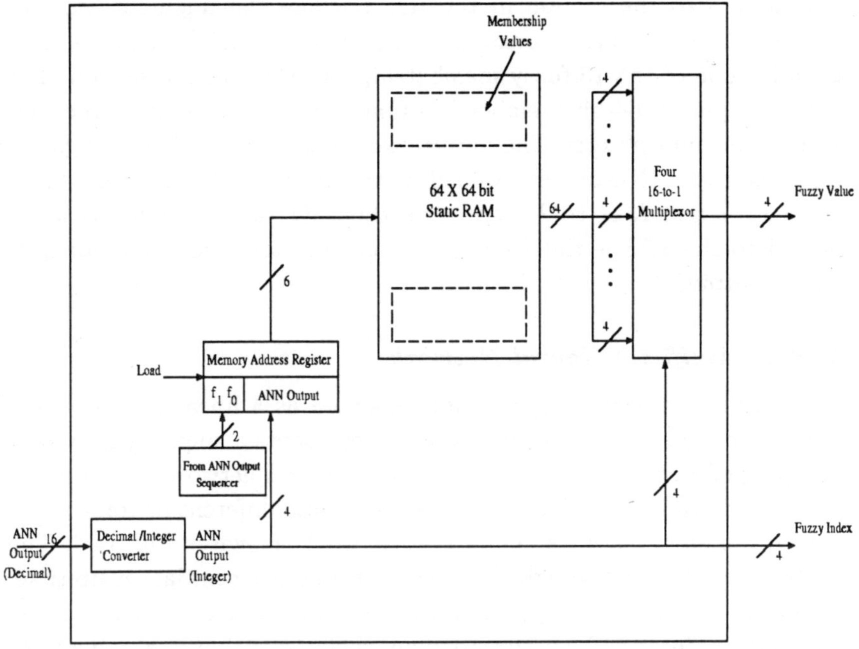

Fig. 14-5. Fuzzification unit.

tions are 64-bits wide. The ANN output sequencer sends a 2-bit address to access one of the 4 sections in the RAM. The 4-bit integer output of the ANN is used to address one of the 16 fuzzy membership functions within a given section.

The ANN output sequencer consists of (i) two 8-to-1 multiplexors, (ii) two 8-bit registers and (iii) a counter. The two 8-bit registers are loaded with a stream of 2-bit values. The 2-bit values correspond to the location of the linguistic variables in one of the four sections in the RAM and are arranged to coincide with each ANN output. In the linear array implementation, the sequence of the ANN outputs is known a priori. At set-up time, the registers are loaded with this sequence. The registers are restricted to a size of 8-bits because each rule in the FES is limited to have 8 antecedents. The registers input data to the multiplexors where the counter is used as a selector. The multiplexors output a 2-bit value that addresses the section in the RAM where the linguistic variable for that ANN output is located. Next, the 4-bit ANN output is used to index one of the 16 fuzzy membership functions in the section. The RAM outputs a 64-bit membership function, and sends it to the four 16-to-1 bit multiplexors. The multiplexors filter out the fuzzy singleton from the other 15 4-bit zero-valued entries from the 64-bit value. Since I_0 to I_{n-1} and I_{n+1} to I_{15} are zero, the only value that needs to be passed to the FES is the 4-bit fuzzy singleton and the 4-bit index, I_n (ANN output).

14.6.2. *Artificial Neural Network*

The artificial neural network is implemented as a linear backpropagation based array of PEs where each PE represents a single neuron. The mapping of a conventional fully connected neural network onto a linear array is shown in Fig. 14-6. It is possible to map different neural topologies to the linear network. The PE structure is shown in Fig. 14-7. The control signals flow through the array of PEs during initial set-up time.

Each neuron contains (i) a RAM to store the interconnection weights and their difference from the previous cycle (Δw^-), (ii) a Z-RAM to store the inputs to the neuron, (iii) an adder, a multiplier and a reciprocal generator, (iv) a two-segment activation function unit [42], (v) a register to store the derivative of the neuron output and 8 general purpose storage registers, (vi) control signal generators, (vii) wait state generators, and (viii) routing multiplexors. The adder and the multiplier have a reduced precision format in order to avoid floating point hardware. The algorithm for further training uses a reciprocal unit to perform divisions, which is implemented as a programmable logic array (PLA). The two-segment activation function unit computes an approximation of the sigmoid activation function. The unit uses two add and a multiplier operation for activation function computation. The control generators sequence each neuron through different steps during the execution, training, and further training modes. The wait state generators, wait1 and wait2, synchronize the neuron with other neurons in the same layer by inserting delays before the data is sent to the adjacent nodes. The routing multiplexors control the data flow between the different layers in the neural network and communicate through three sets of input and output signal lines shown in Fig. 14-7. The first set (DATA1 and DATA2) is used in the execution mode. The second set (DATA3) used in the training and further training modes. The third set is used for neural set-up and control.

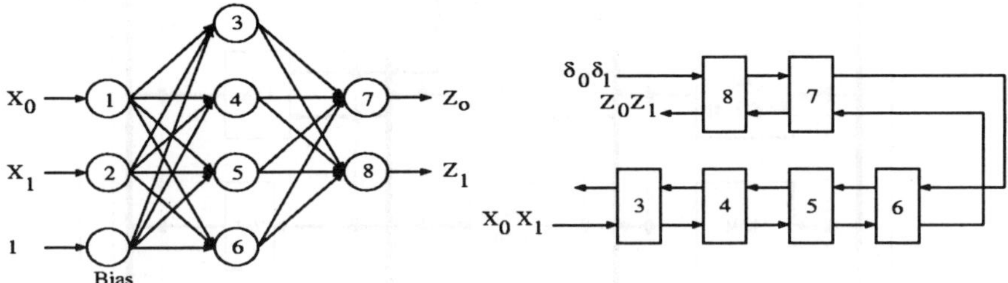

Fig. 14-6. Linear artificial neural network configuration.

In most neural network implementations, floating point arithmetic is commonly used for weight and activation function calculations. The parallel implementations use floating point hardware in each PE and each PE is replicated to form an array, consuming large portions of silicon. Additionally, floating point arithmetic uses extra time for the neural operations. This is due to the aligning of mantissas and exponents for floating point numbers. The floating point operations in the neural network are used during (i) the weight updates and (ii) the y_i activation computation. In order to reduce the amount of hardware, the computational precision was reduced to allow the utilization of integer arithmetic. The precision used to represent the weights and f_{net} calculations was restricted to ± 8 with a variable number of bits to the right of the decimal point. The number of bits to the right of the decimal was determined through experimentation. A series of simulations were done for the neural network using the C programming language including reduced precision arithmetic. Reduced precision was realized by using a truncation function to clip the numbers to within a specified range. The test data for the simulations included 2, 3 and 5-dimensional data as well as data from the XOR problem. The simulations showed that the learning severely degraded when number of bits was reduced to below 13. Any number of bits from 12 and above showed satisfactory learning, where the total sum squared error was below a predefined threshold. Based on the simulations, 12 bits to the right and 4 bits to the left of

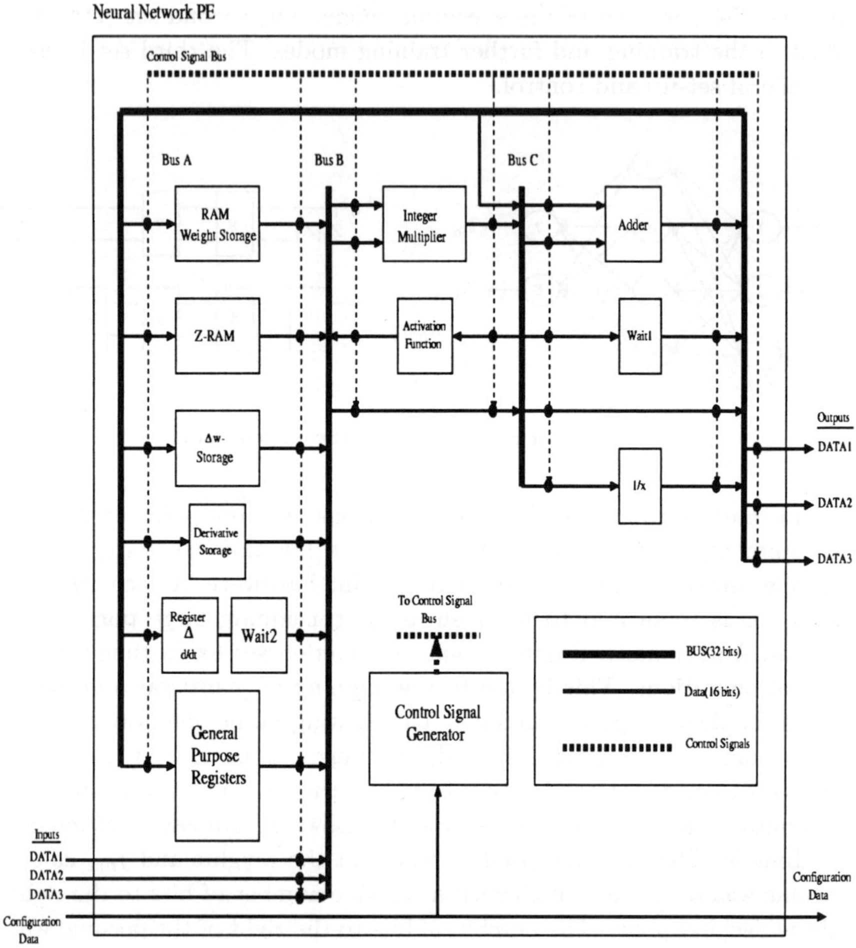

Fig. 14-7. Hardware for a single neuron PE.

the decimal point, adding to 16 bits, was used allowing real values in the range [+7.999755859... − 7.999755859]. Also, the simulations show that additional nodes are needed when using reduced precision arithmetic. The nodes of an ANN can use wide ranges of weight values during training without reduced precision. When reduced precision is implemented, the weight range of the ANN is limited. During training, an extra node or ***auxiliary node*** must empirically be inserted in the corresponding layer when the limit is exceeded or the the current ANN topology does not converge.

Each PE operates in five different modes: (i) initial set-up, (ii) execution, (iii) training, (iv) further training and (v) loading new rescale values. In the set-up mode, the desired neural network is formed into a linear array by propagating the configuration signals through each PE. There are four sets of elements that must be initialized for proper neural operation. These elements are (i) the registers, (ii) the RAM storage, (iii) the step sequencers, and (iv) the routing multiplexors for inter-node connection. The registers within each PE store (i) the neuron position values, (ii) the RAM index values, (iii) the learning rate, (iv) the momentum, (v) the current neuron output and (vi) the derivative of the neuron output. The neuron position values are used in the execution and training modes. They ensure that each neuron is in its correct position in the linear array as shown in Fig. 14-6. The neuron position values also initialize the wait state generators. The RAM index values correspond to the starting addresses for the RAMs.

The RAMs within each PE are initialized either with random or predetermined values using the DATA3 signal lines. Each neuron works in a series of steps for each operational mode and are sequenced using step sequencers. These sequencers are initialized to state 0 in the beginning. Finally, the routing multiplexors control the flow of the data between the neurons within a layer and the neurons that connect one layer to another.

The ANN operates in a linear systolic SIMD fashion during the execution mode and the data flow is shown in Fig. 14-8. In the fig., X is the neuron input, W is the connection weight, $f(net)$ is the neuron activation output. The number in parentheses corresponds to the index position of the neuron in the layer and the wait (wait1) values correspond to the number of cycles each neuron is delayed before transmitting data. In the execution mode, each PE uses the DATA1, DATA2, DVALID1, and DVALID2 signal lines to transmit and receive data. Each neuron operates in a sequence of four steps: (i) idle, (ii) receive, compute and store the value $\sum x_i w_{ij}$, (iii) compute the activation value, y_i, and wait1, and (iv) send the result to the next layer and reset. The neuron in the array receives the input vector and retrieves a weight value stored in the RAM. The neuron multiplies the input with the connection weight and accumulates the results. The neuron repeats the process until all the inputs have been received. Next, the node calculates the activation output and delays (wait1) until the other neurons in the same layer are in their proper states.

After waiting, the neuron outputs the neuron activation using the DATA2 lines to the neurons in the next layer where they are used as inputs. Since the initial input pattern on the DATA1 lines are still being used by the neurons in the current layer, the shifting of the outputs onto the DATA2 lines will not interfere with the current input pattern. The routing of the data is shown in Fig. 14-8 where data subscripts (a and b) refer to the I/O routes.

The computed outputs of each neuron travel through the remaining neurons of the current layer until the last neuron of that layer has been reached. When the output reaches the last neuron, the multiplexors shift the data to the DATA1 lines of the next layer. The neurons in the next layer operate in an identical manner as the previous layer. The performance of a neural network in the execution mode is commonly measured in terms of Connections per Second (CPS) [44]. The performance of the neural network can be derived as follows. Let N be the number of processors and f_{freq} be the clock speed. When the systolic array is full, the performance, in CPS, can be computed by the equation $CPS = N \times f_{freq}$.

The neural network in the training mode operates in a systolic manner similar to the execution mode and the data flow is shown in Fig. 14-9. The computation within a neuron are not described here and can be found from the literature. As input data flows through the network, each neuron propagates the output errors (δs) to the next layer, and learns by adjusting its own connection weights. The neurons in the training mode operate in eight steps: (i) idle, (ii) compute and store the value, $\Delta = \delta \cdot Z(1 - Z)$, and insert wait2 delay cycles (the δ values correspond correspond to the nodes in the output layer, and the nodes in the hidden layer), (iii) compute the error value, $\Delta \cdot w_{ij}$, and transmit to the previous node, (iv) compute and store the value $\delta \cdot Z$, (v) compute and store the value $\alpha \cdot \Delta w_{ij}^{-}$, (vi) reset the RAM index registers, (vii) compute and store the weight adjustment value $\Delta w_{ij} = \Delta Z + \alpha \Delta w_{ij}^{-}$, and (viii) update the connection weights, $w_{ij} = w_{ij} + \Delta w_{ij}$.

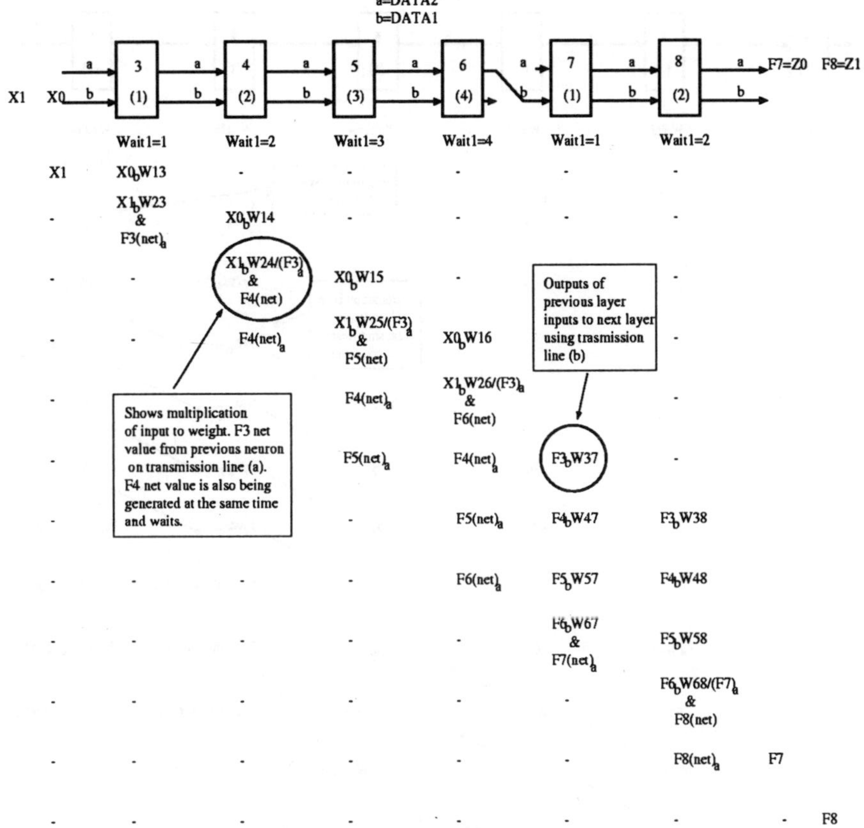

Fig. 14-8. Execution mode data flow (subscripts (a and b) denote route for data).

The performance of a neural network in the training mode is commonly measured in terms of Connection (or Weights) Updates Per Second (CUPS) [44]. The performance of the neural network can be derived as follows: N is the number of processors, N_{instr} is the number of instructions for the update process, and f_{freq} is the clock speed. The performance in the training mode can be computed by the equation:

$$CUPS = \frac{N \times f_{freq}}{N_{intr}} \tag{14-4}$$

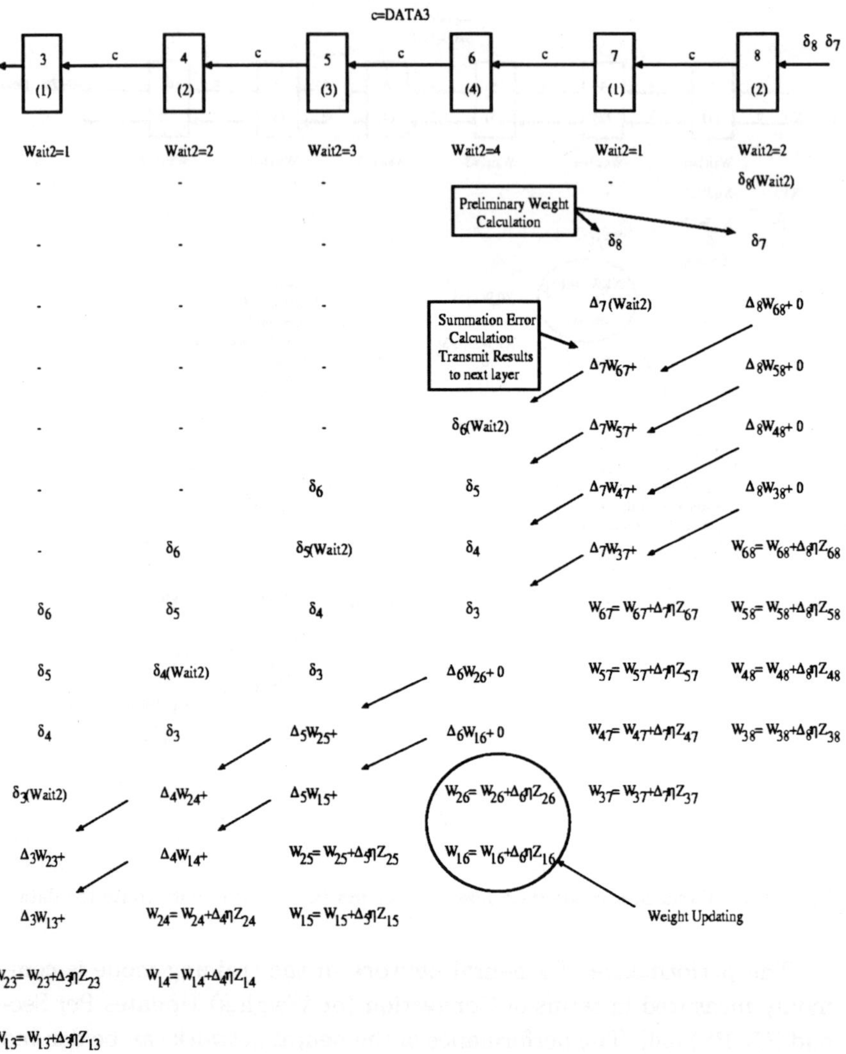

Fig. 14-9. Training mode data flow.

14.6.3. *Pipelined Fuzzy Expert System*

The FES computes a fuzzy decision based on the fuzzy inputs and the rule base using the max min composition rule. The inferencing process is illustrated by an example with n-rules having z-antecedents and one consequent shown in Fig. 14-10. The max min composition rule can be implemented in hardware and performed in 6 steps:

(i) Calculate the minimum of all points in A_1 and A'. Next, compute the maximum of all the minimum values, $\alpha_1^1 = max(min[A_1, A'])$.
(ii) Repeat step 1 for every z-antecedent in rule 1.
(iii) Compute the minimum of all α values in step 2, $\omega_1 = min(\alpha_1^1, \alpha_1^2, \ldots, \alpha_1^z)$.
(iv) Determine the decision of a rule by computing the minimum of the ω value with the consequent of the rule, $C_1' = min(C_1, \omega_1)$.
(v) Repeat for all n-rules.
(vi) Determine the overall decision by computing the maximum of all the decisions from the n rules, $C_{total} = max(C_1', C_2', ..., C_n')$. The fuzzy decision is sent to the defuzzification unit.

Here, α_i^j is the contribution of antecedent j of rule i, ω_i is the contribution of all the antecedents of rule i, $A_i, B_i, .., Z_i$ are the rule antecedents, $A', B', .., Z'$ are the rule inputs and C_i is the computed decision of rule i.

The FES is implemented as a linear array of rule processing elements (RPEs) shown in Fig. 14-11. Each RPE is capable of processing an entire rule. Thus, an n-rule system is constructed with an array of n RPEs.

Each rule processor consists of (i) two 512-bit shift registers, (ii) a composition block for computing the max min rule, (iii) four 16-to-1 bit multiplexors, and (iv) control signal generators all as shown in Fig. 14-12. The shift registers are placed on either side of the composition block and contain a maximum of 8 antecedents and consequents, where each is 64 bits in length. The composition block is composed of two 4-bit minimum units, a group of 16 4-bit minimum units and 16 4-bit maximum units and a 4-bit register. The four 16-to-1 bit multiplexors index the I_n^{th} membership element of the rule antecedent. The control signal generators synchronize the shift registers that feed the compositional block.

The FES data signals are (i) two 4-bit inputs to load the initial rules, (ii) the 4-bit fuzzy singleton input, (iii) the 4-bit fuzzy index, and (iv) the 64-bit max value. The fuzzy singleton is input to the compositional block. The fuzzy index selects 1 of the 16 sets of fuzzy values from the

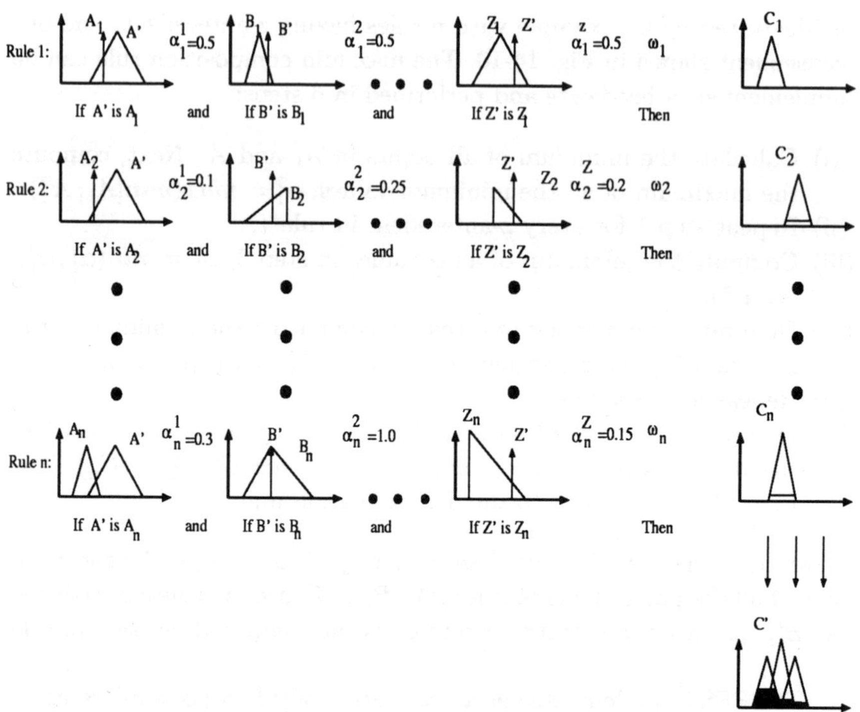

Fig. 14-10. Max Min composition steps

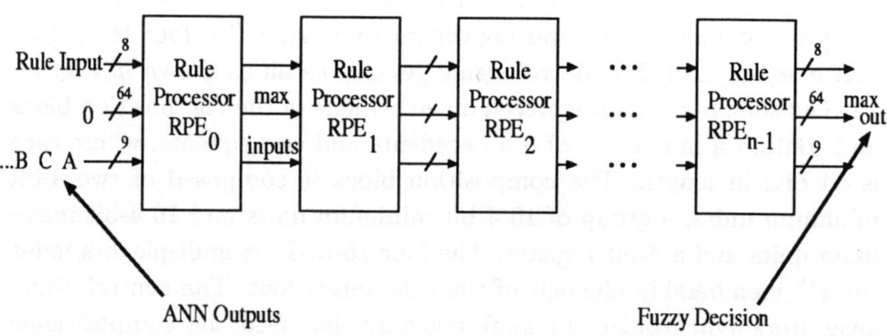

Fig. 14-11. Pipelined FES rule processor.

64-bit membership function using the four 16-to-4 bit multiplexors. The max input represents the cumulative decision of the $i-1$ previous RPEs

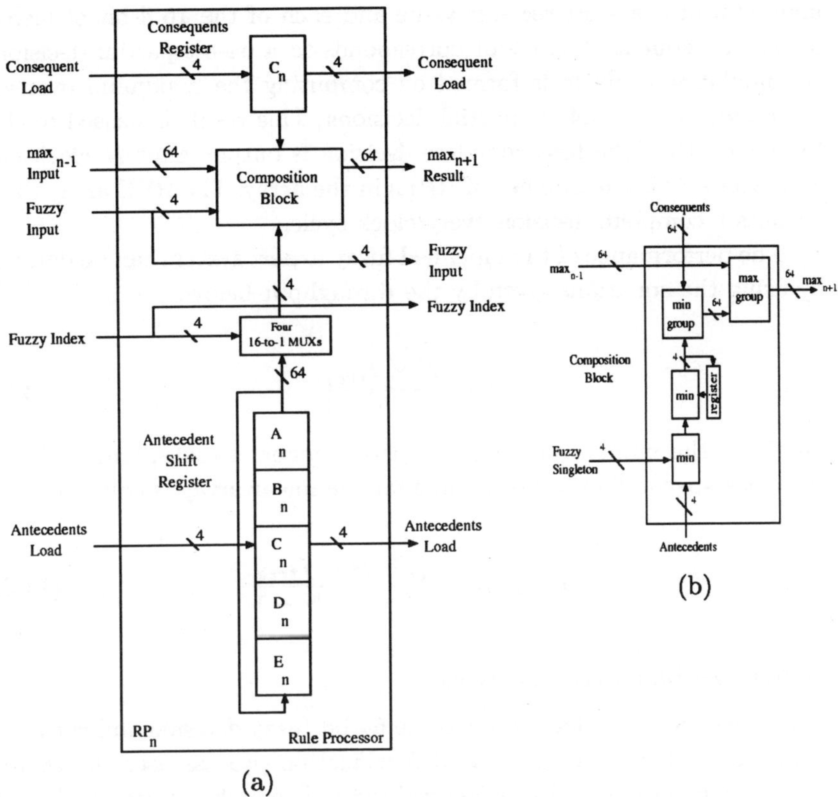

Fig. 14-12. (a) Rule processor (b) and Composition block.

and is passed between processors.

In the inferencing process, each RPE first receives the fuzzy singleton, the fuzzy index, and the max values. Next, the shift register sends the first 64-bit antecedent to the 4 16-to-4 bit multiplexors where the I_n^{th} 4-bit membership element is selected by the fuzzy index. This value is sent with the fuzzy singleton to the composition block for computing the max min rule. The first *min* block computes the minimum between the fuzzy singleton and the membership element of the rule. The minimum of all the antecedents and the fuzzy singletons are computed by the second *min* block and stored in the 4-bit register. After every antecedent has been shifted, the consequent register shifts the 64-bit values to the composition block. The 16 4-bit minimum blocks compute the minimum of the 4-bit register value and each of the 16 4-bit elements of the consequent. The result corresponds to a 64-bit partial decision. A cumulative decision is formed by computing the maximum over all the previous $n-1$ 64-bit partial decisions. This result is passed to the the next RPE. The first complete decision is output after N clock cycles, where N is the number of RPEs in the array. The RPE array then outputs a complete decision every clock cycle.

The performance of the pipelined fuzzy expert system can be derived by using the equations given by the throughput below:

$$\eta = \frac{N \times f_{freq}}{k + N - 1} \tag{14-5}$$

Let N be the number of PEs, k be the number of tasks, and f_{freq} be the clock speed. The response time for the linear array is equal to:

$$t_{response} = N \times \frac{N \times f_{freq}}{k + N - 1} \tag{14-6}$$

14.6.4. *Defuzzification Block*

The defuzzification block receives the 64-bit fuzzy decision and converts it into a 4-bit crisp value. The defuzzification process uses the center of gravity equation in the discretized form. The architecture consists of an adder tree, a multiplier/adder tree and a divider. The x values are index values ranging from 0 to 15, and the membership values (μ-s) are the fuzzy values located at the index positions.

14.7. VLSI Prototype

The main features of the RAPID system hardware architecture are shown in Table 14-1. A prototype VLSI chip implementing the RAPID architecture was designed and simulated using 2-micron CMOS technology. The prototype VLSI chip was designed using a two-phase non-overlapping clocking scheme with the Cadence OPUS design tools. The implementation involved three phases: (i) development of standard library cells, (ii) functional design entry and verification, and (iii) synthesis of the final layout.

Table 14-1. RAPID architecture features.

Issue	Feature
Circuit	Completely Digital
Architecture	On-Chip Memory, SIMD Format
Model	Linear Array
Communication	Nearest Neighbor, No Clock Skew
ANN Performance	$CUPS = \frac{N \times f_{freq}}{N_{intr}}$, $CPS = Nf_{freq}$
FES Performance	$\frac{k+(N-1)}{f_{freq}}$
Neural Training	Backpropagation
Arithmetic	Integer
Neuron Output	16 bits
Weight Size	16 bits

The design was simulated in Verilog-XL at a 50MHz clock rate, using standard cell designs for the data paths, pads, control and other logic. The various components such as the multiplier, the adder, and the RAM were designed and verified [37]. The ANN PE required $56K$ transistors and an array consisting of 40 PEs will require 2.24 million transistors. The rule processor RPE would require $24K$ transistors, and to implement a rule base of 100 rules, an array of 100 PEs will require 2.4 million transistors. Thus, a single system would require 4.64 million transistors. However, the RAPID system cannot be feasibly implemented as a single chip using the standard cell libraries with 2.0 micron technology. The entire system with 4.64 million transistors could be realized on a single chip system using sub-micron technology or as a board level system with multiple chips serially cascaded using 2 micron technology. Again, the system realization depends on the configuration and technology Since the architecture is based on the linear array model, the system can be easily scaled up for larger configurations.

14.8. Performance

The adaptive traffic light application was mapped onto the RAPID architecture for performance study through simulations. While the advantages of a hardware implementation are obvious, it was interesting to see how an integrated model compared versus a purely ANN or FES approach. The details of the simulation study as well as for other appli-

Table 14-2. Comparative performance summary.

System	Correct decision rate	Average Wait times (minutes)	Number of Nodes	Number of Rules
RAPID	95%	2.186	55	40
ANN approach	73%	2.958	83	-
FES approach	95%	2.975	-	40

cations of the integrated system can be found in [37].

A summary of the results for the RAPID system, the ANN and the FES approaches are given in Table 14-2. The results show that the RAPID system provided decisions that relieve intersection congestion better than the ANN approach and was comparable to the FES approach. The results also show that the RAPID system imposed a lower average vehicle wait time than the other two approaches. The ANN approach required more neural nodes than for the ANN in RAPID which led to slower training and a higher implementation cost. The FES approach computed correct decisions well, however the computed decision did not led to a better reduction in the wait times. The FES approach computed decisions based on the current traffic flows only. Additional rules would be required to compute decision using both previous and current values as in in RAPID.

The simulation summary indicates that the RAPID system provided better decisions at a lower cost compared to the other approaches. The summary also shows that the RAPID system provided more effective cycle time adjustments. As more functions are added to the adaptive traffic light problem, the number of rules and the ANN nodes required increases considerably, making the implementation of the ANN and the FES approaches more complex. In the RAPID system, the integration of the two models helps in reducing the number of nodes in the ANN and the number of rules in the FES leading to a better and compact implementation.

14.9. Conclusions

A real time intelligent decision making system called RAPID implementable in VLSI has been presented. The architecture is unique in that the entire system can be implemented as a linear systolic array. The system consists of a backpropagation based ANN that can learn and adapt to a dynamically changing environment and a fuzzy expert system for decision making. It was shown that the integrated system performs better than the ANN and the fuzzy expert system approaches.

The implementation of real time intelligent system architectures will play an increasingly important role in many AI applications. Growth is evident in the application of intelligent systems especially in the fields of machine perception, intelligent signal processing, financial decision making, insurance risk assessment and natural language understanding. The work discussed shows the viability of implementing an integrated intelligent system on silicon and the inherent parallelism in decision making that can be exploited.

References

[1] A.R. Plesckun, M.J. Thazhuthaveetil, "The Architecture of Lisp Machines," *Computer*, IEEE, vol 20, no 3, March 1987.

[2] B. Freisleben, T. Kunkelman, "Combining Fuzzy Logic and Neural Networks to Control an Autonomous Vehicle," *IEEE International Conference on Fuzzy Systems*, 1993,Vol 1, pp 321-326.

[3] B. Kosko, "Neural Networks and Fuzzy Systems," Prentice Hall Inc.,Englewood Cliffs, NJ.

[4] C.A. Cruz, W.A. Hanson, and J.Y. Tam, "Neural Network Emulation Hardware Design Considerations," *Proceedings of IEEE First International Conference Neural Networks*, IEEE Press, June 1987, pp III-427-434.

[5] C. Georgescu, A. Afshari, G. Bornard, "Fuzzy Predictive PID Controllers: A Heating Control Application," *2nd IEEE Conference of Fuzzy Systems*, vol 2, 1993, pp 1091-1098.

[6] C. Pappis, and E. Mamdani, "A Fuzzy Logic Controller for a Traffic Junction," *IEEE Transactions on Systems, Man, and Cybernetics*, Vol SMC-7, no 10, 1977.

[7] D. Hammerstrom, "A VLSI Architecture for High-Performance, Low-Cost, On-Chip Learning," *Proceedings of the International Joint Conference on Neural Networks*, Vol II, 1990, pp 537-543.

[8] D.A. Patterson, "A VLSI RISC," *Computer*, IEEE, September, 1982.

[9] D.E. Rumelhart, "Parallel distributed processing: explorations in the microstructure of cognition," Cambridge, Mass.: MIT Press, 1986.

[10] Dmitriy V. Kirsanov, "Digital Architecture for Neural Networks," *Proceedings of 1993 International Joint Conference on Neural Networks*, 1993, vol 2, Nagoya Congress Center, Nagoya, Japan.

[11] E.H.L. Aarts, "Simulated annealing and Boltzmann machines : a stochastic approach to combinatorial optimization and neural computing ," Wiley Publishing, 1989.

[12] F. Distante, M.G. Sami, R. Stefanelli, G.S. Gajani, "Alternative approaches for Mapping Neural Networks onto Silicon," *Second Italian Workshop on Parallel Architectures and Neural Networks*, Ed. E.R. Cainiello, World Scientific Publishing, 1990, pp. 319-327.

[13] G.F. Luger, "Artificial intelligence and the design of expert systems," Redwood City, Calif. : Benjamin/Cummings Pub. Co., p. 631-644, 1989.

[14] H. Hsin, et. al., "An Adaptive Training Algorithm for Back-Propagation Neural Networks," *IEEE Transactions on Systems, Man, and Cybernetics*, Vol. 25, no. 3, March 1995.

[15] H. Rigas, T. Booth, F. Briggs, at el., "Artificial Intelligence Research in Japan," *Computer*, IEEE, vol 18, no 9, September 1985.

[16] H.J. Zimmerman, "Fuzzy Set Theory and its Applications," 2nd Edition, *Kluwer Academic Publishers*, 1991.

[17] H. Berenji, R.N. Lea, Y. Jani, P. Khedkar, A. Malkani, J. Hoblit, "Space Shuttle Attitude Control by Reinforcement Learning and Fuzzy Logic," *2nd IEEE International Conference of Fuzzy Systems*, vol 2, 1993, pp 1396-1401.

[18] J.J. Hopfield, "Neural Networks and Physical Systems with Emergent Collective Computational Abilities," *Proceedings of National Academy of Science*, vol 79, 1982, pp 2554-2558.

[19] J. Mendel, "Fuzzy Logic Systems for Engineering: A Tutorial," *Proceedings of the IEEE*, vol 83, no 3, March 1995, pp 345-377.

[20] J. Ouali, G. Saucier, "A Flexible Architecture for Neural Networks," *IEEE International Computer Conference On Design: VLSI in Computers*, 1989, pp 483-486.

[21] J. Wawrzyneck, K. Asanovic, and N. Morgan, "The Design of a Neuro-Microprocessor," *IEEE Transactions on Neural Networks*, Vol. 4, no. 3, pp. 394-399, May 1993.

[22] L. Altas, et. al., "A Performance Comparison of Trained Multilayer Preceptrons and Trained Classification Trees," *Proceedings of the IEEE*, Vol. 78, no. 10, 1990, pp. 1614-1619.

[23] Leonid V. Grachev, Eduard J. Kirsov, "The General Purpose Neurocomputer HERCULES," *Proceedings of 1993 International Joint Conference on Neural Networks*, 1993, vol 2, Nagoya Congress Center, Nagoya, Japan.

[24] L.I. Larkin, "A Fuzzy Logic Controller for Aircraft Flight Control," *Proceedings 23rd IEEE Conference on Decision and Control*, December 1984, vol 2, Las Vegas, Nevada.

[25] K. Nishimori, S. Hirakawa, H. Tokutaka, S. Kishida, N. Ishihara, "Fuzzification of Control Timing in Driving Control of a Model Car,"

IEEE International Conference on Fuzzy Systems, vol 1, 1993, pp. 297-302.

[26] M Conti, S. Orocini, F. Piazza, C. Turchetti, "Digital CMOS VLSI Processor Design for the Implementation of Neural Networks using Linear Wavefront Architecture," *Proceedings of 1993 International Joint Conference on Neural Networks*, 1993, vol 2, Nagoya Congress Center, Nagoya, Japan.

[27] M.I. Patel, N. Ranganathan, "A VLSI System Architecture for Real-Time Intelligent Decision Making," *Proc. of International Conference on Application-specific systems Architectures and Processors(ASAP'96)*, 221-230, 1996.

[28] M.I. Patel, N. Ranganathan, "PANTHER: A Parallel Neuro Systolic Architecture for Real-Time Processing," *International Conference on Neural Networks*, ICNN96, Vol 2, 1006-1011, 1996.

[29] M.I. Patel, N. Ranganathan, "A VLSI System for Urban Traffic Control Applications," *IEEE Symposium on Parallel and Distributed Processing (SPDP'96)*, 10-12, 1996.

[30] M. Zubair, B.B. Madan, "Systolic Implementation of Neural networks," *IEEE International Computer Conference on Design: VLSI in Computers*, 1989, pp 479-482.

[31] M. Sugeno, K. Murakami, "Fuzzy parking Control of Model Car," *Proceedings 23rd IEEE Conference on Decision and Control*, December 1984, vol 2, Las Vegas, Nevada.

[32] Marc A. Viredaz, Paolo Ienne, "MANTRA I: A Systolic Neuro-Computer," *Proceedings of 1993 International Joint Conference on Neural Networks*, 1993, vol 3, pp. 3054-3057,Nagoya Congress Center, Nagoya, Japan.

[33] N. Mauduit et al., "Lneuro 1.0: A Piece of Hardware LEGO for Building Neural Network System, *IEEE Transactions of Neural Networks*, Vol. 3, no. 3, 1992, pp 414-421.

[34] N. Morgan, "The Ring Array Processor (RAP): A Multiprocessor Peripheral for Connectionists Applications", *Journal of Parallel and Distributed System*, Vol. 14, no. 3, 1992, pp 248-259.

[35] P. Kotilainen, J. Saarinen, K. Kaski, "Neural Network Computation in a Parallel Multiprocessor Architecture," *Proceedings of 1993 International Joint Conference on Neural Networks*, 1993, no 2.

[36] P. Treleaven, M. Pacheco, M. Vellasco, "VLSI Architectures for Neural Networks," *IEEE MIRCO*, IEEE, December 1989, pp 8-17.

[37] M.I. Patel, "RAPID: A Real Time System Architecture for Intelligent Decision Making", *PhD Dissertation*, University of South Florida, 1997.

[38] R.T. Savely, "The Implementation of Neural Network Technology," Network Systems," *Proceedings of IEEE First International Conference Neural Networks*, IEEE Press, June 1987, pp IV-477-484.

[39] S.M. Barber, J.G. Delgado-Frias, S. Vassiliades, G.G. Pechanek, "SPIN-L: Sequential Pipelined Neuroemulator with Learning Capabilities," *Proceedings of the 1993 International Joint Conference on Neural Networks* vol 2, 1993, pp 1927-1931.

[40] S.M. Smith, G.J.S. Rae, D.T. Anderson, "Applications of Fuzzy Logic to the Control of an Autonomous Underwater Vehicle," *2nd IEEE International Conference on Fuzzy Systems*, vol 2, 1993, pp 1099-1105.

[41] S.Y. Kung, J.N. Hwang, "Parallel Architectures for Artificial Neural Nets," *IEEE International Conference on Neural Networks*, San Diego, CA, July 1988, vol 2, pp 165-172.

[42] S. Vassiliadis, J. Delgado-Frias, M. Zhang, "High Performance with Low Implementation Cost Sigmoid Generator," *Proceedings of 1993 International Joint Conference on Neural Networks*, 1993, Vol 2, pp. 1931-1935, Nagoya Congress Center, Nagoya, Japan.

[43] T. Kohonen, "New Analog Associative Memories," *International Joint Conference on Artificial Intelligence*, IEEE Hong Kong Center, Hong Kong, 1989.

[44] T. Nordstrom, B. Svensson, "Using and Designing Massively Parallel Computers for Artificial Neural Networks," *Journal of Parallel and Distributed Computing*, 1992, Vol 14, pp. 260-285.

[45] W.A. Hanson et al.,"CONE-Computational Network Environment," *Proceedings of IEEE First International Conference Neural Networks*, IEEE Press, June 1987, pp IV-531-538.

[46] U. Ramacher, "SYNAPSE-A Neurocomputer that Synthesizes Neural Algorithms on a Parallel Engine," *Journal of Parallel and Distributed Computing*, Vol. 14, no. 3, 1992, pp 306-318.

[47] Yuji Sato, Katsunari Shibata, Mitsuo Asai, Massaru Ohki, Mamoru Sugie, Takahiro Sakaguchi, Masashi Hashimoto, Yoshihiron Kuwabara, "Development of a High-Performance, General Purpose Neuro-Computer Composed of 512 Digital Neurons," *Proceedings of 1993 International Joint Conference on Neural Networks* ,1993, vol 2, Nagoya Congress Center, Nagoya, Japan.

[48] M.I. Patel, N. Ranganathan, "IDUTC: An Intelligent Decision-Making System for Urban Traffic-Control Applications," *IEEE Transactions on Vehicular Technology*, Vol. 50, No. 3, pp. 816-829, May 2001.

CHAPTER 15

RECONFIGURABLE HARDWARE SYSTOLIC ARRAY FOR REAL-TIME COMPARTMENTAL MODELING OF LARGE-SCALE ARTIFICIAL NERVOUS SYSTEMS

M. Korkin

Genobyte Inc. 1200 Pearl Street, Suite 65
Boulder, Colorado 80302, USA
E-mail: korkin@genobyte.com

This chapter presents a detailed technical description of a large-scale evolvable hardware system for evolving compartmentally modeled neural circuits directly in silicon at high speed. The core of the system is a three-dimensional array of reconfigurable logic with 1.2 million fine-grained function units and 1.2 Gbyte distributed memory. An application example is presented, describing an evolution of compartmentally modeled neural networks, based on a substrate of three-dimensional cellular automata, and a simulation of a hardware-based 75-million neuron artificial brain in real time. The system was developed in 1997-2000 at Genobyte, Inc. (Boulder, Colorado) for ATR HIP (Kyoto, Japan), and is marketed as CAM-Brain Machine (CBM). CBM features a true run-time logic reconfiguration, a hardware implementation of chromosome crossover and mutation, and a hardware-based fitness evaluation. CBM also features a sophisticated genotype-phenotype mapping through the process of embryonic growth.

15.1. Introduction

CAM-Brain Machine (CBM) development [1] pursued three objectives. The main objective was to develop a computationally powerful research tool for evolution of complex digital circuits, such as compartmentally modeled cellular automata based neural networks, directly in reconfigurable hardware. The second objective was to combine the evolutionary capabilities with means for assembling a large-scale circuit from multiple evolved smaller circuits, such as a multi-million neuron artificial brain made of thousands of the evolved neural modules, and running this large circuit at speeds sufficient for real time applications. The third objective was to make the system sufficiently flexible to serve as a general-purpose evolvable hardware research platform. CBM functionality is built around the advanced capabilities of an experimental Xilinx XC6264 FPGA made available to Genobyte by the manufacturer via a special arrangement.

15.2. CBM Hardware Architecture

CBM architecture is subdivided into the following seven blocks:

- Reconfigurable Hardware Core
- Phenotype/Genotype Memory
- Input/Output Unit
- Netlist Unit
- Fitness Evaluation Unit
- Central Processing Unit
- Host Interface

15.2.1. *Reconfigurable Hardware Core*

CBM hardware core is formed by 72 densely interconnected XC6264 FPGAs, packaged in 560-ball plastic ball grid arrays, each containing 16,384 fine-grained reconfigurable function units and 512 reconfigurable I/O blocks. Each of the 72 FPGAs is mounted on a separate printed circuit module board with a 360-pin HDM connector plugged into a single 19" x 19.5" twelve-layer backplane with a total of over 30,000 connections.

The pattern of connectivity forms a three-dimensional (3D) array of 6 x 4 x 3 FPGAs. Each individual FPGA interconnects with six "neighbor" FPGAs in the 3D space using 208 open-drain bi-directional connections: 48 connections north, 48 south, 32 east, 32 west, 24 top, and 24 bottom. The 3D FPGA array is uniform: any cross-section of the array along each of the three axes contains a square matrix of 24 x 24 connections. All FPGA connections on the external faces of the 3D array are fully wrapped around to the corresponding opposite face, so that each FPGA irrespective of its position in the array connects to the six neighbors in 3D. As a result, the array is toroidal on all three dimensions

Fig. 15-1. CAM-Brain Machine.

Each of the 72 FPGAs also has external connections (relative to the 3D array), which are: a dedicated 32-bit data bus, an 18-bit address bus, and a number of control signals used to access both configuration memory space of the FPGA and user-configured logic at run time. A unique feature of the XC6264 FPGA is a uniform access to both configuration memory space and the user-configured logic space.

In order to ensure clock integrity in the large hardware array, each individual module board receives a dedicated differential clock signal. On the receiving end each clock signal is individually multiplied to 66 MHz by a digital frequency multiplier. In addition, groups of six clocks can be individually skew-adjusted via programmable delay lines by the host computer software. Skew adjustment is used to manage ground bounce effects caused by thousands of simultaneously switching signals.

CBM was designed to support two different kinds of evolvable hardware, known as "gate-level" and "function-level" evolvable hardware

[2]. Evolution of gate-level circuits is supported by loading chromosome bitstrings directly into the FPGA configuration memory to configure an arbitrary digital circuit made up by the interconnected elementary Boolean gates and the flip-flops. Unlike any other FPGA, including Xilinx Virtex series FPGA, the XC6200 series is uniquely suited for this approach – random logic or random routing configuration cannot physically damage the device.

An example of the function-level evolvable hardware applications is a cellular automata based compartmentally modeled neural networks. In this particular application, each FPGA's internal logic is configured as a 3D cellular automata block of 4 x 6 x 8 cells. Each cell connects bi-directionally to six neighbor cells in three dimensions. External connections of the "block" reach to corresponding cells inside "neighbor" FPGAs in the 3D FPGA array. As a result, a uniform toroidal cube of 24 x 24 x 24 cells is formed.

A manually pre-designed digital circuit implementing CA rules is repeated for each cell of the array. This circuit utilizes multiple elementary Boolean gates and flip-flops, a total of 64-128 FPGA function units. Writing specific short bitstrings into this user-defined cell logic can modify cell's behavior. This is an example of CBM-based implementation of function-level evolvable hardware with user-designed "functions" or building blocks at a level higher than the gate-level. A great variety of function-level evolvable hardware applications can be attempted using a total of 1.2 million fine-grained logic units, configurable as either an elementary Boolean function, or a flip-flop, or a combination thereof.

15.2.2. *Phenotype/Genotype Memory*

Each of the 72 FPGA module boards also contains a dedicated 16 Mbyte EDO DRAM tightly coupled with the FPGA via a 32-bit data bus. A data exchange rate during FPGA write access is 66 Mbyte/s; while read access rate is 22 Mbyte/s. All 72 DRAMs form a large distributed 1.2 Gbyte memory array.

Overall control of the CBM hardware core is organized as a "Single Instruction Multiple Data" (SIMD), so that each FPGA receives an identical address and control signals, while the data flow between the FPGA and the DRAM is unique on each module board. During evolutionary runs, DRAM data can be streamed into FPGA into either configuration memory space, or into user-designed logic space. In either case, the loaded bitstrings can serve as circuit chromosomes, which determine circuit functionality during subsequent circuit execution and fitness evaluation.

When CBM is running the cellular automata neural network applications, the evolutionary process employs a sophisticated genotype-phenotype mapping through an intermediate process of embryonic growth.

The ultimate purpose of growing neural circuits on a substrate of three-dimensional cellular automata is to form dendritic and axonic branches from multiple small compartments linked together. Before the growth process starts, the cellular array logic is pre-configured to implement the "growth phase" cellular rules. Each cell is then loaded with a growth instruction, such as "turn south", "grow straight", "split west, north and top", "block growth", etc. A complete chromosome is a string of growth instructions for each cell of the array. Next, a few cells in the 3D space are seeded as future "neurons" in the otherwise "blank" cellular automata space.

Then, a neural network is gradually grown clock by clock, originating at the seeded neurons, guided and constrained by the loaded chromosome. Each neuron grows a mathematically correct set of branching dendritic and axonic trees in the 3D space.

After growing is completed, the grown network is saved back into the DRAM as a resultant "phenotype", which encodes the 3D topology of the grown network.

The phenotype can be later loaded back into the cellular array after the array logic is reconfigured with a different set of rules for the purpose of exercising the grown neural network in order to evaluate the network's performance, or fitness. This exercising phase is called "signaling phase".

During the signaling phase signals propagate through the grown dendritic branches toward their respective neurons, which can be excited or inhibited by the incoming signal activity. Sufficiently excited neurons can fire and further propagate signals through their outgoing axonic branches. According to the grown network topology, the axonic branches of some neurons may terminate on the other neurons' dendritic branches resulting in a axono-dendritic connection made.

Thus, depending on the phase of the process, DRAM data encode either a genotype (for the growth phase) or a phenotype (for the signaling phase). Alternative CBM applications may not necessarily separate genotype memory from phenotype memory in cases where a direct genotype-phenotype mapping is utilized.

In neural network applications the phenotype/genotype memory capacity is 64,640 complete neural module phenotypes, each encoding a 24 x 24 x 24 cellular cube circuit. Each cube can contain up to 1,152 neurons endowed with a set of branching dendritic and axonic trees. Loading a particular phenotype from memory into the cellular array logic instantiates this neural module in hardware for the purpose of exercising it at a hardware speed by propagating signals through it. Typical instantiation time is 12-16 microseconds.

In a "brain" simulation mode, CBM sequentially instantiates one evolved phenotype after another into the hardware core, and propagates certain stimuli signals through the instantiated neural module. The intermediate results are saved back into the DRAM. These intermediate results are the status of the signals propagating through dendritic and axonic branches, as well as neuronal accumulator values, and other variable data. To minimize any idle time for the hardware array, a pipelined operation is implemented – while one neural module is being exercised, the next module is pre-loaded into the FPGA hardware, while simultaneously the previous module's results are saved back into the DRAM. After the current module run time is over, the next module is instantiated in a matter of nanoseconds, and the process continues. A module's run time is matched to module's pre-loading time (12-16 microseconds), so that hardware core is exercised without considerable downtime between the successive modules.

CBM's computational power in implementing cellular automata based neural networks is on the order of 10,000 Pentium III 500 MHz computers running identical algorithms in software.

A similar pipelined streaming process takes place during an evolution mode. First, the hardware array is pre-reconfigured for the growth phase cellular rules. Then, a population of up to 1024 neural networks is grown in hardware by loading growth chromosomes and growing phenotypes one after another. The resultant phenotypes are saved back into the phenotype/genotype memory. Then, the cellular array is reconfigured again, this time for signaling phase, and each of the grown phenotypes is instantiated one after another to run identical stimuli patterns through the modules. A fitness evaluation unit (see below) evaluates the response of each neural module during the signaling phase. After one generation has been fully evaluated, the relatively better performers are selected for reproduction, and their chromosomes are crossed over and mutated to create the next generation of genotypes.

The crossover and mutation operations are performed also in the hardware cellular array for speed. The array is completely reconfigured one more time for implementing a special set of cellular automata rules needed for this third "genetic" phase. During this phase, CBM loads two parents' chromosomes into the hardware array, and then loads a crossover mask and a mutation mask randomly generated by the host computer for each bit of the chromosome. An offspring chromosome is created in a few nanoseconds, and saved back into the same phenotype/genotype memory to be further used for growing and evaluating of the next generation of neural networks. Mask generation is performed prior to the run time.

The evolutionary process continues until a specified number of generations is reached. Because all of the time-consuming stages of the evolutionary process are implemented in hardware, the evolution time typically takes on the order of seconds for a population of 100 evolving over 1000 generations.

To recoup, during an evolutionary run the FPGA hardware core is repeatedly reconfigured to change the cellular automata rules: first for the growth phase, next for the signaling phase, and finally for the genetic phase. Reconfiguration takes place in all of the 72 FPGAs in parallel by loading an identical pre-compiled logic configuration file (CAL file) from the phenotype/genotype memory into the FPGA configuration memory. Full reconfiguration of the FPGA array takes less than 10 ms.

The short logic reconfiguration time in the XC6264 FPGA yields a number of computational advantages. First, it opens up a possibility of reduced complexity of the instantiated cellular automata rules limited to a particular subset needed in each phase of the evolutionary run.

Consequently, it allows making cellular automata logic much simpler and thus faster, and, finally, it allows fitting a larger number of cells in the same amount of logic. If the reconfiguration time were prohibitively long, the cells would have to be more complex to include a full set of CA rules for all phases, their logic would have been accordingly slower, and a much smaller number of cells would be fittable in the array.

Future improvements in CBM performance will include further reduction of FPGA reconfiguration time by exploiting a unique partial reconfiguration capability of XC6264 – the only true partially reconfigurable logic array ever built.

Fig. 15-2. A fully grown compartmentally modeled neural circuit on a substrate of three-dimensional cellular automata.

15.2.3. *Input/Output Unit*

When a neural module is exercised, a set of input stimuli is presented to the input ports of a neural module, and the response signal activity is recorded. In CA neural networks applications signals are implemented as serial sequences of ones and zeroes called spiketrains, by analogy with biological neural networks. A real-valued information is encoded in the spiketrains using an interspike interval (temporal) coding [3], similar to the type of encoding discovered in animal nervous systems. This type of encoding is very efficient – each spike on average encodes 3-5 bits of information. To convert a spiketrain into the corresponding real-valued waveform, the spiketrain is convolved with a 24-tap convolution filter, which shape is borrowed from a recent neuroscientific research [4]. A reverse transform from spiketrains into waveforms is also available.

A neural module instantiated in the cellular cube of 24 x 24 x 24 cells receives up to 188 input stimuli spiketrains evenly distributed along the external surfaces of the cube. Four additional connections are designated as the module's outputs. Each spiketrain is a 96-bit long sequence of ones and zeroes, which encodes a segment of a real-valued waveform.

Spiketrains are stored in a Spiketrain Memory, a 2 Mbyte SRAM, sufficient to store up to 131K 96-bit spiketrains. Each spiketrain has a unique identifier, and is individually accessible using this identifier as an address. Because all of the afferent spiketrains must be fed into an instantiated neural module in parallel at hardware speeds, they are first retrieved from the Spiketrain Memory and loaded into a set of 188 spiketrains buffers. These buffers are implemented using three additional XC6264 FPGAs. Each FPGA houses 64 dual-pipelined buffers, and is coupled with its own dedicated Spiketrain Memory SRAM section, each containing a copy of all available spiketrains. This redundant architecture allows high-speed parallel retrieval and storage of spiketrains, three spiketrains at a time. A pipelined process similar to the one implemented in CBM core is used in spiketrain buffers: while one set of spiketrains is being fed into the neural module, the next one is preloaded from the Spiketrain Memory. At the same time, the four response spiketrains generated by the neural module are recorded by a set of response buffers to be later stored back into the Spiketrain Memory.

In the "brain" mode, when CBM sequentially streams one phenotype after another into the hardware core, sets of corresponding spiketrains are sequentially streamed into the neural modules, while the response spiketrains are saved back into the Spiketrain Memory.

15.2.4. *Netlist Unit*

In order to connect specific spiketrains to specific ports on the external surfaces of the cellular cube, each neural module is associated with a netlist. The netlist is a list of 188 spiketrain identifiers corresponding to the module's input port numbers, followed by the four spiketrain identifiers for module's outputs. Netlists are stored in a dedicated Netlist Memory, a 64 Mbyte DRAM, capable of storing up to 65K different netlists.

In "brain" mode, when CBM sequentially streams one phenotype after another into the hardware core, the corresponding netlists are retrieved from the Netlist Memory -- each neural module has its own unique netlist describing its input and output "connections". Each spiketrain identifier listed in the netlist is used as an address to retrieve/save the spiketrain from/into the Spiketrain Memory. This mechanism gives the user an unlimited flexibility in module interconnection in a multi-thousand-module brain architecture. Any response spiketrain (with a unique identifier) originating at any neural module can be connected to any input port of any other neural module in the brain. Additionally, any number of spiketrains can be designated as inputs or outputs from the "external world" relative to the "brain" in a variety of applications such as robotics.

In CBM's "evolution" mode netlists are used to specify a set of stimuli and their input port connectivity. Each individual in the evolving population receives the same set of stimuli. The length of a stimulus is not limited to the 96-bit spiketrain size. Successive netlists grouped together provide up to 6.5-Mbit long stimuli capability, especially needed when multiple cases of the same stimulus pattern are presented during evolution.

Alternative CBM applications which don't involve cellular automata based neural networks still require the above netlist mechanism for "interconnecting" multiple sub-circuits into a larger circuit.

15.2.5. *Fitness Evaluation Unit*

In order to guide the evolutionary process, each circuit's fitness must be evaluated. In CBM, circuit's fitness is evaluated by comparing a response from an individual circuit with a targeted response, specified by the user, given a set of stimuli. In cellular automata based neural network applications this comparison is performed between a set of four target waveforms and a set of four response waveforms. In CBM, a (partial) measure of fitness is a sum of absolute deviations between a target waveform and a response waveform. A full measure of fitness is a sum of four partial fitnesses.

Because the actual waveforms to be compared must be first decoded from the corresponding spiketrains, the fitness evaluation unit contains a set of eight parallel convolution filters of 24 7-bit taps each, followed by four arithmetic units and the 32-bit accumulators. This fitness evaluation unit is implemented using one XC6264 FPGA to carry out the large amount of computations at a high speed. The filter tap values are pre-loaded directly into the FPGA by the host computer – the user can readily redefine the shape of the filter at any time. During each candidate neural module evaluation, the corresponding fitness values are computed at the rate of response spiketrain generation by the neural network. The target spiketrains are fed into the fitness evaluation unit from the target spiketrain buffers in the manner similar to the input spiketrain stimuli. Fitness values for each member of the evolving population are accumulated in a dedicated Fitness Memory (512 Kbyte SRAM) during the course of the signaling phase. At the end of each generation the host computer retrieves the fitness values to perform selection of breeding parents in software.

15.2.6. *Central Processing Unit*

In order to ensure that CBM high-speed hardware resources are utilized at their capacity, a dedicated custom central processor unit (CPU) was developed with a specialized custom 50-instruction code set using 32-bit instructions. The CPU is coupled with 2 Mbyte SRAM instruction memory, and is implemented using one Xilinx XC95288 CPLD.

At any given time the CPU executes two completely independent processes, at 33 MHz clock rate. The main process controls all hardware resources in the FPGA hardware core and the phenotype/genotype memory, while the secondary process controls the hardware resources of the input/output unit, the netlist unit, and the fitness evaluation unit. Both main and secondary programs share the same instruction memory.

For each of the two processes the tandem CPU contains a program counter (PC), a stack pointer, an instruction fetch register, and a program flow control state machine. The CPU instructions include a 16-bit address field, an 8-bit opcode field, a 2-bit instruction count field, and a 2-bit program flow control field. The instructions can be up to 4 execution cycles long. The program flow control field determines operations of the program counter, which can be incremented, loaded with an address from the instruction address field, popped from stack, or pushed onto stack. The stack is implemented in the CPU instruction memory using two separate stack areas for each of the two processes. The instruction fetch register can be loaded not only from the instruction memory, but also directly from the host computer. This mechanism is used to launch CPU program execution by pre-loading an appropriate jump instruction.

The tandem CPU instruction set contains a special SYNC instruction, which allows synchronization of the two processes. When either one of the two processes fetches this instruction, it suspends execution until the other process also fetches a SYNC instruction, after which both processes continue. Additionally, both processes can be suspended when a real-time refresh timer asserts a memory refresh request. The CPU grants refresh request when the main process fetches a RFRSH instruction. CBM applications use this instruction to make sure that a refresh request is granted only when appropriate, depending on the currently executed operations. For example, when page mode DRAM access is in progress, no refresh should be allowed. Both the Genotype/Phenotype memory (1.2 Gbyte DRAM) and the Netlist memory (64 Mbyte DRAM) are refreshed every 32 ms.

The opcode field and the address field in the majority of the instructions are not intended for the CPU itself, but instead are decoded and executed by the local state machines in the underlying hardware resources, such as FPGA module boards in the hardware core, or the fitness evaluation unit. A typical opcode may instruct the local state machine to transfer one 32-bit data (double) word from DRAM into FPGA using the address contained in the instruction's address field. The DRAM addressing in most cases is implicit using autoincremented page mode streaming.

Other CPU instructions in the code set are used for a variety of hardware functions, such as loading DRAM page row and column addresses into the DRAM controller, clearing FPGA internal registers, resetting local state machines, issuing control signals, and other functions. Certain instructions are recognized by some hardware resources and ignored by others. In many instances an instruction which starts a very long execution sequence is executed in parallel with multiple subsequent instructions responded to by a different state machine.

To facilitate CBM machine code development, a custom machine code assembler, compiler, and linker were developed. Frequently used machine code routines make up the CBM Operating System (CBMOS). CBMOS is loaded into instruction memory during the system initialization.

15.3. Host Interface

The CBM host computer is a Windows NT Pentium III 500 MHz computer connected to CBM with a 10' high-speed differential cable via a PCI interface board. At any given moment CBM is operating in one of the two modes: a "host mode" or an "auto mode".

In the host mode, CBM host computer performs access to any hardware resources of the machine under software control. Host access is implemented using a custom developed Windows NT kernel level PCI driver based on a pass-through mechanism of data transfer with I/O mapped address space. The host mode access is primarily used to burst data into/from the genotype/phenotype memory, the netlist memory, the spiketrain memory, or the CPU instruction memory. Additionally, the host can directly access FPGAs in the hardware core and those implementing spiketrain buffers. A variety of other CBM hardware resources are also directly accessible by the host, each resource mapped onto PCI pass-through address space.

Before initiating any access, the host computer polls the status of the refresh timer to compare the time needed to transfer the requested amount of data with the time left till the next memory refresh. A safety margin is included in the transfer time computation to guarantee that any preemption, which might occur in Windows NT, would not result in refresh period overrun. While in the host mode, the host computer can invoke CBM auto mode by setting certain control registers in the CBM hardware.

In the auto mode the system is operating at high speed under its own tandem CPU control according to the machine code preloaded into CBM instruction memory (see Section 15.2.6). No data transfer with the host computer takes place during the auto mode. The only host access allowed in the auto mode is polling of the main program counter contents in the tandem CPU. The auto mode yields back to the host computer by implementing a jump to a designated memory location, which contains a BREAK instruction suspending any further program execution. When the host computer reads the expected program counter value it relinquishes CBM control in host mode.

In particular, during an evolutionary run, each generation evaluation is performed at high speed in the auto mode. After each generation evaluation is over, the auto mode yields back to the host, which then reads the resultant fitness values from the CBM fitness memory and proceeds with parent selection routine in software. Because selection algorithms don't yield themselves to parallel execution to any considerable degree, selection of breeding pairs is performed by the host computer in software.

After the pairs of breeding parents are selected, the host uploads the modified machine code fragments containing the selected parent genotypes' memory pointers, and then invokes the genetic phase, followed by the growth and the signaling phase, all three phases in auto mode.

15.4. CBM Software Tools

User interaction with CBM is implemented through a number of graphical software tools, and a custom high-level language allowing complete control over the specification of the various evolutionary parameters and data structures. While tailored for use with cellular automata based neural networks, they could be readily adapted to support other evolvable circuits.

These parameters and data structures are specified by the user in the form of project data files using a high-level language, which are compiled and loaded into CBM. Some of the parameters, such as the genotypes, are loaded directly into CBM phenotype/genotype memory. Other control parameters, such as the population size, the number of generations, are pre-compiled into a machine-code "templates" which are executed by the CBM's tandem CPU to control the evolutionary and the brain mode runs at high speed.

In the evolutionary mode CBM first grows a population of genotypes into their corresponding phenotypes (the growth phase), then performs fitness measurements of the resulting phenotypes (the signaling phase), and finally performs reproduction to arrive at the next generation of genotypes (the genetic phase).

Evolutionary mode is the more complex of the two. It consists of three phases: growth, signaling, and genetic, each phase further subdivided into multiple sub-steps. The brain mode configures a set of phenotypes interconnected with the user defined topology and then continuously cycles through the phenotypes, exercising them in signaling mode, as described in Section 15.2.

As described above in Section 15.2, all of the algorithmic steps except one are handled by the CBM hardware at high speed. Selection of breeding pairs is implemented in software because the algorithmic steps necessary in selection (such as sorting and Roulette Wheel sampling) cannot take considerable advantage of the hardware parallelism in CBM.

For faster execution, the selection algorithm is performed by the host directly at the Windows NT driver level to avoid the relatively lengthy times switching control from NT driver level to NT application level, and then back to driver level to load the results into CBM. The typical selection execution time is of the same order as double propagation time between NT driver level and NT application level. In general, all of the CBM operations requiring top execution speed are implemented at the driver level to reduce CBM hardware idle time.

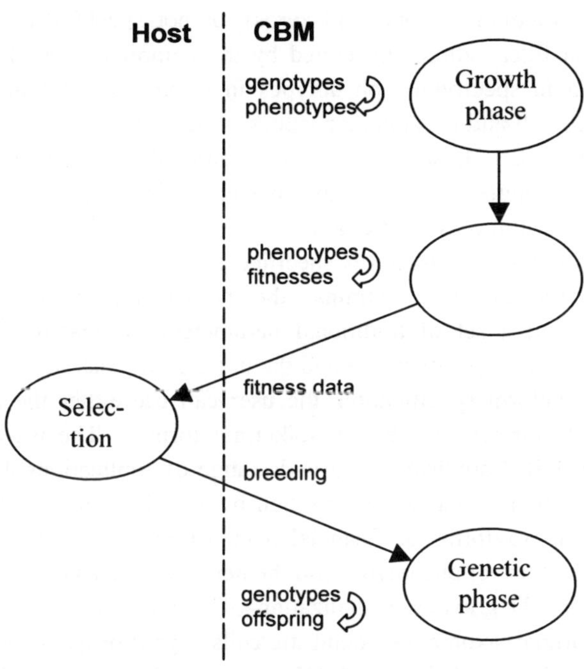

Fig. 15-3. Flow control over one generation.

The implemented selection process is the Rank selection algorithm in combination with Elitism (user- controllable). Alternative selection algorithms, such as fitness-proportionate reproduction, are avaiable to the user. Rank selection was initially chosen to keep the genetic algorithm from converging too quickly on (possibly localized) winners, since rank selection chooses its breeding members proportionate to their relative fitness instead of their absolute fitness as is the case in fitness-proportionate breeding. This tends to keep the population from becoming too homogenous too quickly.

Depending on which mode is desired, the appropriate data files are constructed by the user and compiled by software tools. Users have the flexibility to specify data to any level of relevant detail, down to a single data bit. A genotype may be specified in full detail for each of the 13,824 cells in a module, partially specified using default filler patterns for the unspecified portions (e.g. random, all ones, all zeros), or not specified at all, relying completely on filler patterns generated by the compiler. This flexibility is also provided in specifying phenotypes in brain mode (including signal states for each cell and accumulator values in neurons), as well as crossover and mutation masks. The user can control additional evolutionary parameters for the starting population, such as network branching rate, neuronal density, and excitation/inhibition synaptic ratio.

The additional data structures specified for an evolutionary run are the stimuli and the target spiketrains, the netlist specifying input/output connectivity, and a set of traditional parameters associated with genetic algorithms such as population size and the number of generations. In stimuli and target waveform specification, the user can use either the real-valued waveform data format, or a binary spiketrain format. The waveform data format makes CBM applicable in a wide range of applications, because the users can prepare and analyze data in their own application-specific domain, such as speech wavforms or financial data. Brain mode data is primarily specified in the form of phenotypes and the netlist connectivity.

Interactive 3D graphics visualization of a neural network phenotype, including neuronal, axonic, and dendritic cells, signal propagation direction, and cell states, is provided by a CBMServer tool. Visualization is available only before or after high-speed execution completes in hardware. Phenotypes can be displayed as a three-dimensional structure in both growth and signalling phases. A portion of a phenotype undergoing signalling is shown in Fig. 15-2. Cells types (neuronal, axonic, dendritic) are indicated by coloring, as are signal states of cells (active, inactive, neuron-firing). The user can interactively navigate in 3D through the network, select individual cells to display their status, manually advance execution process in growth or signaling phase, and observe input/output signal activity. The CBMServer application acts as the central interface between the user and CBM, controlling all load and run operations. The 3D rendering in CBMServer is performed with OpenGL.

Another tool, called Wavetool, provides visualization of the spiketrains, either as "spikes" separated by intervals of no activity, or as the equivalent real-valued waveform. The user can assign labels and select colors of the individual spiketrains displayed, scale waveforms vertically and horizontally, overlay waveforms, switch the waveforms on and off. The Wavetool utility executes as a separate application providing a COM Automation server interface allowing for control by CBMServer acting as an Automation client. Multiple instances of the Wavetool can be invoked to display large numbers of spiketrains and waveforms. The users can view not only CBM-generated signals, but their own application-specific data as waveforms or as spiketrains.

At the end of each evolutionary run the CBMServer instantly displays the run history in the form of average fitness plot and the best fitness plot over generations. Also, the Wavetool is automatically invoked to display the target versus evolved response waveforms of the best evolved individual. The user can further select other individuals from the last generation to display their response waveforms versus the target.

In addition to CBM application software, a completely separate set of tools is provided for the user, called "Access". These are CBM test and diagnostic routines for in-depth verification of CBM proper functioning and repair.

Finally, a set of C libraries (wrapped C++ classes) is provided for the advanced users to allow direct access to all of the CBM resources in user-developed software applications. Using the libraries, a wide range of evolvable hardware research applications can be readily developed.

15.5. Conclusions

CBM is a unique dedicated research platform for large-scale evolvable hardware applications. A considerable effort was made during CBM development to implement time-consuming functions directly in hardware to ensure the shortest execution time possible. A large parallel array of reconfigurable logic exercised at high-speed yields a supercomputer level computational performance at a fraction of the cost. CBM opens up a possibility of realizing research ideas currently unfeasible, such as meta-level evolution. As this new tool is becoming available to the research community, the authors expect CBM to produce results unattainable or impractical with any other available technology.

Acknowledgments

The author gratefully acknowledges an invaluable contribution of Gary Fehr and Gregory Jeffery, as well as the pioneering effort of Prof. Hugo de Garis, and an unwavering support of Katsunori Shimohara.

References

[1] Michael Korkin, Hugo de Garis, Felix Gers, and Hitoshi Hemmi. CBM (CAM-Brain Machine): A hardware tool which evolves a neural net module in a fraction of a second and runs a million neuron artificial brain in real time. In John R. Koza, Kalyanmoy Deb, Marco Dorigo, David B. Fogel, Max Garzon, Hitoshi Iba, and Rick L. Riolo, editors, *Genetic Programming 1997: Proceedings of the Second Annual Conference*, July 1997.

[2] T. Higuchi, M. Iwata, and W. Liu, (eds.). Evolvable Systems: from Biology to Hardware. Springer-Verlag, 1997. *Lecture Notes in Computer Science* No 1259.

[3] Michael Korkin, Norberto Eiji Nawa, and Hugo de Garis. A"spike interval information coding" representation for ATR's CAM-Brain Machine (CBM). In Proceedings of the Second International Conference on Evolvable Systems: From Biology to Hardware (ICES'98). Springer-Verlag, September 1998.

[4] Fred Rieke, David Warland, Rob de Ruyter van Steveninck and William Bialek. *Spikes: exploring the neural code.* MIT Press/Bradford Books, Cambridge, MA, 1997.

[5] Hugo de Garis, Felix Gers, Michael Korkin, Arvin Agah, and Norberto Eiji Nawa. Building an artificial brain using an FPGA based "CAM-Brain Machine". *Artificial Life and Robotics Journal*, 1999.

CHAPTER 16

IMPLEMENTING AND MAPPING ANNS ON RECONFIGURABLE MESH MASSIVELY PARALLEL ARCHITECTURES

W.N. Li

Department of Computer Science and Computer Engineering
University of Arkansas, Fayetteville, AR 72701, USA
E-mail: wingning@mail.uark.edu

J.J. Jenq

Computer Science Department, Montclair State University
Upper Montclair, NJ 07043, USA
E-mail: jenq@pegasus.montclair.edu

Artificial neural networks (ANNs) have been used successfully for solving hard problems in many disciplines. Multi-layer feed-forward ANNs with back-propagation learning is one of the most widely used models. Methods for efficiently implementing and mapping this model on general-purpose, massively parallel reconfigurable mesh architectures are presented.

16.1. Introduction

Parallel implementation of artificial neural networks (ANNs) involves mapping of neural network models onto parallel architectures. Due to the large, diverse, and growing number of learning algorithms and architectures in the ANN literature, it is impossible to come up with a single classification scheme that can categorize all the essential features of all these paradigms. Nonetheless, ANN models are frequently characterized by their learning rules (supervised, un-supervised, hybrid), their network topology (feed-forward, recurrent, single layer, multi-layer), and neuron characteristic. Some of the most commonly used and discussed ANN models include multi-layer feed-forward networks with supervised learning by error back-propagation [27], recurrent networks [9] (also referred to as feedback network), adaptive resonance theory (ART) networks [3], self-organizing map networks [13]. The mapping methods presented in this chapter focus on multi-layer feed-forward networks with supervised learning by error back-propagation ANN model.

Back-propagation [26] is one of the most widely used training algorithms for multi-layer feed-forward artificial neural networks, which gives the learning ability to an ANN. The ability to learn is crucial in most ANN applications such as pattern recognition, image processing, automation and control. However, the Back-propagation algorithm is computationally intensive and learning times (training times) on the order of days and even weeks are not uncommon when the algorithm is executed on sequential machines [6,7,28]. The problem of long training time can be overcome either by devising faster learning algorithms or by implementing the existing algorithms on parallel computer architectures. Improving the learning algorithm is an active area of research [25], however, here we shall instead focus on the later approach of parallel implementation. An additional benefit of this approach is that an improved, more efficient learning algorithm can be further speeded up by parallel implementation.

Using parallel processing to implement ANNs has attracted much interest during the past years. Research and development efforts fall into three general categories: mapping and implementing ANNs directly in hardware [2,5], mapping and implementing ANNs on special-purpose processor architectures and computing devices that are designed for ANN simulation [21,22], mapping and implementing ANNs on general-purpose parallel architectures [4,8,10,14,16,17].

Implementing ANNs directly in hardware has the advantage of fully exploiting the inherent parallelism in the neural networks and fully utilizing the strength of VLSI technology. However, the approach faces some tough challenges due to different hardware requirements for the recall phase, which requires low precision and less computing power because the weights and the structure of the network are fixed, and the training phase that demands high precision and a lot of computing power to adjust the weights and sometimes the structure of the neural network. Sometimes it is appropriate to study several different neural network structures and learning algorithms for the same problem to obtain the optimal solution. The degree of freedom needed makes it infeasible to build the special hardware in certain cases.

Parallel computers designed with special hardware architectures customized for mapping ANNs are good for both the training and recall phases. The approach provides the flexibility that the direct hardware approach cannot offer. This flexibility permits the mapping of different ANN models of varying sizes and complexity and realizing them basically in VLSI. Currently the number of processing elements (PEs) in the special machines makes the machines fall into the moderately parallel [24] and highly parallel [24] categories.

General-purpose parallel architectures provide the continuously parallel [24] computational models for studying parallel computation as well as the basis for developing massively parallel [24] machines solving computational intensive applications. Using general-purpose massively parallel architectures as the target machines for ANN implementation allows one to see the maximum parallelism achievable and helps one to evaluate the architecture models. In practice, the massively parallel machines are extremely useful for parallelization of the learning phase [23] for a wide range of applications.

The mapping methods to be described in this chapter focus on the last approach. The particular massively parallel architecture targeted is reconfigurable mesh.

The rest of this chapter is organized as follows. Section 16.2 describes the multi-layer feed-forward ANN model and the back-propagation algorithm. The description concentrates on the computational aspect of the model so that its parallel implementation can be better understood. Section 16.3 introduces reconfigurable mesh model (RMESH), the massively parallel architecture on which the ANN model is mapped. Several algorithms for the fundamental data manipulation operations are also presented in the section. These basic algorithms are useful building blocks for the development of efficient mapping methods in later sections. Another purpose of describing these fundamental algorithms is using them as examples to illustrate how RMESH model operates for the reader who is less experienced with the model. Section 16.4 presents the first mapping algorithm which has optimal asymptotic time complexity for RMESH architecture and conceptually simpler to develop. Section 16.5 presents the second mapping algorithms which by employing pipelined approach further reduces the space complexity of the first algorithm while at the same time maintaining the same time complexity as the first algorithm. Section 16.6 discuss some of the issues related to the two mapping algorithms that are not yet addressed and review related work that maps the ANN model to other massively parallel architectures. Section 16.7 summarizes the overall conclusions from this study.

16.2. Feed-Forward Networks: Back-propagation learning

A feed-forward artificial neural network consists of three types of layers in general, i.e., input layer, hidden layers and output layer. The first layer is the input layer, whose only role is to feed input patterns into the rest of the network. The last layer is the output layer off which the result of computation is read. The intermediate layers are called the hidden layers. The total number of layers can be different for different applications. The network connection is such that each node (neuron) in a layer receives input from every node of the previous layer and produces output to every node of the next layer. Each node computes a weighted sum of all its inputs. Then it applies a nonlinear activation function, F, to the sum, resulting in an activation value (or output) of the neuron. A sigmoid function, with a smooth threshold-like curve is the most frequently used activation function, though hard limiters are also used.

An ANN has two distinct operational states: recall state and learning state. The computation of the learning state can be further break down into two phases: forward phase and backward phase. Since the computation of the recall state is identical to that of the forward phase of the learning state, here we shall only describe the algorithm of the learning state and its parallel implementation in later sections.

In the forward phase of the algorithm the input to the network is provided and values propagate forward from one layer to another layer through the network to produce the output vector. The output vector is then compared with an expected output vector, provided by a teacher, resulting in an error vector. Let the layers be numbered from 0 to $L-1$ and the network has L layers. Let $O[L-1]$ be the output vector, T the target vector, and E the error vector. We have $E = T - O[L-1]$.

Let $W_{i,j}[l]$ denote the connection weight between node i of layer l and node j of layer $l+1$. Let $O_j[l]$ denote the output of node j of layer l, F the activation function. Also let the number of nodes of layer l be n_l and the nodes are numbered from 0 to $n_l - 1$. Then the output of a node is determined as follows.

$$O_i[l] = F(\sum_{0 \leq j \leq n_{l-1}-1} W_{j,i}[l-1]O_j[l-1]) \tag{16-1}$$

In the backward phase the values of the error vector are propagated back from the last layer towards the first layer through the network. The error signals for the hidden layers are determined recursively: Error values of layer l are determined from the error values of layer $l+1$. First the error values of layer $l+1$ are multiplied by the corresponding weights between the two layers. Then the products are propagated "backward" and get added at each node in layer l. The Let $\delta_i[l]$ denote the error of node i of layer l. Then, we have

$$\delta_i[l] = F'(O_i[l]) \sum_{0 \leq j \leq n_{l+1}-1} \delta_j[l+1]W_{i,j}[l] \tag{16-2}$$

F' stands for the derivative of F. Initially, in the last layer, $\delta_i[L-1]$ is defined as $F'(O_i[L-1]) \times E_i$, where E_i is the the ith component of the error vector E. For the sigmoid activation function, $F' = F(1-F)$, so we have

$$\delta_i[l] = O_i[l](1 - O_i[l]) \sum_{0 \leq j \leq n-1} \delta_j[l+1]W_{i,j}[l] \tag{16-3}$$

Once the error values are computed, appropriate changes of weights can be made. The weight change $\Delta W_{i,j}[l]$ in the connection from node i of layer l to node j of layer $l+1$ is proportional to the product of the the output value, $O_i[l]$, and the error value, $\delta_j[l+1]$. The proportional factor, η, is called the learning rate. The algorithm is summarized in Fig. 16-1.

16.3. Reconfigurable Mesh with Buses

The particular reconfigurable mesh architecture that we use in this chapter is due to Miller, Kumar, Resis and Stout [18,20,19]. Related recon-

Algorithm 1

1 Apply input, $O[0] = I$

2 Propagate the signal forward to compute the output $O[L-1]$ using $O_i[l] = F\left(\sum_{0\leq j\leq n_{l-1}-1} W_{j,i}[l-1]O_j[l-1]\right)$, for l from 1 to $L-1$

3 Compute the deltas for the output layer using, $\delta_i[L-1] = F'(O_i[L-1])(T_i - O_i[L-1]) = F'(O_i[L-1])E_i$

4 Compute the deltas for the preceding layers by propagating the error backwards, $\delta_i[l] = F'(O_i[l])\sum_{0\leq j\leq n_{l+1}-1}\delta_j[l+1]W_{i,j}[l]$, for l from $L-2$ to 1

5 Adjust weights using,
$\Delta W_{i,j}[l] = \eta O_i[l]\delta_j[l+1]$
$W_{i,j}^{new}[l] = W_{i,j}^{old}[l] + \Delta W_{i,j}[l]$

6 Repeat from 1 for the next input pattern

Fig. 16-1. Backpropagation learning algorithm.

figurable mesh architectures have also been proposed [1,15]. It has been demonstrated that these architectures are often very easy to program and that in many situations it is possible to obtain constant time algorithms that use a polynomial number of processors for problems that are not so solvable using PRAM model [18,12]. The fundamental data manipulation algorithms that we cover in this section have been considered in [11].

16.3.1. *Computational Model*

Fig. 16-2 shows a 4×4 RMESH. By opening some of the switches, the bus may be reconfigured into smaller buses that connect only a subset of the processors.

The important features of a RMESH are:

(1) An $N \times M$ RMESH is a 2-dimensional mesh connected array of processing elements (PEs). Each PE in the RMESH is connected to a broadcast bus which is itself constructed as an $N \times M$ grid. The PEs are connected to the bus at the intersection of the grid. Each

processor has up to four bus switches (Fig. 16-2) that are software controlled and that can be used to reconfigure the bus into sub-buses. The ID of each PE is a pair (i, j) where i is the row index and j is the column index. The ID of the upper left corner PE is $(0, 0)$ and that of the lower right one is $(N - 1, M - 1)$.

(2) There are up to four switches associated with a PE and are labeled E (east), W (west), S (south), and N (north). Notice that the east (west, north, south) switch of a PE is also the west (east, south, north) switch of the PE on its right (left, top, bottom). Two PEs can simultaneously set (connect, close) or unset (disconnect, open) a particular switch as long as the settings do not conflict. The broadcast bus can be subdivided into sub-buses by opening

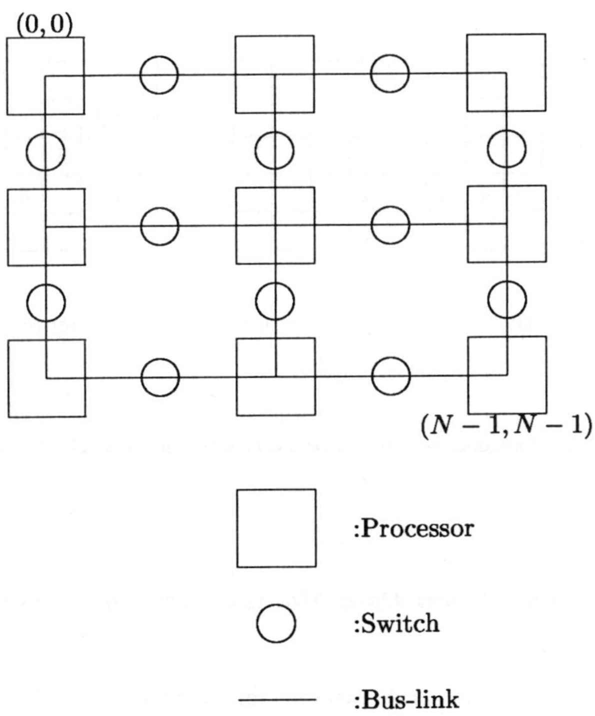

Fig. 16-2. 3 × 3 RMESH.

(disconnecting) some of the switches.

(3) Only one processor can put data onto a given sub-bus at any time.

(4) In unit time, data put on a sub-bus can be read by every PE connected to it. If a PE is to broadcast a value in register I to all the PEs on its sub-bus, then it uses the command broadcast(I).

(5) To read the content of the broadcast bus into a register R, the statement R = content(bus) is used.

(6) Row buses are formed if each PE disconnects (opens) its S switch and connects (closes) its E switch. Column buses are formed by disconnecting the E switches and connecting the S switches.

(7) Diagonalize a row (column) of elements is a command to move the specific row (column) elements to the diagonal position of a specified window. This is illustrated in Fig. 16-3.

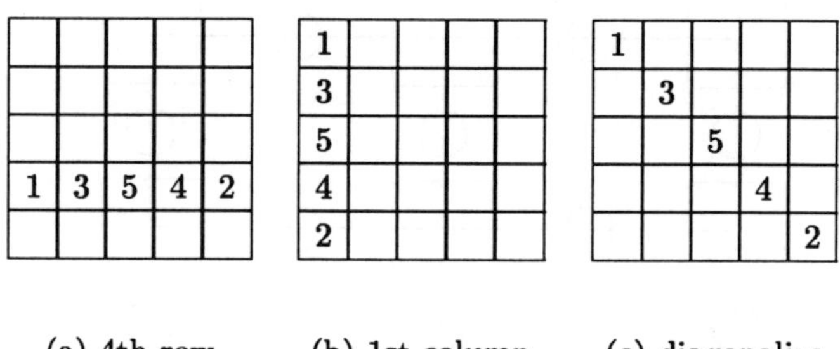

Fig. 16-3. Diagonalize 4th row or 1st column elements of a 5 × 5 window.

16.3.2. *Fundamental Data Manipulation Operations*

16.3.2.1. *Shift*

Each PE has data in its A variable that is to be shifted to the B variable of a PE that is s, $s > 0$, units to the right but on the same row. Following the shift, we have

$$B(i,j) = \begin{cases} null, & j < s \\ A(i,j-s), & j \geq s \end{cases} \tag{16-4}$$

A circular shift variant of the above shift requires

$$B(i,j) = A(i,(j-s) \bmod N) \tag{16-5}$$

Let us examine the first variant. This can be done in $O(s)$ time by dividing the sender and receiver processor pairs $((i,j-s),(i,j))$ into $s+1$ equivalent classes as below:

$$class\ k = \{((i,j-s),(i,j)) \mid (j-s) \bmod (s+1) - k\} \tag{16-6}$$

The sender and receiver pairs in each class can be connected by disjoint buses and so we can accomplish the shift of the data in the sender PEs of each class in $O(1)$ time. The algorithm is given in Fig. 16-4. The number of broadcasts is $s+1$. The procedure is easily extended to handle the case of left shift. Assume that $s < 0$ denotes a left shift by s units on the same row. This can also be done in $O(s)$ time. Noting the symmetry between rows and columns, the procedure is also easily extended to column shift. Let ShiftDown(s,A,B) be the procedure that shifts $A(i,j)$ to $B(i+s,j)$, $s > 0$.

A circular shift of s can be done in $O(s)$ time by first performing a ordinary shift of s and then shifting $A(i,N-s)$,...,$A(i,N-1)$ left by $N-s$. The latter shift can be done by first shifting $A(i,N-s)$, then $A(i,N-s+1)$,... and finally $A(i,N-1)$ using one broadcast for each shift.

The circular shift operation can be extended to shift in $1 \times W$ row window or $W \times 1$ column window. Let RowCircularShift(A,s,W) and ColumnCircularShift(A,s,W), respectively, be the procedures that shift A values by s units in windows of size $1 \times W$ and $W \times 1$. Let A^{in} and A^f, respectively, denote the initial and final values of A. Then, for RowCircularShift(A,s,W) we have the following relation

$$A^{in}(i,j) = A^f(i,q) \tag{16-7}$$

where PEs (i,j) and (i,q) are, respectively, the $a = j \bmod W$'th and $b = q \bmod W$'th PEs in the same $1 \times W$ row window and $b = (a-s) \bmod W$.

```
Procedure Shift(s,A,B)
/* shift from A(i,j) to B(i,j+s), s > 0 */
begin
    All PEs disconnect there N and S switches;
    for k=0 to s do
    begin
    /* shift class k */
        PE(i,j) disconnects its E switches if (j - s) mod (s + 1) = k;
        PE(i,j) disconnects its W switches if j mod (s + 1) = k;
        Broadcast(A(i,j)) if j mod (s + 1) = k;
        B(i,j)=content(bus) for every PE with (j - s) mod (s + 1) = k;
    end;
end;
```

Fig. 16-4. Shifting by s, $s > 0$.

The strategy of Fig. 16-4 is easily extended so that RowCircularShift and ColumnCircularShift are done in $O(s)$ time and using $2s+1$ broadcasts. The RowCircularShift is used to develop the ConsecutiveSum procedure to be described next.

16.3.2.2. *Consecutive Sum*

Assume that an $N \times N$ RMESH is tiled by $1 \times M$ blocks (M divides N) in a natural manner with no blocks overlapping. So, PE(i,j) is the j mod M'th processor in its block. Each PE(i,j) of the RMESH has an array X[0,$M-1$](i,j) of values. If j mod $M = q$, then PE(i,j) is to compute S(i,j) such that

$$S(i,j) = \sum_{r=0}^{M-1} X[q](i,(j\ div\ M) * M + r) \tag{16-8}$$

That is, the q'th processor in each block sums the q'th X value of the processors in its block. The consecutive sum operation is performed by

```
Procedure ConsecutiveSum(X,S,M)
/* Consecutive Sum of X in 1 × M blocks */
begin
    S(i,j) = X[((j mod M) + 1) mod M](i,j), 0 ≤ i,j < N
    for k=2 to M do
    begin
    /* circularly shift S in 1 × M blocks and add terms */
        RowCircularShift(S,M,-1);
        S(i,j) = S(i,j) + X[((j mod M) + k) mod M](i,j), 0 ≤ i,j < N
    end;
end;
```

Fig. 16-5. Consecutive sums in $1 \times M$ blocks.

having each PE in a $1 \times M$ block initiates a token that will accumulate the desired sum from the processor to its right end in the block. More specifically, the token generated by the q'th PE in a block will compute the sum for the $q+1$ mod M'th PE in the block, $0 \leq q < M$. The token are shifted left circularly within their $1 \times M$ block until each token has visited each PE in its block and arrived at its destination PE. The algorithm is given in Fig. 16-5. The consecutive sum operation can be done in a similar way for an $M \times 1$ block and in either case the time complexity of the operation is $O(M)$.

16.4. First Mapping Method

We assume that there are L layers numbered from 0 to $L - 1$ in the multiple layer feed-forward neural network just as in Section 16.2 and borrow the notation and terminology introduced in that section. Furthermore, it is assumed that each layer consists of n nodes, the size of RMESH is $N \times N$, and $n = N$. The validity of the assumption as well as the techniques to remove them is addressed in the discussion section later.

Since each layer has n nodes and the nodes between any adjacent layers are completely connected (the connection forms a complete bipartite graph), the number of weights between two adjacent layers is $n \times n$. These $n \times n$ weights are mapped to the $N \times N$ PEs one each. For example, considering the weights between layer 0 and layer 1, $W_{i,j}[0]$ is mapped to PE(i,j), $0 \leq i,j \leq N-1$, as illustrated in Fig. 16-6. The mapping is carried out for each layer l, $0 \leq l \leq L-1$, in a preprocessing step. After preprocessing, each PE(i,j) has L weights, $W_{i,j}[0]$, $W_{i,j}[1]$, ..., $W_{i,j}[L-1]$.

$W_{0,0}$	$W_{0,1}$	$W_{0,2}$
$W_{1,0}$	$W_{1,1}$	$W_{1,2}$
$W_{2,0}$	$W_{2,1}$	$W_{2,2}$

Fig. 16-6. An illustration of the initial weight distribution for both methods on a 3×3 RMESH.

Now, with the initial setting up of the weights, let us examine how the rest of parallel implementation are done on RMESH. Let us assume PEs with indices (0,0) to $(n-1,0)$ (first column) are conected to the input interface of the system and can retrieve the pattern input and its expected (desired) output from this interface. Let input(p,i) be the ith input value of pattern p, and output(p,i) the ith component value of the output expected from pattern p.

Let us first consider the mapping of the forward phase where a group of n patterns are simultaneously processed. The procedure is shown in Fig. 16-7.

Initially we distribute the n patterns to the PEs in such a way that the ith input value of each pattern is stored at the PEs on the ith row (i.e., each pattern is stored at every column). Hence, the ith input value of pattern p is stored at variable $O[p,0](i,j)$ of $PE(i,j)$, $0 \leq j < N$. The

```
begin
S1  for p=0 to n-1 do
    begin
    S1.0  PE(i,0) read input(p,i) and output(p,i), for 0 ≤ i < N
    S1.1  Setup row buses;
    S1.2  PE(i,0) broadcast(input(p,i)), 0 ≤ i < N;
    S1.3  O[p,0](i,j)=content(bus), 0 ≤ i,j < N;
    S1.4  PE(i,0) broadcast(output(p,i)), 0 ≤ i < N;
    S1.5  T[p](i,j)=content(bus), i = j;
    end
S2  for k=0 to L-2 do
    begin
    S2.0  for p=0 to n-1 do
                    t[p](i,j)=O[p,k](i,j) × W[k](i,j), 0 ≤ i,j < N
    S2.1  ConsecutiveSum(t,S,N) for tiled columns;
    S2.2  A(i,j) = F(S(i,j)), 0 ≤ i,j < N;
    S2.3  for p=0 to n-1 do
          begin
          S2.3.0 PE(i,j) diagonalize A(i,j) to D(j,j), for i = p;
          S2.3.1 Setup row buses;
          S2.3.2 Broadcast(D(i,j)), for PEs with i = j;
          S2.3.3 O[p,k+1](i,j)=content(bus), 0 ≤ i,j < N;
          end
    end
S3  for p=0 to n-1 do
          δ[p,L-1](i,j)=F'(O[p,L-1](i,j))(T[p](i,j)-O[p,L-1](i,j)), for i = j;
end
```

Fig. 16-7. ANN forward phase computation of the first method.

first index of variable O identifies the training pattern and the second index identifies the layer of the network that produces this output. The meaning of this variable is similar to the O variable in Section 16.2, except for the augmentation of another pattern index. The expected output of each pattern is distributed to the diagonal PEs and stored at variable T. $T[k](i,i)$ gives the value of the ith component of the expected output resulted from pattern k. The reason why diagonal PEs are used will become apparent later. The details of accomplishing the above are in S1 portion of the algorithm. Step S1.0 reads in the input data and the desired output data of the pattern. The values will then be distributed to the processors through row buses and broadcast operations, steps S1.1 to S1.5. The operation will be done n times since we have n patterns.

Once the input patterns are distributed, the outputs of layer 0 are now available and need to be used to generate the outputs of the next layer. This process is repeated for each layer according to equation (1), except that now we need to take care of n patterns at once.

Let us take a close look at how the outputs of layer 1 are generated. To generate the n outputs, each $PE(i,j)$ first multiplies its $O[p,0](i,j)$ variable with its $W[0](i,j)$ variable for each pattern p, $0 \leq p < n$. The resutls are stored at a temperary varaible $t[0..n-1](i,j)$. Notice that $\sum_{0 \leq i < N} t[k](i,j)$ gives the weighted input sum to node j of layer 1 when pattern k is applied. Since there are n input patterns and each pattern produces n weighted input sums for layer 1, we have a total $n \times n$ such weighted input sums to compute. These sums are determined by ConsecutiveSum procedure on tiled columns of RMESH. Each sum is stored at variable S of each PE. Notice that $S(i,j)$ gives the weighted input sum to node j resulted from pattern i. After activation function F is applied to $S(i,j)$ by each $PE(i,j)$, each row i contains the output of layer 1 corresponding to pattern i. More specifically, $(F(S(i,1)), F(S(i,2)),...,F(S(i,N-1)))$ is the output vector resulted from input pattern i. Now each rwo r needs to move its S data to each column just as the initial input patterns are distributed so that the above process can be repeated.

To move its data to each column, row r first diagonalizes its data, then the system sets up the row buses, and then each diagonal PE broadcasts the data, finally $PE(i,j)$ in each column receives the data in $O[r,1](i,j)$. Once every row has done that, we have the needed output in variable O to be fed to the next layer.

The details of accomplishing the above for every layer are in S2 portion of the algorithm. Step S2.0 carries out the mutiplication of weights with the corresponding outputs or activation values. Step S2.1 sums up the results of multiplying the activation values by the weights towards the nodes of the next layer. Step S2.2 computs the activation values of the nodes in the next layer. Step S2.3 redistributes the new activation values to variables O in preparation of the next iteration. The loop at step S2 makes sure the above steps are repeated for each layer.

Notice that when we leave step S2, the outputs of the last layer have already been redistributed to the columns. In particular, one copy of the outputs is contained in the diagonal. More specifically, $(O[k, L-1](0,0), O[k, L-1](1,1), ..., O[k, L-1](n-1, n-1))$ is the output vector produced from input pattern k at the last layer. Since the expected output of each pattern is also stored at the diagonal, the diagonal PEs are ready to compute the error values, δ, for the last layer and to redistribute the δ values for the backward phase. The computation of δ values is done in S3 portion of the algorithm.

Notice the symmetry in the summation parts of equations (1) and (2). The mapping of the backward phase is very similar to that of the forward phase. Instread of moving the output forward through the network, we need to propagate the error backward. In terms of implementation, instead of distributing output vectors to the columns, we move error vectors to the rows; instead of using ConsecutiveSum on the columns, we apply it for the rows; instead of making each row to diagonalize the results for the next layer computation, we let each column diagonalize. The procedure is shown in Fig. 16-8. We leave the details of how this algorithm works to the reader.

We have discussed the mapping of the network to RMESH for a group of n patterns. The long network training time is partly due to enormous training patterns (the total number of patterns involved in the training, including the repeated use of the same pattern). So it is not unusual to assume that $P \gg n$ and $P \gg l \cdot n$, where P is the total number of patterns used in the learning. In such a case, we can partition the P patterns in $\lceil P/n \rceil$ groups where each group holds a maximum of n patterns and use the above method for each group.

Now let us consider the time and space complexities of the parallel algorithm. The space used by each PE is dominated by two 2-dimensional arrays $O[0..n-1, 0..L-1]$ and $\delta[0..n-1, 0..L-1]$. Hence, the space complexity is $O(nL)$ for n input patterns each having n components and L

```
begin
S4  Setup column buses;
    for p=0 to n-1 do
    begin /* redistribute errors to rows */
        PE(i,j) broadcast(δ[p, L − 1](i, j)), for i = j;
        δ[p, L − 1](i, j)=content(bus), 0 ≤ i, j < N;
    end
S5  for k=L-2 downto 1 do
    begin
            /* compute the errors of the previous layer */
            for p=0 to n-1 do
                    t[p](i,j)=δ[p,k+1](i,j) × W[k](i,j), 0 ≤ i, j < N
            ConsecutiveSum(t,S,N) for tiled rows;
            Setup column buses;
            PE(i,j) broadcast(O[i,k](i,j)), for i = j;
            tO(i,j)=content(bus), 0 ≤ i, j < N;
            A(i,j) = F'(tO(i, j)) × S(i, j), 0 ≤ i, j < N;
            /* move the new errors into position for the next iteration */
            for p=0 to n-1 do
            begin
                    PE(i,j) diagonalize A(i,j) to D(j,j), for j = p;
                    Setup column buses;
                    Broadcast(D(i,j)), for PEs with i = j;
                    δ[p,k](i,j)=content(bus), 0 ≤ i, j < N;
            end
    end
end
```

Fig. 16-8. ANN backward phase computation of the first method.

layer networks. S1 part of the algorithm takes $O(n)$ time, because each operation inside the loop takes $O(1)$ time. So is S3 part of the algorithm. For S2 part, each top level step takes $O(n)$ time except for S2.2 which takes $O(1)$. Therefore, S2 part of the algorithm takes $O(nL)$ time. Since the computational resources needed for the backward phase is similar to that of the forward phase, the overall time and space complexities are both $O(nL)$. If we consider the network being fixed, then the time and space complexities become $O(n)$ for n input patterns. For P patterns, the complexity changes to $O(LP)$ and the space complexity remains the same.

The next mapping method that we will present reduces the space complexity to $O(L)$ while at the same time maintaining $O(nL)$ time complexity for n patterns.

16.5. Second Mapping Method

We shall make the same sef of assumptions for this method as we did for the first method. Identical to the first method, $W_{i,j}[l]$ are mapped to $PE(i,j)$, $0 \leq i,j < N$, $0 \leq l < L-1$.

Recall that in the first mapping, each input pattern and its outputs from every layer are stored at each column. Thus, for n patterns, each PE takes $O(nL)$ space. To reduce the space to $O(L)$, different input patterns are stored at different columns, as illustrated in Fig. 16-9. The output from each layer is stored at the same column where the input pattern that produces it is stored, just as the first mapping does. Note that the initial input pattern distribution can done using shift operation in $O(n)$ time.

Now, in order to compute the weighted input sum of each node, we have to 'move' the weights, assigned to the other PEs during the preprocessing step, around to meet its corresponding inputs. Then we 'move' the products to be added together for the final sum.

The preceding description is actually accomplished by a technique in which the weight associated with each PE is broadcasted using a pipelined method as illustrated in Fig. 16-10. The arrows indicate the set of PEs participating in the broadcast of their weights for one iteration of the pipeline. The order in which the set of PEs participates is from the upper-left corner arrow towards the the lower-right corner arrow.

The basic idea is as follows. Each PE located at the current arrow broadcasts its weight to every PEs in the same row. Then each PE in those rows participating in the broadcast computes the product of the weight just received and the output value that it has. The product is added to the previously accumulated sum, if any, sent from the PEs in the row immediate above and the total is stored at a variable T. The the value of the T variables in each row participating in the broadcast, except for the last row, are shifted down to the next row. Note that the value shifted down is the previously accumulated sum mentioned earlier.

The above process is repeated for each arrow. When the last row starts participating in the broadcast, the weighted input sums begin

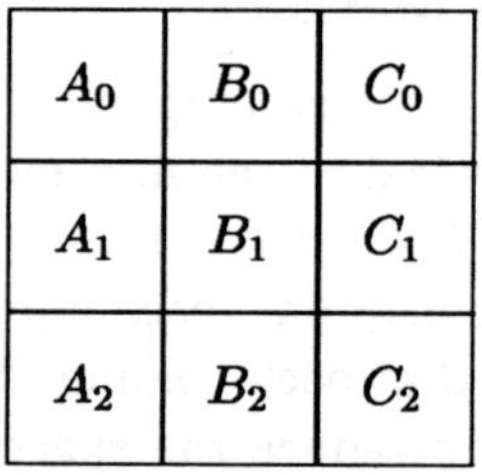

A_0	B_0	C_0
A_1	B_1	C_1
A_2	B_2	C_2

Fig. 16-9. An illustration of the distribution of three input pattern A, B, C on a 3×3 RMESH for method 2.

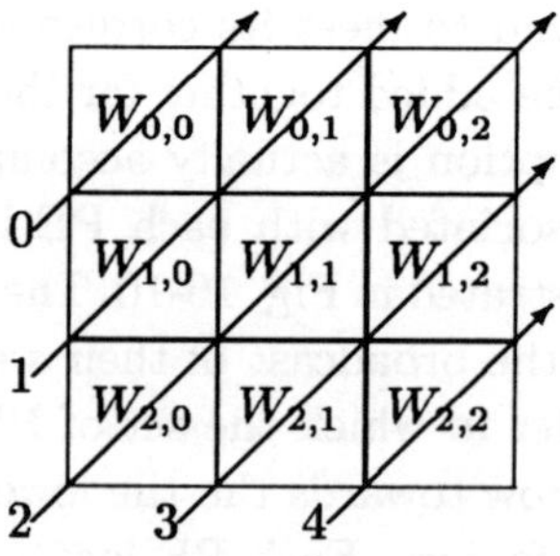

Fig. 16-10. An illustration of the broadcasting pattern for sending all the weights through row bus in the pipeline approach of method 2.

to emerge at the last row. Now the last row broadcasts just completed sums. The appropriate reads in the data to its O variable and is ready for moving forward to the next layer. The procedure shown in Fig. 16-11 is based on this idea and shows the computation for one layer. The procedure assumes that the outputs from the previous layer are set up in the O variable as illustrated in Fig. 16-9 and stores the outputs from the current layer in the same fashion.

Procedure ForwardComputation(layer)
begin
1 D(i,j)=0, $0 \leq i, j < N$;
2 for k=0 to 2(n-1) do
  begin;
  2.0 Setup row bus;
  2.1 PE(i,j) broadcast(W[layer](i,j)), for $i + j = k$;
  2.2 C(i,j)=content(bus),
     for ($k < n$ *and* $i \leq k$) or ($k \geq n$ *and* $i > (k \bmod n)$);
  2.3 T(i,j)=C(i,j)×O[layer](i,j)+D(i,j),
     for ($i \leq k < n$) or ($k \geq n$ *and* $i > (k \bmod n)$);
  2.4 ShiftDown(1,T,D),
     for (($i \leq k < n$) or ($k \geq n$ *and* $i > (k \bmod n)$)) and $i \neq n - 1$;
  2.5 Setup column bus, if $k \geq n - 1$;
  2.6 PE(n-1,j) broadcast(F(T(n-1,j))), for $0 \leq j < N$, if $k \geq n - 1$;
  2.7 O[layer+1](i,j)=content(bus), for $i = (k \bmod n)$, if $k \geq n - 1$;
  end
end

Fig. 16-11. ANN one layer forward phase computation of the second method.

Steps 2.1 and 2.2 carry out the broadcast of the weights contained in the PEs along the current arrow, see Fig. 16-10. In step 2.1, the condition $i+j = k$ identifies the PEs participating in the broadcast. There are two sets of disjoint conditions for the rows receiving the broadcasted data.

One is for $k < n$ and the other is for $k \geq n$, as described in the two clauses for the logical or operator in step 2.2. Steps 2.3 and 2.4 compute the partial sums for the weighted inputs and move the results to the next row. Notice that the condition, $i \neq n-1$, in step 2.4 forbids the last row from shifting. Steps 2.5 to 2.7 apply the activation function to the completed sum and transfer the results to the approciate row. These steps are executed only after the last row joins in the set of broadcasting rows. The condition, $k \geq n-1$, in each step guarantees that. Fig. 16-12 shows how weighted inputs are accumulated in variable T in the algorithm. It uses the weights of Fig. 16-6 and the input patterns of Fig. 16-9 as an example.

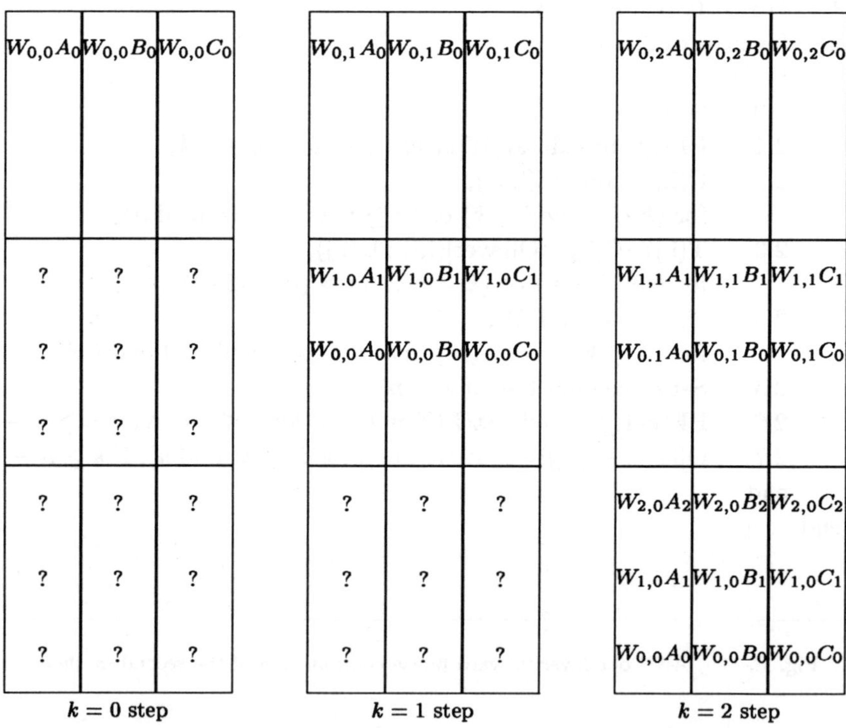

Fig. 16-12. An example of weighted input acumulation in variable T.

It is not difficult to see how this procedure can be applied $L-$

1 times to obtain the output of each pattern at the last layer. The time complexity of the procedure is $O(n)$ because each step in the loop takes $O(1)$ time. Hence, the overall time complexity is (nL), and for P patterns $O(PL)$. The time complexity matches the first method. The space complexity is $O(L)$, mainly due to variable $O[0..L-1]$. Notice that variable O is a 1-dimensional array. Thus, an improvement in space complexity is made over the first method.

For the backward phase, we can make similar adjustment to this procedure as we did to the procedure in Fig. 16-7 to obtain that in Fig. 16-8, in the preceding section. We shall leave it to the reader to come up with the exact code for RMESH.

16.6. Discussions

Let us first consider some of the assumptions we made in the preceding two sections. We know for a fact that many very successful applications are developed using multi-layer feed-forward with backpropagation learning networks that do not have the same number of nodes (or neurons) for every layer. In those cases, let n be the number of nodes in the largest layer. We then augment those layers whose node number is less than n by introducing dummy nodes. The dummy nodes are connected to the nodes (real or dummy) in the adjacent layers with zero weight. Clearly, the argmented network is functionally equivalent to the pre-argmented one and satisfies our assumption. Now, the methods can be applied.

The other assumption we made is $n = N$. That is, the number of nodes in each layer is the same as the number of rows or columns of a RMESH. Clearly when $n < N$, our methods can still be applied. We can simply load a special program that does nothing to each redundant PE in a preprocessing step and use the rest of the PEs for our method. Hence, the true rationale for the assumption is the desire for studying the maximum fine-grain parallelism achievable. However, the assumption does pose a problem for the methods when a real RMESH machine does not have enough PEs. One way to solve the problem is relying on a somewhat coarse-grain approach. The nodes in each layer are divided into super nodes such that the number of super nodes is equal to N. Then, we apply the methods to the super node network, with the understand that the computation within the super node is carried out by a single PE sequentially.

Widely considered massively parallel architectures that are similar to RMESH include mesh and hypercubes. The ANN model considered in this chapter has been mapped to both architectures [14,16,17]. The mapping methods on RMESH compare well with the work cited above. For single pattern learning, the time complexity of the proposed methods is $(logn)$ which matches the best known in the literature [17]. For a group of n patterns, the time complexity of the proposed methods is $O(n)$, asympototic optimal for RMESH, while the best known in the literature is $O(nlogn)$. Notice that we have dropped L in the asympototic time complexity analysis. The reason is that, in the literature, complexity results of the mapping methods are usually given for the computation between two layers.

16.6.1. *Conclusion*

In this chapter we discussed the need for implementing ANNs on parallel architectures, in particular the backpropagation learning algorithm. We categoried the parallel mapping approaches into three general schemes: mapping and implementing ANNs directly in hardware, mapping and implementing ANNs on special-purpose processor architectures and computing devices that are designed for ANN simulation, mapping and implementing ANNs on general-purpose parallel architectures, with the advantage of each approach briefly commented . We reviewed the computational aspect of feedforward network with backpropagation learning and described in detail the massively parallel reconfigurable mesh architectures, as well as some algorithms for fundamental data manipulation operations. Based on the ANN model and the parallel architecture, we then developed two different methods for mapping the network model and its learning algorithm to the architecture. The first method has space complexity of $O(nL)$, where n is the number of nodes in one layer and L the number of layers, but is conceptually simpler to develop. The second method uses a pipeline approach which reduces the memory requirement to $O(L)$. Both methods achieve $O(PL)$ time complexity for P number of input patterns which are asymptotic optimal for RMESH architecture.

References

[1] Y. Ben-Asher et al. The power of reconfiguration. Technical report, The Hebrew University, Israel, 1990.

[2] B. Boser et al. An analog neural network processor with programmable topology. *IEEE Journal of Solid-State Circuits*, 26(12), 1991.

[3] G. Carpenter and S. Grossberg. A massively parallel architecture for a self-organizing neural pattern recognition machine. *Computer Vision, Graphics and Image Processing*, 37:54–115, 1987.

[4] L. Chu and W. Wah. Optimal mapping of neural network learning on message-passing multicomputers. *¿Journal of Parallel and Distributed Computing*, 14:319–339, 1992.

[5] T. Clarkson, C. Ng, and Y. Guan. The pram: an adaptive vlsi chip. *IEEE Transactions on Neural Networks*, 4(3):408–411, 1993.

[6] Y. Le Cun et al. Backpropagation applied to handwritten zip code recognition. *Neural Computation*, 1(4):541–551, 1989.

[7] H. Drucker and Y. Le Cun. Improving generalization performance using double backpropagation. *IEEE Transaction on Neural Networks*, 3(2):991–997, 1992.

[8] J Ghosh and K. Hwang. Mapping neural networks onto message-passing multicomputers. *Journal of Parallel and Distributed Computing*, 6:291–330, 1989.

[9] J Hopfield. Neural networks and physical systems with emergent collective computational abilities. In *Proceedings of National Academy of Science*, pages 2554–2558, 1982.

[10] J. Jenq and W. Li. Feedforward backpropagation artificial neural networks on reconfigurable meshes. *Future Generation Computer Systems*, 14:313–319, 1998.

[11] J. Jenq and S. Sahni. Reconfigurable mesh algorithms for fundamental data manipulation operations. In F. Ozguner, editor, *Computing on Distributed Memory Multiprocessors, NATO Series F*, pages 27–46. Springer.

[12] J. Jenq and S. Sahni. Reconfigurable mesh algorithms for the hough transform. In *Proceedings of 1991 International Conference on Parallel Processing*, pages 34–41, 1991.

[13] T. Kohonen. *Self-Organization and Associative Memory*. Springer-Verlag, Berlin, third edition, 1989.

[14] V. Kumar et al. A scalable parallel formulation of the backpropagation algorithm for hypercube and related architectures. *IEEE Transactions on Parallel and Distributed Systems*, 5(10):1073–1090, 1994.

[15] H. Li and M. Maresca. Polymorphic-torus network. *IEEE Transactions on Computers*, 38(9):1345–1351, 1989.

[16] W. Lin et al. Algorithmic mapping of neural network models onto parallel simd machines. *IEEE Transactions on Computers*, 40(12):1390–

1401, 1991.

[17] Q. Malluhi et al. Efficient mapping of anns on hypercube massively parallel machines. *IEEE Transactions on Computers*, 44(6):769–779, 1995.

[18] R. Miller et al. Data movement operations and applications on reconfigurable vlsi arrays. In *Proceedings of 1988 International Conference on Parallel Processing*, pages 205–208, 1988.

[19] R. Miller et al. Image computations on reconfigurable vlsi arrays. In *Proceedings IEEE Conference on Computer Vision and Pattern Recognition*, pages 925–930, 1988.

[20] R. Miller et al. Meshes with reconfigurable buses. In *Proceedings 8th MIT Conference on Advanced Research in VLSI*, pages 163–178, 1988.

[21] N. Morgan et al. The ring array processor(rap): A multiprocessing peripheral for connectionist applications. *Journal of Paralel and Distributed Computing*, 14, 1992.

[22] U. Muller et al. Fast neural net simulation with a dsp processor array. *IEEE Transactions on Neural Networks*, 6(1):203–213, 1995.

[23] T. Nordstrom and B Svensson. Massively parallel architecture for large scale neural network simulations. *IEEE Transaction on Neural Networks*, 3:876–887, 1992.

[24] T. Nordstrom and B Svensson. Using and designing massively parallel computers for artificial neural networks. *Journal of Paralel and Distributed Computing*, 14:260–285, 1992.

[25] P. Osowski and M. Stodolski. Fast second order learning algorithms for feedforward neural networks and its applications. *Neural Networks*, 9(9):1583–1596, 1996.

[26] D. E. Rumelhart, G. E. Hinton, and R. J. Williams. Learning internal representations by error propagation. *Nature*, 323:533–536, 1986.

[27] D.E. Rumelhart and J.L. McClelland. *Paralle Distributed Processing: Explorations in the Microstructure of Cognition.* MIT Press, Cambridge, MA, 1986.

[28] A Waibel. Consonant recognition by modular construction of large phonetic time-delay neural networks. *Advances in Neural Information Processing Systems*, pages 215–223, 1989.

Index

About the Editors

David Zhang graduated in Computer Science from Peking University and received his M.Sc. and Ph.D. degrees in Computer Science and Engineering from the Harbin Institute of Technology (HIT) in 1983 and 1985, respectively. From 1986 to 1988 he was a post-doctoral fellow at Tsinghua University and became an associate professor at Academia Sinica, Beijing, China. In 1988, he joined the University of Windsor, Ontario, Canada, as a visiting research fellow in Electrical Engineering. He received his second Ph.D. degree in Electrical and Computer Engineering at the University of Waterloo, Ontario, Canada, in 1994. After that, he was an associate professor at City University of Hong Kong. Currently, he is a professor at Hong Kong Polytechnic University. He is founder and director of both Biometrics Technology Centers supported by UGC/CRC, the Hong Kong Government, and the National Nature Scientific Foundation (NSFC) of China, respectively. He is also a guest professor at Tsinghua University, Shanghai Jiao Tong University and HIT. In addition, he is founder and editor-in-chief of *Int. Journal of Image and Graphics*, and an associate editor for *IEEE Trans. on Systems, Man and Cybernetics, Pattern Recognition*, *Int. Journal of Pattern Recognition and Artificial Intelligence*, *Int. Journal of Robotics and Automation,* and *Neural, Parallel and Scientific Computations.* He has given many keynotes, invited talks and tutorial lectures, as well as served as a member of many organizing committees at international conferences in recent years. He has published over 170 papers including six books.

Sankar K. Pal is a professor and distinguished scientist at the Indian Statistical Institute, Calcutta. He is also the founding head of the Machine Intelligence Unit. He received his M.Tech. and Ph.D. degrees in Radio physics and Electronics in 1974 and 1979, respectively, from the University of Calcutta. In 1982 he received his second Ph.D. in Electrical Engineering along with DIC from Imperial College, University of London. He worked at the University of California, Berkeley and the University of Maryland, College Park during 1986–87 as a Fulbright postdoctoral visiting fellow; at the NASA Johnson Space Center, Houston, Texas during 1990–92 and 1994 as a guest investigator under the NRC-NASA Senior Research Associateship program; and at the Hong Kong Polytechnic University, Hong Kong in 1999 as a visiting professor. He served as a distinguished visitor of the IEEE Computer Society (USA) for the Asia-Pacific Region during 1997–99.

Professor Pal is a fellow of the IEEE, USA, Third World Academy of Sciences, Italy, and all the four National Academies for Science/ Engineering in India. His research interests include Pattern Recognition, Image Processing, Data Mining, Soft Computing, Neural Networks, Genetic Algorithms, and Fuzzy Systems. He is a co-author/co-editor of nine books including *Fuzzy Mathematical Approach to Pattern Recognition*, John Wiley (Halsted), N.Y., 1986, *Neuro-Fuzzy Pattern Recognition: Methods in Soft Computing*, John Wiley, N.Y., 1999; and of about three hundred research publications.

He received the *1990 S. S. Bhatnagar Prize* (which is the most coveted award for a scientist in India), *1993 Jawaharlal Nehru Fellowship, 1993 Vikram Sarabhai Research Award, 1993 NASA Tech Brief Award, 1994 IEEE Trans. Neural Networks Outstanding Paper Award, 1995 NASA Patent Application Award, 1997 IETE-Ram Lal Wadhwa Gold Medal, 1998 Om Bhasin Foundation Award, 1999 G. D. Birla Award for Scientific Research, 2000 Khwarizmi International Award (1st winner) from the Islamic Republic of Iran, 2001 INSA-Syed Husain Zaheer Medal*, and the *FICCI-2001 Award in Engineering and Technology.*

He is an associate editor for *IEEE Trans. Neural Networks* (1994–98), *Pattern Recognition Letters, Int. Journal of Pattern Recognition and*

Artificial Intelligence, *Neurocomputing*, *Applied Intelligence*, *Information Sciences*, *Fuzzy Sets and Systems*, and *Fundamenta Informaticae*; a member of the executive advisory editorial board of *IEEE Trans. Fuzzy Systems*, *Int. Journal on Image and Graphics*, and *Int. Journal of Approximate Reasoning*; and a guest editor of many journals including the *IEEE Computer*.